CALCULUS
A Modern Approach

Karl Menger

With a new Preface and
Guide to Further Reading by

B. Schweizer
Professor Emeritus of Mathematics
University of Massachusetts

A. Sklar
Professor Emeritus of Mathematics
Illinois Institute of Technology

DOVER PUBLICATIONS, INC.
Mineola, New York

Bibliographical Note

This Dover edition, first published in 2007, is an unabridged republication of the work originally published by Ginn and Company, Boston, in 1955. The present edition adds a new Preface and Guide to Further Reading by B. Schweizer and A. Sklar.

Library of Congress Cataloging-in-Publication Data

Menger, Karl, 1902–
Calculus : a modern approach / Karl Menger ; with a new preface and guide to further reading by B. Schweizer, A. Sklar. — Dover ed.
p. cm.
Originally published: Boston : Ginn, 1955.
Includes bibliographical references and index.
ISBN 0-486-45771-0
1. Calculus. I. Schweizer, B. (Berthold) II. Sklar, A. (Abe) III. Title.

QA303.M39 2007
515—dc22

2006103547

Manufactured in the United States of America
Dover Publications, Inc., 31 East 2nd Street, Mineola, N.Y. 11501

Preface to the Dover edition

What makes Karl Menger's "Calculus: a Modern Approach" different from all other calculus books? And what should this difference mean to the student, to the instructor, to the user of calculus, or to the philosopher of science? The short answer to the second question is that a person in any of these four overlapping categories will benefit from the effort of looking at calculus through the eyes of one of the outstanding mathematicians of the 20th century who yet took a great interest and pleasure in "elementary" teaching—and who, in his early years, was closely allied to the Vienna Circle and delighted in "the explication of concepts".

For someone approaching calculus for the first time, the way to benefit from this book is to skip over all the preliminary material, go directly to page 1 and proceed from there. For those with previous experience with calculus, on the other hand, some amplification of our 'short answer' is in order. This involves a somewhat detailed answer to the first question, as follows:

Menger begins his exposition with the development of a simple yet elegant miniature version of calculus in which the functions to be integrated are constant functions and the functions to be differentiated are linear functions[1]. Since each set of functions is closed under addition (the constant functions are also closed under multiplication), since any (indefinite) integral of a constant function is a linear function, and since the derivative of a linear function is a constant function, this allows him to develop virtually all the basic results of calculus, including the so-called "Fundamental Theorem of Calculus"[2], in this simple setting, all without any appeal to the notions of limits or continuity.

This mini-calculus is then extended by enlarging the set of constant functions to the set of step-functions ("step-lines" in Menger's terminology) and the set of linear functions to the set of (continuous) piecewise-linear functions ("polygons"). The basic results already obtained carry over: this is crucial, for, in calculus, functions are integrated by approximating them by step-functions and differentiated by approximating them by piecewise-linear functions. This is what Menger does in going from his mini-calculus to "full" calculus.

In effect, Menger's method is to get the basic concepts across in a setting in which one need not worry about the

[1] Actually, since Menger identifies a real function with its Cartesian graph (both being, in his view, the same set of ordered pairs of numbers) and uses the same terminology for both, he speaks here rather of "horizontal lines" and "non-vertical lines", respectively. This has the advantage of allowing him, at this stage, to keep the exposition completely geometrical.

[2] It is interesting to note that nowhere in Menger's book do the words "Fundamental Theorem of Calculus" actually occur.

difficult[3] notions of *limit*, *continuity*, etc. Does Menger's method work? Those of us who have tried it in actual classroom practice think so: see, e.g., [5].

The other major difference between Menger's *Calculus* and other calculus books is that Menger provides a name for every function mentioned. This allows him to consistently distinguish between a function and its value at a given point[4]. In particular, he calls the identity function on the reals j, so that $j(x) = x$ for all real numbers x. This has given rise to a considerable amount of misunderstanding, especially since Menger then uses the name j^2 for the square function, $j^{1/3}$ for the cube root function, etc. To someone who is used to speaking of "the function x", "the function x^2", etc., it looks as though Menger is simply replacing the letter x by the letter j, which is not the case at all. To Menger, the letter j designates a specific function and nothing else, whereas to the "x-ist", at each occurrence, the meaning of the letter x has to be inferred from the context. This is often far from straightforward. At a higher level than the x-ist's, one finds two figures on the same page of a generally excellent textbook, one figure captioned "Graph of the identity function: $f(x) = x$", and the other captioned "Graph of g when $g(x) = x^2$". Thus the square function has the name g, at least on this page, but the identity function has no name

[3]That these notions are difficult is substantiated, not only by the reactions of students, but also by the fact that it took the mathematical community some 300 years—say from Cavalieri to Weierstrass to A. Robinson—to straighten them out to the satisfaction of most, if not all, present-day mathematicians. Moreover, by introducing the very intuitive phrases "arbitrarily close" and "sufficiently close", Menger cleverly avoids all reference to the usual $\varepsilon - \delta$ arguments that confound and plague the beginning student.

[4]And to avoid such wondrously nonsensical statements as "$f(a)$ is the value of $f(x)$ when x has the value a"—simply interchange a and x.

(unless one jumps to the unlikely conclusion that "$f(x) = x$" is a name). In addition, the introduction of j allows Menger to explicitly express the fact that it is the identity[5] (hence its name) for the binary operation of composition of functions, viz., $j \circ f = f \circ j = f$, for every function f.

The seemingly small innovation of providing each function with a name is vital, for as noted in [12], "at one stroke it removes most if not all of the ambiguities that plagued, and still plague, the standard presentation of calculus, and converts the subject into a pure and unambiguous calculus of functions." It also has a similar dramatic effect on the introduction and development of the Laplace Transform (see pages 271–272).

The mini-calculus and the provision of proper names for all functions are Menger's major innovations. Besides these, there are many small felicities in the book. One such is the technique for graphical composition of real functions on pages 89–90. Again quoting [12], "this technique is well-known in the special case of iteration, i.e., composition of a function with itself, but seems to be still quite unknown for the general case". Another is the discussion of the standard exponential function on pages 31 and 123–124. Here Menger gives simple geometric arguments which show why the base e is different from all other bases, motivates the standard limit definition of e, and shows that the slope of the exponential function at 0 is 1.

Another feature of Menger's book is his approach to the applications of calculus to problems in the physical sciences. Here the first issue that Menger faces and explicitly enun-

[5]For two-place functions, Menger introduces the *selector* functions I and J which are defined by $I(x, y) = x$ and $J(x, y) = y$ for any pair of real numbers (x, y).

ciates is the fact that in real-world situations the "variables" are generally not the numerical and function variables that play the central role in "pure" calculus. Consider one of his favorite examples: In the classical "equation", $s = s(t) = 16t^2$ (see page 194), where t denotes the elapsed time and s the distance travelled by a freely falling body, s and t by themselves are what he calls "consistent classes of quantities", or, following Newton, "fluents", while $s(t)$, if it is to have any consistent meaning at all, can only mean (as on page 195) the function connecting s with t. Thus $s(t)$ is definitely not the same as s, and to state that "$s = s(t)$" is simply incorrect. On the other hand, armed with specific symbols for the identity and square function, one can explicitly make the correct statement: $s(t) = 16\jmath^2$.

For another example, consider the plight of a student who is told in a calculus course that $\int f(x)dx$ and $\int f(y)dy$ have exactly the same meaning, but at the same time in a physics course (or later in the same calculus course) that $\int p(t)dt$ and $\int p(s)ds$ (p force, t time, s distance) not only involve different functions, $p(t)$ and $p(s)$, but yield completely different "physical quantities", namely "impulse" and "work". Small wonder that such a student is likely to wind up not a little confused!

Menger, unlike virtually every other author of a calculus text, confronts such issues head-on. In doing so, he eliminates much of the usual confusion, e.g., in the preceding example, by deleting the superfluous "$(x)dx$" and "$(y)dy$" in integrals of functions in "pure" calculus, and by carefully defining, on pages 202–210, the "cumulation" of one fluent with respect to another to obtain a new fluent.

Of course, Menger's book is not perfect. There are features in it that will tend to annoy readers. Many of these annoyances could have been avoided if Menger had been *more* rather than *less* "Mengerian". For example, consider his insistence on distinguishing between roman type (for numbers and numerical variables, page 62) and italic type (for functions and function variables, page 75). This convention is utterly impractical in a classroom situation or for applications (and made the book devilishly hard to proofread). Its only real achievement is to enable one to distinguish between say, 3, the number, and *3*, the constant function of value 3. But this could have been achieved by, say, naming that constant function k_3—and likewise for all other constant functions[6].

Another source of such annoyances was Menger's desire, at the time, to put composition of functions ("substitution" in his terminology) on a par with addition and multiplication[7]. Thus in analogy with the notation $1/f$ for the reciprocal of f, we find the ghastly expression $j//f$ for the inverse of f. In an earlier mimeographed version of the book Menger had used f^*. This symbol or $f^{(-1)}$ or, in accordance with current usage in iteration theory, f^{-1}, would have been far better—all the more so since he almost always denotes the reciprocal of f by $1/f$. Menger never returned to this nota-

[6]Menger had great difficulty finding a publisher for his book; and when Ginn and Company finally agreed to put it out, they did so in a miserly way. The most irritating outcome of this "pulp fiction" production is the fact that it is very difficult to distinguish roman and italic type. Fortunately, in almost all cases the intended meaning of the symbol is clear from the context. (Indeed, Menger could have made the point that there is a difference between a number and the constant function whose value is that number and left it up to the reader to decide from the context which is which—but that would have been decidedly "un-Mengerian".)

[7]This is a view he later changed. For example, in [M8] he says: "Beyond any question, in the realm of functions substitution is the operation par excellence."

tion and in some of this later writings he used the symbol "INV*f*" [M8].

But such caveats do not detract from the fact that the book, on the whole, is a stunning achievement. This was recognized by several of the early reviewers. Thus I.T. Adamson writes in [1]:

> "Every teacher of the Calculus must read [Menger's book]. And he will be thick-skinned indeed who, having read it, does not seriously rethink his lectures: the thinner-skinned reader will rewrite his—and the reviewer confidently believes he will rewrite them á la Menger."

And, A. Rapoport, in a review [8] directed at non-mathematicians writes:

> "Whether Menger's reforms will revolutionize the teaching of the calculus and the writing of calculus textbooks depends on the outcome of the battle between semantic awareness and loyalty to established routines. In the long run, semantic awareness usually wins, but many short-term victories fall to the entrenched forces of conformity[8]. Pending the final victory, Menger's *Cal-*

[8]This is indeed what happened, but in an unexpected way. The appearance of Menger's book coincided with the beginning of the era of "The New Math" and many other post-Sputnik attempts to modernize the teaching of mathematics and science. So, not only was Menger opposed by the usual conservative establishment, but also by a cadre of "new revolutionaries"—many intent on infusing elementary mathematics with the spirit of the "Bourbaki" group of research mathematicians. Faced with this opposition, lack of support, and widespread rumors to the effect that here was an outstanding mathematician who had now

culus is strongly recommended to all brave teachers willing to blaze new paths, to confront raised eyebrows of department chairmen, and to defend their convictions of what constitutes good teaching of mathematics."

B. Schweizer
Professor Emeritus of Mathematics
University of Massachusetts

A. Sklar
Professor Emeritus of Mathematics
Illinois Institute of Technology

"lost it" and thought that replacing x by j was a major achievement, Menger had as much chance as today's small shopkeeper versus Walmart. Thus, in spite of many favorable, indeed enthusiastic reviews—based on the earlier mimeographed version—e.g., [1, 3, 8], the book soon disappeared from the mathematical horizon.

Guide to Further Reading

Two collections of Menger's papers contain material of interest to readers of this book:

The first is the book "Selected Papers in Logic and Foundations, Didactics, Economics" [M1]. This contains many of Menger's expository and philosophical papers, in particular, the paper [M12] which, in his words, "supercedes all preceding papers on variables", and the second part of [M15] which presents his last published thoughts on the application of the function concept to science. In addition, there are the three Gulliver satires [M9, M10, M11] and the popular exposition [M7]. The mordant comments in these four papers are as valid today—or even more so—as they were half a century ago.

The second is the two-volume "Karl Menger: Selecta Mathematica" [M2] which contains both papers that led up to this book, e.g., [M3, M4], and papers that were motivated by questions raised in this book, e.g., [M5, M6, M8, M13, M14]. Before turning to these works, the reader should consult the biographical sketches [9, 14] and the commentaries [6, 12, 13, 15]. These present not only surveys and analyses of Menger's own work, but also discuss the influence of his work on the work of others, e.g., in [2, 4, 5, 7, 10, 11]. One detail: in [15] it is pointed out that [10] introduces genuinely *functional* versions of Menger's set-theoretic identities (13), (14) on page 80, (16), (17) on page 88 and (5) on page 235 of this book. The functional versions are much easier to handle—particularly in an axiomatic development of the algebra of functions [M8, M13, 10, 11]—so more useful than the set-theoretic ones.

References To Menger

M1. Selected Papers in Logic and Foundations, Didactics, Economics, Reidel, Dordrecht 1979.

M2. Karl Menger, Selecta Mathematica, B. Schweizer, K. Sigmund, A. Sklar et al., eds., Springer, Wien, New York, vol. 1, 2002, vol. 2, 2003.

M3. Algebra of Analysis, Notre Dame Mathematical Lectures, vol. 3, 1944. Reprinted in M2, vol. 2, pp. 135–180.

M4. The Mathematics of Elementary Thermodynamics, American Journal of Physics 18(1950)89–103. Reprinted in M2, vol. 2, pp. 351–365.

M5. Random Variables from the Point of View of a General Theory of Variables, Proceedings of the Third Berkeley Symposium on Mathematical Statistics and Probability, L. M. LeCam and J. Neyman, eds., University of California Press, Berkeley 1956, vol. 2, pp. 215–229. Reprinted in M2, vol. 2, pp. 367–381.

M6. What are x and y?, The Mathematical Gazette 40(1956) 246–255. Reprinted in M2, vol. 2, pp. 383–392.

M7. Why Johnny Hates Math, The Mathematics Teacher 49(1956) 578–584. Reprinted in M1, pp. 174–184.

M8. An Axiomatic Theory of Functions and Fluents, in The Axiomatic Method with Special Reference to Geometry and Physics, L. Henkin et al., eds., Studies in Logic and the Foundations of Mathematics, North-Holland, Amsterdam, 1959, pp. 454–473. Reprinted in M2, vol. 2, pp. 205–224.

M9. Gulliver in the Land Without One, Two, Three, The Mathematical Gazette 43(1959)241–250. Reprinted in M1, pp. 305–314.

M10. Gulliver's Return to the Land Without One, Two, Three, American Mathematical Monthly, 67(1960) 641–648. Reprinted in M1, pp. 315– 319.

M11. Gulliver in Applyland, Eureka 23(1960) 5–8. Reprinted in M1, pp. 320–323.

M12. Variables, Constants, Fluents, in Current Issues in the Philosophy of Science, H. Feigl and G. Maxwell, eds., New York 1961, pp. 304–313 and 316–318. Reprinted in M1, pp. 144–152.

M13. The Algebra of Functions: Past, Present, Future, Rendiconti di Matematica 20(1961)409–430. Reprinted in M2, vol. 2, pp. 225–246.

M14. On Substitutive Algebra and Its Syntax, Zeitschrift für mathematische Logik und Grundlagen der Mathematik 10(1964) 81–104. Reprinted in M2, vol. 2, pp. 247–270.

M15. Mathematical Implications of Mach's Ideas: Positivistic Geometry, The Clarification of Functional Connections, in Ernst Mach, Physicist and Philosopher, Boston Studies in the Philosophy of Science VI, R. S. Cohen and R. J. Seeger, eds., Reidel, Dordrecht 1970, pp. 107–125. Reprinted in M2, vol. 1, pp. 575–593.

References To Others

1. I. T. Adamson. Review of K. Menger: Calculus, a Modern Approach, The Mathematical Gazette 39(1956) 245–250.

2. I. T. Adamson. Transformation of Integrals, American Mathematical Monthly 65(1958)590–596.

3. H. E. Bray. Review of K. Menger: Calculus, a Modern Approach, American Mathematical Monthly 61(1954) 483–492.

4. T. Erber. Modern Calculus Notation Applied to Physics, European Journal of Physics 15(1994)111–118.

5. J. B. Harkin. The Pre-limit Calculus of K. Menger, Mathematics Teachers Journal 23(1973)71–75.

6. H. Kaiser. The Influence of Menger's Ideas on the Work of Nöbauer and His School, M2, vol. 2, pp. 127–134.

7. M. A. McKiernan. On the n-th Derivative of Composite Functions, American Mathematical Monthly 63 (1956) 331–333.

8. A. Rapoport. Review of K. Menger: Calculus, a Modern Approach, ETC 12(1955)137–144.

9. B. Schweizer. Introduction, M2, vol. 1, pp. 1–5.

10. B. Schweizer and A. Sklar. Function Systems, Mathematische Annalen 172(1967)1–16.

11. B. Schweizer and A. Sklar. A Grammar of Functions, Aequationes Mathematicae 2(1969)62–85 and 3(1969) 15–43.

12. B. Schweizer and A. Sklar. Commentary on Didactics, Variables, Fluents, M2, vol. 2, pp. 111–126.

13. L. Senechal and B. Schweizer. A Mengerian Tour Along Caratheodory's Royal Road, M2, vol. 2, pp. 317–323.

14. K. Sigmund. Karl Menger and Vienna's Golden Autumn, M2, vol. 1, pp. 7–21.

15. A. Sklar. Commentary on the Algebra of Analysis and Algebra of Functions, M2, vol. 2, pp. 111-126.

FOREWORD

The primary aim of this text is the presentation of pure and applied calculus in a clarified conceptual frame, to the end that students may acquire as quickly and as easily as possible a thorough understanding of the theory as well as the ability to use calculus as a tool. The importance of this undertaking is underlined by the increasing demand for mathematicians, scientists, and engineers with a training in calculus beyond manipulative facility.

The need for such a book became clear to the author during the Second World War, when he had the privilege of directing the mathematical instruction in the Navy Training Center at the University of Notre Dame, one of the largest V-12 units in the country. It was in connection with this large-scale accelerated program that he began serious study of how some of the stultifying routine drill in mathematics might be replaced by instruction that would lead to better understanding. In 1944 he published the booklet "Algebra of Analysis" (see the Bibliography on p. 346), stressing substitution and the identity function, without, of course, incorporating this material in the student's program at that time. But many exercises, used in the present book, were then first presented and aroused great interest.

After the war, more pronounced efforts in the direction of the modern approach were made in teaching large classes, mainly attended by former members of the Armed Forces, at the Illinois Institute of Technology. Mimeographed sheets and, later, pamphlets supplementing the text in use were distributed – the first set of notes prepared with the valuable help of Dr. Burton D. Fried.

But the modern approach to calculus presented in this text was first taught in 1952, and has been taught ever since, in the evening school of I.I.T.* There, surrounded by railroads and factories, and in the center of the most extended slum-clearing program ever undertaken, engineers after a day's work in industry and in laboratories gather together, sometimes slightly tired but always eager to learn and always appreciative of aids to clear thinking. The reaction of most of them to the new approach to calculus has been enthusiastic.

In this text, the attempt has been made to present the subject with all the directness of word-of-mouth teaching. Moreover, the author has tried to anticipate the questions that may arise in the student's mind and to quote them in the form in which they are usually asked by beginners.

---

* The author covers the material in 30 weekly 2-hour lectures – the first three chapters in the first three lectures. For beginners without previous training it is easy to grasp the bulk of the first seven chapters in 30 semiweekly 2-hour classes and the second half of the book, after a review of Chapters VI and VII, in another series of 30 double-classes.

The answers are accompanied by motivations of the concepts introduced and by warnings against wayward lines of reasoning. Various original pedagogical devices are used to achieve lucidity of presentation.

But the most outstanding feature of this book is the thoroughness and care with which basic ideas are explained, a feature which makes the book equally suitable for young beginners and for mature readers. Before a symbol appears in statements or formulas, its meaning is unambiguously defined and rules are formulated according to which the symbol is to be consistently used. This method makes it possible to reduce the number of drill questions in favor of more interesting problems, of which the more difficult ones are marked by a star.

The theory is applied to various branches of science, and the illustrations, which have been formulated with particular care, follow actual scientific procedures. Students of science and engineering, eager to apply calculus to their respective fields, will find in Chapter VII the very technique of that application treated extensively and explicitly. Articulate rules and clear instructions for the use of calculus as a tool are provided.

Teachers of mathematics are offered in this text a wealth of suggestions concerning the presentation of mathematical material, which are the result of the author's extensive teaching experience. Particular attention is called to the new theory of variables presented in Chapters IV, VII, and XI; the scrutiny (at the end of Chapter VII) of the scientific significance of rates of change – the idea that dominated Newton's research; and the clarification and correction (in Section 2 of Chapter VI) of formulas such as $\int \cos x\, dx = \sin x + c$ in Leibniz' notation.

In presenting the conceptual frame of calculus the author has found it necessary to introduce some new symbols for the sake of clarity. Questions of notation, however, play a merely subsidiary role and are important only as they aid in an understanding of concepts. Not only is each symbol, old and new, clearly defined before it is used, but sufficient discussion of the various traditional notations is given so that the student will experience no difficulty in reading other literature in the field.

KARL MENGER

CONTENTS

ACKNOWLEDGMENTS

In preparing the manuscript and reading the proofsheets the author enjoyed extensive help from Professor H.J. Curtis (University of Illinois, Chicago) who made many valuable suggestions and also prepared the Index. Sister Felice (Mount Mary College, Milwaukee) and Messrs. M.A. McKiernan, R.J. Mihalek, B. Schweizer, and D. Sachs (all at I.I.T.) read parts of the proofsheets, and all of them offered very helpful remarks. Excellent editorial assistance was provided by Ginn and Company. The author has also utilized suggestions made in former years by Messrs. J.M. Elkin and B.D. Fried and, especially, by Professor Alexander Wundheiler, as well as remarks contained in two letters by Professors W.V. Quine and N.E. Steenrod. To all of them and, especially, to his wife, Hilda, for her constant help, the author wishes to express his sincere gratitude.

TO THE INSTRUCTOR AND GENERAL READER*

What Is A Variable?

The conceptual and semantic clarification of calculus is centered on the analysis of the hitherto obscure general term "variable," which is resolved into an extensive spectrum of well-defined meanings.

The only clear (if one-sided) definition heretofore formulated goes back to Weierstrass who, in his celebrated lectures in the 1880's, defined it as a symbol that stands for any number or any element belonging to a certain class of numbers. Bertrand Russell, who at the turn of the century investigated the various aspects of variables probably more thoroughly than anyone before him, said: "Variable is perhaps the most distinctly mathematical of all notions; it is certainly also one of the most difficult to understand ... and in the present work [*The Principles of Mathematics*, 1903] a satisfactory theory as to its nature, in spite of much discussion, will hardly be found." In fifty years this situation has not been improved.

In this book a solution of the problem is attempted by distinguishing and making precise various equally important uses of the term "variable" in pure and applied calculus. Some of these variables differ from each other as profoundly as do trigonometric and geometric tangents. But whereas no one has, on account of a flimsy equivocation, confused tangents of angles and tangents of curves, this book seems to be the first to maintain clear distinctions between the following three concepts:

I. Variables according to Weierstrass, herein called *numerical variables*, as x and y in

(1) $\mathrm{x}^2 - 9\mathrm{y}^2 = (\mathrm{x} + 3\mathrm{y}) \cdot (\mathrm{x} - 3\mathrm{y})$ for any two numbers x and y.

Here, as throughout this book, the numerical variables are printed in roman type. Without any change of the meaning, x and y may be replaced by any two non-identical letters, e.g., by a and b or by y and x; that is to say, two numerical variables may be interchanged:

$$\mathrm{y}^2 - 9\mathrm{x}^2 = (\mathrm{y} + 3\mathrm{x}) \cdot (\mathrm{y} - 3\mathrm{x}) \text{ for any two numbers y and x}$$

is tantamount to (1). In calculus, numerical variables may be used or they may, as will be shown herein, be dispensed with.

II. Variables or variable quantities in the sense in which scientists use these terms; for instance, *t*, the time; *s*, the distance traveled (in chosen units); *x* and *y*, the abscissa and ordinate in a physical or postulational plane (relative to a chosen frame of reference); etc. These "variables" are defined and thoroughly discussed in Chapter VII under

*The beginner should start on page 1 or the one facing it.

the names of *consistent classes of quantities* and – reviving Newton's terminology – of *fluents*. They are herein consistently denoted by letters in italic type. Fluents cannot be dispensed with in formulas expressing scientific laws, such as Galileo's

(2) $$s = 16t^2.$$

Nor can they be interchanged:

$$2x + 3y = 5 \quad \text{and} \quad 2y + 3x = 5$$

are different lines. (If, on the other hand, in pure analytic geometry, the first of these two lines is defined as

the class of all pairs (x, y) of numbers such that 2x + 3y = 5,

where x and y are numerical variables, then

the class of all pairs (y, x) of numbers such that 2y + 3x = 5

is an equivalent definition.)

III. Variables in the sense of u and w in statements such as

(3) If $w = 16u^2$, then $\frac{dw}{du} = 32u$ for any two fluents, u and w.

These "variables" belong to a third type, first explicitly introduced by the author in 1952 (see Bibliography). They are herein referred to as *fluent variables*, since they partake in characteristics of numerical variables as well as of fluents. In (3), u and w may be replaced by any two elements of a certain class – but not by two numbers. If u were replaced by 3, and w by 144, then the antecedent $144 = 16 \cdot 3^2$ would be valid, and yet the consequent $\frac{d\,144}{d\,3} = 32 \cdot 3$, utterly nonsensical. What u and w in (3) may be replaced by are fluents, such as t and s regarding a motion, or x and y along a plane curve:

(3′) If $s = 16t^2$, then $\frac{ds}{dt} = 32t$;

(3″) If $y = 16x^2$, then $\frac{dy}{dx} = 32x$.

About the Notation

Obviously, the preceding clarifications have nothing to do with notation. But it may be pointed out that, in the literature, the equivocations of the term "variable" are accentuated by the use of the same letters (mainly, x and y in italic type): (1) as numerical variables; (2) to denote specific consistent classes of quantities or fluents, namely abscissa and ordinate; (3) as fluent variables; indeed, (3″) often stands for (3).

In formulas such as

$$(4)\qquad \mathbf{D}\, sin\ x = cos\ x \quad \text{and} \quad \int_0^{\pi/2} cos\ x\ dx = [sin\ x]_0^{\pi/2},$$

x can be simply shed; and the formulas may be replaced by

$$(4')\qquad \mathbf{D}\, sin = cos \quad \text{and} \quad \int_0^{\pi/2} cos = [sin]_0^{\pi/2}.$$

In fact, there is a trend toward actually shedding x after the letters f and g – symbols that stand for any function or any element of a certain class of functions and which, therefore, have been called *function variables*. In several more advanced books on analysis one can read

$$(5)\qquad \text{If } \mathbf{D}f = g, \text{ then } \int_a^b g = [f]_a^b.$$

Only the second formula in (5) contains numerical variables (namely, a and b) and g may be replaced by any function that is continuous between a and b.

Why then is x in (4) usually retained,and, even in more advanced books, reintroduced in applications of (5) to specific functions? The reason is one of the great curiosities in the history of modern analysis – comparable only to the lack of a symbol for the number zero in ancient arithmetic. The function that is perhaps more important than any other – the *identity function* assuming the value x for any x – lacks a traditional symbol. Therefore, in contrast to (4), in

$$(6)\qquad \mathbf{D}\, log\ x = 1/x \text{ for } x > 0, \quad \mathbf{D}x^3 = 3x^2, \quad \mathbf{D}(\tfrac{1}{2}x^2) = x,$$

x cannot be shed. One obviously would not write

$$\mathbf{D}\, log \quad = 1/\quad, \qquad \mathbf{D}\ {}^3 = 3\ {}^2, \quad \mathbf{D}(\tfrac{1}{2}\ {}^2) = \quad,$$

and this is why, for the sake of uniformity, x is also retained in (4). Traditionally, the identity function is referred to by its value for x as "the function x" – incidentally, a fourth current meaning of the italic letter x.

There is, however, another way to preserve uniformity, namely, by rescuing the identity function from anonymity. If one denotes it by j (so that $j\,x = x$ and $j^n x = x^n$ for any x), then (6) reads

$$(6^*)\qquad \mathbf{D}\, log\ x = j^{-1}x \text{ for } x > 0, \quad \mathbf{D}j^3x = 3j^2x, \quad \mathbf{D}(\tfrac{1}{2}j^2)x = j\,x;$$

and in (6*), x can be shed as in (4):

$$(6')\qquad \mathbf{D}\, log = j^{-1} \text{ (on the class P of all positive numbers)},$$

$$\mathbf{D}j^3 = 3j^2, \quad \mathbf{D}(\tfrac{1}{2}j^2) = j.$$

The formulas thus obtained (or, if one pleases, the results of calculus "in this notation") connect the functions themselves rather than their values and are of algebraic beauty – as it were, *streamlined.* Yet it is a matter of taste which one prefers: (4′) and (6′) or (4) and (6). But it is a fact (proved at the end of Chapter IV) and not a mere matter of taste that only if the identity function is granted the same status that *log* and *cos* have enjoyed for centuries can calculus be (to use another term that is in vogue) completely *automatized.* Automation is impossible "in the classical notation." In the automatized calculus one obtains specific results from general theorems by replacing function variables by the designation of specific functions, and numerical variables by the designations of specific numbers. For instance, from (5) one obtains:

$$\text{If } \mathbf{D}\, log = j^{-1}, \text{ then } \int_a^b j^{-1} = [log]_a^b.$$

Since, according to (6′), actually $\mathbf{D}\, log = j^{-1}$ on the class P, one concludes that $\int_a^b j^{-1} = [log]_a^b$, where the numerical variables a and b may be replaced by any two numbers in the class P. For instance,

$$\int_1^2 j^{-1} = [log]_1^2 \ (= log\ 2 - log\ 1 = log\ 2).$$

A symbol for the identity function was introduced, as early as 1924, by the Russian logician Schönfinkel and has been adopted by H. B. Curry in his work on combinatoric logic. But whereas in topology extensive use has been made of a symbol for the identity mapping, that for the identity function had not, before the publication of the booklet "Algebra of Analysis" (see Bibliography), found its way into mathematical analysis – in particular not even into Curry's own mathematical papers.

To the specialist, perhaps more interesting is the automation of the theory of what, traditionally, is called the theory of functions of two variables. Most important are the functions assuming for every pair (x, y) the values x and y. Traditionally, they are referred to by their values as "the function x" – a fifth meaning of x – and "the function *y*." In an automatized calculus, they cannot remain anonymous. In Chapter XI they are called the *selector functions* and are denoted by *I* and *J* so that

$$I(x, y) = x \text{ and } J(x, y) = y \text{ for any x and y.}$$

Chapter XI contains the theories of partial derivatives and partial integrals. Calculus thus developed as an algebra of functions synthesizes the spirit of the oldest logic and that of the most modern machine age.

Extensions of the ideas here expounded to *complex functions*, to the theory of *operators*, to *random variables*, etc. will be given in other publications.

Applications to Science

Perhaps the most important automation is that of the applications of calculus to science. An example is the application of derivatives of functions to rates of change of one fluent with respect to another. The scheme of these applications is the following statement: If w is connected with u by the function f, then the rate of change $\frac{\mathbf{d}\,w}{\mathbf{d}\,u}$ is connected with u by the derivative $\mathbf{D}\,f$; or

(7) $$\text{If } w = f(u), \text{ then } \frac{\mathbf{d}\,w}{\mathbf{d}\,u} = \mathbf{D}\,f(u).$$

Traditionally, "derivative" and "rate of change" are considered as synonyms. But the former operator associates a function with every function of a certain kind; the latter, a fluent with certain pairs of fluents. If, in (7), the function variable f is replaced by *sin*, the result is

$$\text{If } w = \mathit{sin}\ u, \text{ then } \frac{\mathbf{d}\,w}{\mathbf{d}\,u} = \mathbf{D}\,\mathit{sin}\ u.$$

Hence, by virtue of $\mathbf{D}\,\mathit{sin} = \mathit{cos}$,

(8) $$\text{If } w = \mathit{sin}\ u, \text{ then } \frac{\mathbf{d}\,w}{\mathbf{d}\,u} = \mathit{cos}\ u.$$

(Incidentally, (3) results from (7) by virtue of $\mathbf{D}(16\,j^2) = 32\,j$.) Replacing the fluent variables u and w by the time and the position of a linear oscillator and calling v its velocity one obtains

(8′) $$\text{If } s = \mathit{sin}\ t, \text{ then } \frac{\mathbf{d}\,s}{\mathbf{d}\,t} = \mathit{cos}\ t \text{ or } v = \mathit{cos}\ t.$$

If in physics one ascertains the validity of $s = \mathit{sin}\ t$ for an oscillator, then he can conclude that for this oscillator $v = \mathit{cos}\ t$.

The contributions of mathematics ($\mathbf{D}\,\mathit{sin} = \mathit{cos}$) and of physics ($v = \frac{\mathbf{d}\,s}{\mathbf{d}\,t}$ and $s = \mathit{sin}\ t$) are completely separated, whereby the roles of these two sources of insight into nature are greatly clarified. (7) – it may be repeated – is the general scheme for applying the calculus of derivatives to science. There is an analogous and, in fact, more important scheme for the application of integral calculus to consistent classes of quantities. In Chapter XI, these ideas are extended to partial rates of change and partial derivatives, and are applied to thermodynamics thereby preparing the basic mathematical tools needed in physical chemistry.

On the basis of pure observations alone, all that physics can claim about a falling object is that s equals $16t^2$ *within certain limits* and that v equals $32t$ *within certain limits*. What is the logical connection between these two statements? In Chapter VII it will be shown that, if $\sim$ denotes "is equal within certain limits," then

$$v \sim 32t \text{ (in presence of proper initial conditions) implies } s \sim 16t^2;$$

but that $s \sim 16t^2$ does not, conversely, imply $v \sim 32t$; in fact, $s \sim 16t^2$ *does not permit any inference whatever concerning* v. In the realm of strictly observational statements, inferences "by integration" are highly significant, whereas inferences "by differentiation," while often extremely important heuristically, are not of logical character.

Calculus is a theory of pure functions just as arithmetic is a theory of pure numbers. There are analogies between the two in the realms of applications as well as of pure theory, both on the specific and on the general levels.

	CALCULUS	LEVEL	ARITHMETIC
APPLIED	If $s = \sin t$, then $\frac{ds}{dt} = \cos t$	Specific	If 1 ft. = 12 in., then 3 ft. = 36 in.
	$\frac{d \sin u}{d u} = \cos u$ for any fluent u	General	$3 \cdot 12 a = 36a$ for any object a
PURE	$\mathbf{D} \sin = \cos$	Specific	$3 \cdot 12 = 36$
	$\mathbf{D} f^2 = 2f \cdot \mathbf{D} f$ for any function f	General	$(a + 1)^2 = a^2 + 2a + 1$ for any number a.

But there is a difference in attitude. No arithmetician stoops to references to objects, whereas some analysts refuse to raise their eyes above the level of fluent variables. Yet the transition to the pure level does not in any way jeopardize the applicability of arithmetic to feet and inches or of calculus to time and position, nor does the shedding of numerical variables in calculus. The formulas

$$\mathbf{D} \sin x = \cos x \text{ for any } x \text{ and } \frac{d \sin u}{d u} = \cos u \text{ for any } u,$$

which are often considered as having the same meaning, really belong to altogether different realms (one using a numerical variable, the other a fluent variable), but both follow from $\mathbf{D} \sin = \cos$, – the first by evaluation, the second by virtue of (7). Nothing is lost by writing simply $\mathbf{D} \sin = \cos$ or $\sec = 1/\cos$. In fact, as will be shown in Chapter VII, it is the latter formulas (and not those obtained by inserting numerical variables) that are the perfect analogue of the traditional formulas $\frac{d s}{d t} = \cos t$ and $v = 1/p$ in physics.

Students' Questions

Within the traditional conceptual frame of calculus it is exceedingly difficult to answer questions pertaining to some classical definitions, statements, and formulas such as the following:

The function x is the function assuming for any x the value x; or the class of all pairs (x, x).

The line $2x + 3y = 5$ is the class of all pairs (x, y) such that $2x + 3y = 5$, which is the same as the class of all pairs (y, x) such that $2y + 3x = 5$, but quite different from the line $2y + 3x = 5$.

In algebra, $\frac{x^2 - 1}{x - 1} = x + 1$, whereas in calculus, the functions $\frac{x^2 - 1}{x - 1}$ and $x + 1$ are not identical.

For falling objects, $s = s(t) = s(v)$.

The expressions

$$\int_a^b f(x)\,dx \quad \text{and} \quad \int_a^b f(t)\,dt$$

are not only numerically equal but have precisely the same meaning. On the other hand, if p denotes force, then

$$\int_a^b p\,ds \quad \text{and} \quad \int_a^b p\,dt,$$

and, if one writes $p = p(s) = p(t)$, then even

$$\int_a^b p(s)\,ds \quad \text{and} \quad \int_a^b p(t)\,dt$$

have not only different meanings (work and impulse) but are, in general, numerically different.

Questions concerning such statements naturally arise in the minds of the very best students. Some of them may be not only disappointed but discouraged in the pursuit of mathematical studies if they do not receive clear answers. Following the approach described in this book, teachers can readily answer these and similar questions.

TO THE STUDENT WITH PREVIOUS TRAINING IN CALCULUS ALONG TRADITIONAL LINES

Note. The beginner should omit a reading of this page and start directly with page 1.

1. Instead of a Cartesian cross of axes to which the reader may be accustomed, in Fig. 1 on page 1 he finds (since the Y-axis plays no role in calculus) only a horizontal line with a scale. Above it, there is a curve that intersects every vertical line either in one point or not at all. The curve is (as herein are all such curves) denoted by a symbol in italic type, f; the basic horizontal line, by italic O. Analytic geometry, where the curve is described by $y = f(x)$, and the line by $y = 0$, will be discussed in Chapter VII.

2. Also on page 1, the symbols $\int_2^{4.5} f(x)\,dx$ and $\int_a^b f(x)\,dx$ to which the reader may be accustomed are abbreviated to $\int_2^{4.5} f$ and $\int_a^b f$. These short symbols render, term by term, the words "the area (indicated by $\int$) from 2 to 4.5 under the curve f" and "the area from a to b under f." Since these unambiguous verbal descriptions are perfectly intelligible, the reader will also get used to the parallel abbreviated symbols whose advantages (besides brevity) he will realize as he progresses in the study of this book.

3. The slope of the curve f at 7 is denoted by **D** f 7, which again follows, term by term, the verbal description (**D** indicating slope). The reader may be accustomed to $\left(\frac{df}{dx}\right)_{x=7}$ or $f'(7)$. Applications to rates of change will be discussed in Chapter VII.

4. On page 4, there is a discussion of the line described in analytic geometry by $y = x$. This extremely important line is herein given a name and a symbol, just as two important curves are called the logarithmic curve and the cosine curve or the curves *log* and *cos*. The line having the altitude x above any point x on the basic horizontal line is called the *identity line* or the line j.

CHAPTER I

THE TWO BASIC PROBLEMS OF CALCULUS AND THEIR SOLUTIONS FOR STRAIGHT LINES

In Fig. 1 there appears a horizontal line *O* and a simple curve *f*. A curve is said to be *simple* if it intersects no vertical line (i.e., no line perpendicular to *O*) in more than one point. The letter S is not a simple curve – a fact illustrated by any $-sign; neither is a circle. The upper half and the lower half of a circle, taken separately, are simple curves.

On the line *O*, a point 0 and a point 1 have been chosen. The segment between them is called a linear unit. The point which is 2 (or any number, a) units to the right of 0 is called the point 2 (the point a). Vertical distances are measured in the linear unit between 0 and 1 turned through a right angle; areas, in square units such as that marked above the segment from 0 to 1.

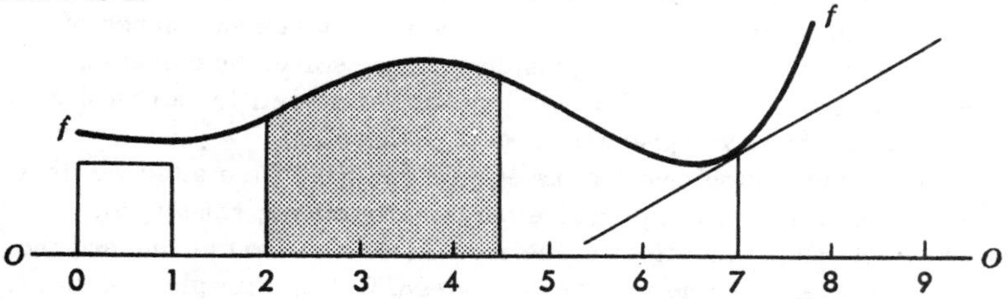

Fig. 1

A domain has been shaded. It is bounded by a portion of the curve *f*, the segment from 2 to 4.5 on *O*, and two lateral vertical segments. WHAT IS THE AREA OF THIS DOMAIN? In other words, HOW MANY SQUARE UNITS ARE CONTAINED IN THIS DOMAIN? This question is a typical example of the first basic problem to which this book is devoted. The number in question will be denoted by the symbol $\int_2^{4.5} f$ (read: integral of *f* from 2 to 4.5). If a and b are any two points on *O* belonging to the projection of *f*, then

$$\int_a^b f \quad \text{(read: the } \textit{integral} \text{ of } f \text{ from a to b)}$$

denotes the area of the domain above *O* and below *f* from a to b – briefly, *the area under f from* a *to* b.

In Fig. 1, at the point of *f* above 7, the tangent to that curve has been drawn. HOW STEEP IS THAT TANGENT? In other words, HOW MANY

UNITS DOES THE TANGENT RISE VERTICALLY FOR EACH HORIZONTAL UNIT? This question is an example of the second basic problem to be treated in this book. The number in question is called the derivative of f at 7 and will be denoted by $\mathbf{D}f7$. If x is any point on the projection of f on O, then

$$\mathbf{D}f\,x \quad \text{(read: the } \textit{derivative} \text{ of } f \text{ at } x\text{)}$$

denotes the steepness or slope of the tangent to f above the point x – briefly, *the slope of f at* x.

Approximate solutions of area problems are as old as geometry. For a few special curves, Archimedes (about 250 B.C.) found precise areas and slopes, but in a cumbersome way. The systematic study of the two problems began about 1700, when Newton in England and Leibniz in Germany discovered an exact method applicable to many curves. Their ingenious theory has become known as CALCULUS.

In the attempt to understand and control Nature man is faced with innumerable scientific questions and engineering problems that can be reduced to the determination of areas under simple curves and slopes of tangents. Since these geometric problems can be solved by calculus, calculus has become one of the most powerful of scientific tools and a major factor in the development of modern civilization.

Before the basic problems for curves like f in Fig. 1 are attacked, they will be solved for the most primitive curves imaginable, namely, for straight lines; more specifically, the area under a horizontal line and the slope of a nonvertical line will be determined. In these simple situations, first numerical and then graphical solutions of the problems are presented; finally the solution is given that is based on the idea of Newton and Leibniz.

1. NUMERICAL SOLUTIONS

a. Area. In Fig. 2(a), the horizontal line is drawn that is .5 units above the line O; it is denoted by $.5$; in Fig. 2(b), the horizontal line $-.5$ is .5 units below O. More generally, if c is any number, the horizontal line that is c units above the line O will be called *the line* c – briefly, c. Clearly, O is the line at the altitude 0. ($.5$, $-.5$, c for lines are italic type.)

The domain below the line $.5$ and above O from 2 to 5 is shaded. The domain is a rectangle of height .5 and of base 5 – 2. Thus $\int_2^5 .5$ (which denotes the area of this domain) is $.5\,(5-2) = 1.5$ square units. More generally,

If c is the horizontal line of the altitude c, then $\int_a^b c = c \cdot (b - a)$, for any a, b, and c. For instance,

$$\int_1^3 2.5 = 2.5 \cdot (3 - 1) = 5; \quad \int_1^3 O = 0; \quad \int_2^5 (-.5) = (-.5) \cdot (5 - 2) = -1.5.$$

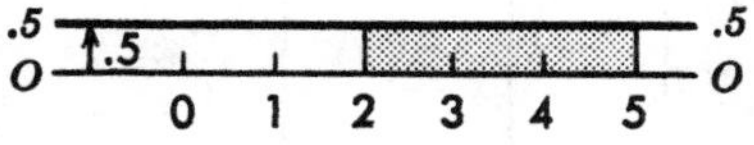

Fig. 2a

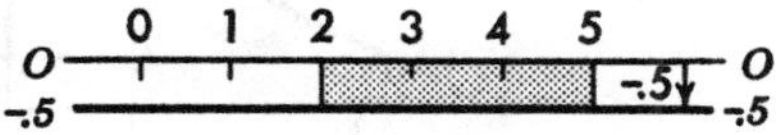

Fig. 2b

Fig. 2(b) illustrates the last equality. The shaded rectangle of 1.5 square units below O and above the line $-.5$ is said to be -1.5 square units above O and below the line $-.5$.

That the area of this rectangle is negative cannot be proved (nor, of course, disproved) just as it cannot be proved (or disproved) that its altitude is negative. Both statements express conventions concerning the use of the words area and altitude. The conventions are universally adopted because they are useful in distinguishing between the altitudes of the shaded rectangles in Fig. 2(a) and (b), and between the areas of these two rectangles. Without these conventions both rectangles would have the altitude .5 and the area 1.5.

b. Slope. In Fig. 3, l denotes the straight line that has the altitude 1 above the point 2, and the altitude 2.5 above the point 5 on O. These facts are conveniently expressed in the formulas

$$l\,2 = 1 \quad \text{and} \quad l\,5 = 2.5.$$

More generally, the altitude of l above any point a on O – briefly, the *altitude of l at* a – will be denoted by l a.

The difference $l\,5 - l\,2$ is called the *rise* of l from 2 to 5, and is denoted by $[l]_2^5$. More generally, if a and b are any two numbers, then l a and l b are numbers and $[l]_a^b = l\,b - l\,a$.

By the slope of l at 2 (which, according to what was said on p. 2, will be denoted by $\mathbf{D}\,l\,2$) is meant the ratio of the rise of l from 2 to any other point, say 5, to the distance on O from 2 to that point; in a formula,

$$\mathbf{D}\,l\,2 = \frac{[l]_2^5}{5-2}.$$

Clearly, $\frac{[l]_2^5}{5-2} = \frac{1.5}{3} = .5$. Hence $\mathbf{D}\,l\,2 = .5$.

From similar triangles in Fig. 3 it is clear that the same ratio is obtained if 5 is replaced by 4 or 6; $\frac{[l]_2^4}{4-2} = \frac{[l]_2^6}{6-2} = .5$.

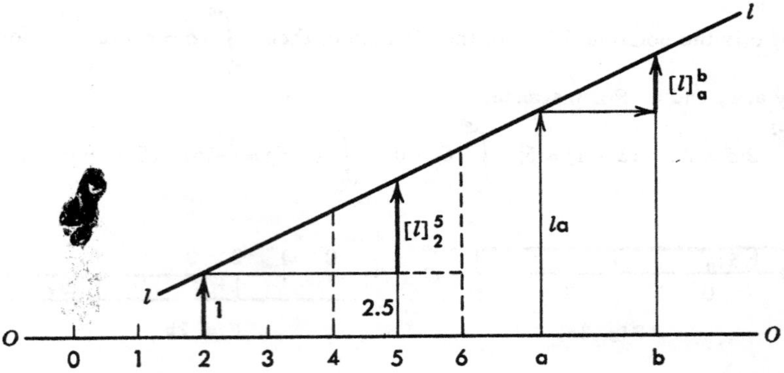

Fig. 3

Moreover, it is clear that, for any two different points a and b, $\dfrac{[l]_a^b}{b-a} = \dfrac{[l]_2^b}{b-2} = \dfrac{[l]_2^5}{5-2} = .5$. Hence, for any a on O and for any $b \neq a$, $\mathbf{D}\, l\, a = \dfrac{[l]_a^b}{b-a} = .5$. Similarly one sees:

If m is any nonvertical straight line and a is any point on O, then

$$\mathbf{D}\, m\, a = \frac{[m]_a^b}{b-a} = \frac{m\,b - m\,a}{b-a}, \quad \text{for any } b \neq a, \text{ and}$$

$$\mathbf{D}\, m\, x = \mathbf{D}\, m\, a, \quad \text{for any point x on } O.$$

For instance, let j denote (*as it will throughout this book*) the line having the altitude 0 at 0, the altitude 2 at 2, the altitude −3 at −3 – more generally, the line which, for any number x has the altitude x above the point x on O; in a formula, j is the line for which

$$j\,x = x \quad \text{for any x.}$$

Then

$$\mathbf{D}\, j\, a = \frac{[j]_a^b}{b-a} = \frac{j\,b - j\,a}{b-a} = \frac{b-a}{b-a} = 1 \quad \text{for any a and any } b \neq a.$$

Next consider the line having, for any number x, the altitude $\frac{1}{2}$x above the point x. This line (which, incidentally, is the line l in Fig. 3) may be denoted by $\frac{1}{2}\, j$. Its slope at any a is $\mathbf{D}\,(\frac{1}{2}\, j)\, a = \dfrac{[\frac{1}{2}\, j]_a^b}{b-a} = \dfrac{\frac{1}{2}\, b - \frac{1}{2}\, a}{b-a} = \dfrac{1}{2}$.

Clearly, $\mathbf{D}\,(\frac{1}{2}\, j)\, x = \dfrac{1}{2}$ at any x.

Then consider the line whose altitude at any point x is $\frac{1}{2}$x + 1, that is, the sum of the altitudes of the lines $\frac{1}{2}j$ and *1*, wherefore that line may be conveniently denoted by $\frac{1}{2}j + 1$. Its slope at a is

$$\mathbf{D}(\tfrac{1}{2}j + 1)\mathrm{a} = \frac{[\frac{1}{2}j + 1]_{\mathrm{a}}^{\mathrm{b}}}{\mathrm{b} - \mathrm{a}} = \frac{(\frac{1}{2}\mathrm{b} + 1) - (\frac{1}{2}\mathrm{a} + 1)}{\mathrm{b} - \mathrm{a}} = \frac{1}{2}.$$

Similarly, $-3j + 2$ would denote the line having at any x the altitude -3x + 2 (the sum of the altitudes of the lines $-3j$ and *2*). Its slope at any x is $\mathbf{D}(-3j + 2)\mathrm{x} = -3$.

It is left to the reader to verify that, if *l* is any nonvertical straight line, there are two numbers c and c′ such that, for any number x, the altitude of *l* above the point x on *O* is cx + c′. If this line is denoted by $cj + c'$, then

$$\mathbf{D}(cj + c')\mathrm{x} = \mathrm{c} \quad \text{for any x on } O.$$

EXERCISES

INTRODUCTORY NOTE. On the preceding pages, numerals and letters in roman type (0, 1, 7, a, c, x, etc.) designate numbers or points on the line *O*; whereas numerals and letters in *italic* type designate simple curves (the horizontal lines *O*, *.5*, *c*, etc., the nonvertical lines *l*, *j*, etc., the curve *f*). *This typographical distinction will be maintained throughout this book.* If in longhand (on paper or on a blackboard) the distinction between roman and italic type should be difficult, one may denote

numbers and points on O by *0, 1, 7, a, c, x*;

simple curves by $\underline{0}$, $\underline{.5}$, $\underline{c}$, $\underline{l}$, $\underline{j}$, $\underline{f}$ (underlined).

If one avoids j and f as designations of numbers, then he may omit the underlining in the designation of the line *j* and the curve *f*. But one must not omit it in the symbols for the horizontal lines *O*, *.5*, and *c*, since 0, .5, and c designate numbers.

The letters x and y will be avoided as designations of altitudes of horizontal lines. The reason is that lines with the altitudes x and y would have to be denoted by *x* and *y*, whereas we wish to reserve the letters *x* and *y* (in italics) for concepts of analytic geometry discussed in Chapter VII.

Finally, it should be noted that, in this book, the symbols

$$\mathbf{D}3, \quad \mathbf{D}\mathrm{c}, \quad \int_2^4 3, \quad \int_{\mathrm{a}}^{\mathrm{b}} \mathrm{c}$$

remain undefined, and therefore meaningless. A number has neither a slope, nor an area, nor a weight, nor a color, ...

1. On cross-section paper, draw figures illustrating

(a) $\int_2^5 .5 = \int_5^8 .5 = \int_{-1}^{2} .5 = \int_{-4}^{-1} .5 = 1.5$;

(b) $\int_0^3 (-4) = \int_{-2}^{1} (-4) = \int_{-4}^{-1} (-4) = -12$;

(c) $\int_1^3 O = 0$ and $\int_1^1 3 = 0$.

2. Show that $\int_a^b c = \int_{a+k}^{b+k} c$ and $\int_a^b c + \int_b^{b'} c = \int_a^{b'} c$.

★3. Apply the formula $\int_a^b c = c(b-a)$ to cases where b is less than a (that is, the point b on O is to the left of the point a). For instance, what are $\int_5^2 .5$, $\int_3^1 2.5$, $\int_3^1 O$, $\int_5^2 (-.5)$? When is $\int_a^b c$ positive and when negative?

4. Determine the numbers c_1 and c_0 in such a way that $c_1 j + c_0$ is the line having

(a) the altitude 1 at -1, and the altitude 5 at 1;

(b) the altitudes a′ at a and b′ at b, where a, b, a′, b′ are any four given numbers ($a \neq b$);

(c) the slope 3 and the altitude 2 at 1;

(d) the altitude 4 at both 1 and 3;

(e) the altitude 0 at a, and the slope c;

(f) the altitude c at a, and the rise r from a to b.

5. On cross-section paper, draw figures illustrating $\mathbf{D}(2j+3)a = 2$ for (a) $a = 1.5$ using $b = 3$; (b) $a = -3$ using $b = -1$; (c) $a = -1$ using $b = 2$. Then illustrate $\mathbf{D}(2j+3)(-1) = 2$ using any point b.

6. Show that $\mathbf{D}(-2j+3)2 = \mathbf{D}(-2j+4)2 = \mathbf{D}(-2j-1)2$ and illustrate this equality on cross-section paper.

2. GRAPHICAL SOLUTIONS

a. Slope and Slope Line. The slope of a nonvertical line can be determined graphically. No knowledge of arithmetical division (used in the numerical solution of the slope problem) is presupposed. However, parallel lines have to be drawn (an operation not needed in the numerical solution).

In Fig. 4, the slope of the line l at a is determined as follows: Through the point a on O, the segment m parallel to l is drawn, whose projection on O has unit length, that is, extends from a to a + 1. The triangles in Fig. 4 are similar. In the large triangle, the height is $[l]_a^b$, and the base is b − a. In the small triangle, the height is m(a + 1) (that is, the altitude of m at a + 1), and the base is (a + 1) − a = 1. Hence the proportion

$$\frac{[l]_a^b}{b - a} = \frac{m(a + 1)}{1}.$$

The quotient on the right side is equal to m(a + 1). The quotient on the left side is equal to the slope **D** l a. Consequently

$$\mathbf{D}\, l\, a = m(a + 1).$$

Hence the following

Graphical Solution of the Slope Problem for Nonvertical Lines. *To find the slope of l at* a, *construct m as indicated, and measure the altitude of m at* a + 1.

In Fig. 4 there appears also the horizontal line of altitude **D** l a, which passes through the terminal point of m. This line will be called

the *slope line* of l or the *derivative* of l – briefly, **D** l.

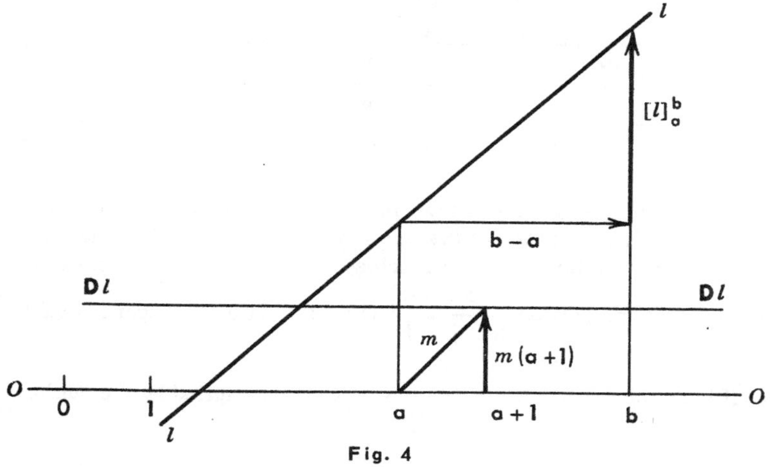

Fig. 4

The horizontal line **D** l has the same altitude at any point of O, just as the line l has the same slope at any point. The slope line of the horizontal line *3* is the line O; in a formula, **D** *3* = O. More generally, **D** c = O for any horizontal line c.

Since, if l is a line, $l3$ and la (for any a) are numbers, $D(l3)$ and $D(la)$ are undefined symbols. Therefore $Dl3$ cannot mean anything but the slope of l at 3 or the altitude, $(Dl)3$, at 3 of the slope line Dl. This is why, from the outset, symbols such as $Dl2, D(\frac{1}{2}j)x$, etc. have been used and could be used without any ambiguity.

b. Area and Area Lines. $\int_a^b c$, the area of the shaded domain under the horizontal line c from a to b in Fig. 5, can also be determined graphically. Draw the line k joining the point a on O to the point on the line c above the point a + 1 on O. The right triangles in Fig. 5 whose hypotenuses lie on the line k are similar. In the small triangle, the height is c linear units,

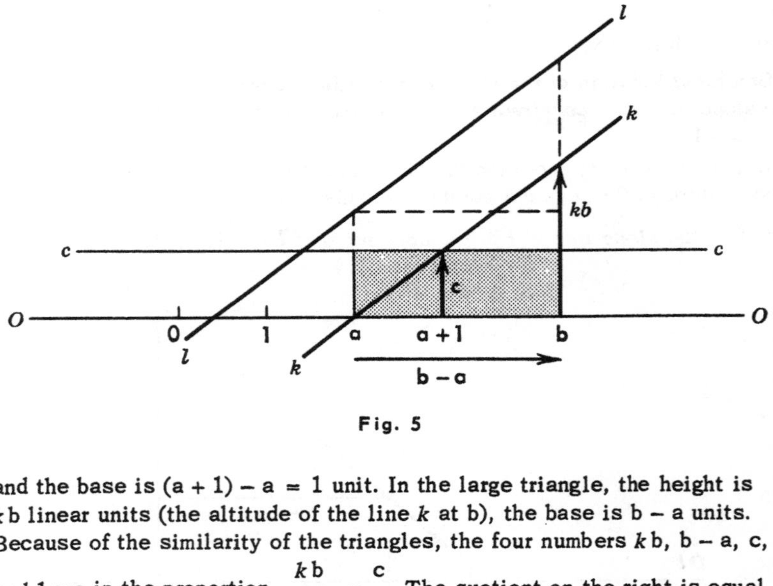

Fig. 5

and the base is $(a + 1) - a = 1$ unit. In the large triangle, the height is kb linear units (the altitude of the line k at b), the base is $b - a$ units. Because of the similarity of the triangles, the four numbers kb, $b - a$, c, and 1 are in the proportion $\frac{kb}{b-a} = \frac{c}{1}$. The quotient on the right is equal to c. If both sides of the equality $\frac{kb}{b-a} = c$ are multiplied by the number $b - a$, the result is $kb = c \cdot (b - a)$. Since $c \cdot (b - a) = \int_a^b c$, it follows that

$$\int_a^b c = kb;$$

that is to say, the number of square units in the shaded rectangle is equal to the number of linear units in the altitude of k above the point b. Thus, to find $\int_a^b c$, construct the line k as indicated and measure its altitude at b. The situation is particularly clear on cross-section paper. E.g., in

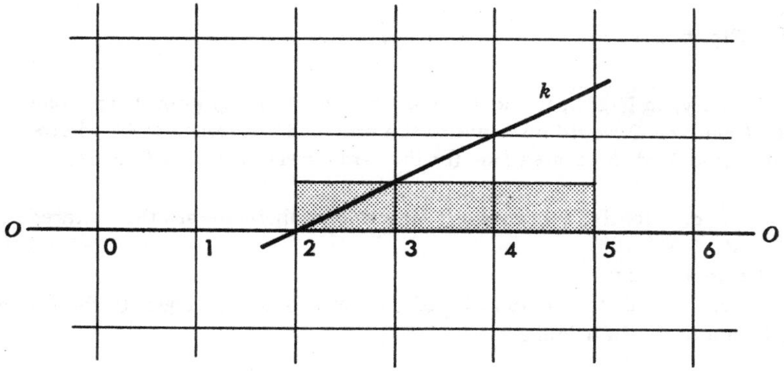

Fig. 6

Fig. 6 the shaded rectangle under the line $.5$ from 2 to 5 obviously consists of three rectangles, each of one half of a square unit. Thus $\int_2^5 .5 = 1.5$. The construction yields a line k of the altitude 1.5 above the point 5 on O. Thus, in the vertical segment from that point to the line k there are as many linear units as there are square units in the shaded rectangle. In other words, the number of linear units in that segment is equal to the product of the numbers of linear units in the two perpendicular sides of the shaded rectangle.

The age in years of a father, say 35, may be equal to the product of the ages in years, 7 and 5, of his children, and there is nothing paradoxical about this fact. Of course, it must not be expressed by saying that the father is the product of the two children. Nor should one say that the vertical segment from 5 to k is the product of two perpendicular sides of the rectangle.

The line k in Fig. 6 has not only the altitude $\int_2^5 .5$ at 5, but also the altitudes $\int_2^4 .5$ and $\int_2^3 .5$ at 4 and 3 respectively. In fact, for any number b between 0 and 5, the altitude of k at b equals the area under $.5$ from

2 to b. (Even at 2, its altitude is 0, as is $\int_2^2 .5$, the area of a rectangle whose base is 0.) For this reason, the line k in Fig. 6 will be called *the 2-area line* of the horizontal line *.5* or *the 2-integral* of *.5*, denoted by $\int_2 .5$; in a formula,

in Fig. 6, $$k = \int_2 .5.$$

Similarly, in Fig. 5, the altitude of the line k at any point b′ that one might mark on O would be equal to the area under c from a to b′, wherefore k is called the a-area line (or the a-integral) of c; in a formula, $k = \int_a c$. Clearly, k is the *only* line whose altitudes are thus related to the areas under c.

Since $k\text{a} = 0$, the altitude $k\text{b}'$ of k at any point b′ is equal to the rise of k from a to b′. In a formula,

$$[k]_a^{b'} = \int_a^{b'} c \text{ and, more generally, } [k]_x^{x'} = \int_x^{x'} c \text{ for any x and x}'.$$

But the last equality holds also for some lines other than k, for instance for the line l (parallel to k) in Fig. 5, and, in fact, for any line that is parallel to k. Indeed any such line has the same rise from x to x′ as has k. Any such line is therefore called *an area line* or *an integral* of c. The a-area line of c is the one that has the altitude 0 at a (or crosses the line O at a). The line l in Fig. 5 is the .5-area line of c or $\int_{.5} c$. For any number h, there is an area line of c having the altitude h at a.

The preceding results yield the following rule:

Graphical Solution of the Area Problem for Horizontal Lines. *To find the area under the horizontal line* c *from* a *to* b, *construct an area line as indicated in Fig. 5 (either k or any line parallel to k) and measure its rise from* a *to* b.

The reader may wonder why, besides k, area lines parallel to k are even considered. Those lines will play an important role in Chapter II.

EXERCISES

7. Which of the following are numbers and which are lines?

$$\int_1 3, \quad \mathbf{D}c2, \quad \int_1^2 5, \quad \mathbf{D}j3, \quad \mathbf{D}3, \quad \sqrt[3]{5}, \quad \mathbf{D}(2j), \quad 4, \quad [2j]_3^5 .$$

8. On cross-section paper construct the slope and the slope line of l, if:

(a) $l1 = 1.5$ and $l1.5 = 2.75$ (b) $l4 = 1$ and $l6.5 = -2.75$

(c) $l1 = -2$ and $l1.5 = -.75$ (d) $l1 = -2$ and $l3 = .5$

(e) $l2 = 1$ and $l3.5 = 1$ (f) $l = 2j + 1$

(g) $l = -3j + 2$ (h) $l = .5$.

Check the results by the numerical method. Then draw nonvertical lines and determine their slopes graphically without measuring the altitudes of the lines anywhere. Only one mensuration is indispensable in each case: that of the altitude of the slope line.

9. What can be said about the slope line of a line which *(a)* has the rise 0 from a to b? *(b)* has the rise r from a to a + 1? *(c)* has the rise 1 from a to b? Show that a straight line l is a simple curve if and only if l is not vertical.

10. For what lines l is *(a)* $\mathbf{D}l = l$; *(b)* $\mathbf{D}l$ parallel to l? *(c)* Prove that if $l = cj + c'$ (see p. 5), then $c = \mathbf{D}l0$ and $c' = l0$.

11. On cross-section paper construct

(a) $\int_2^{3.5} 2$ (Count the number of unit squares or halves of unit squares under the line 2 from 2 to 3.5, and measure the rise of the area line from 2 to 3.5.)

(b) $\int_2^{2+\sqrt{2}} \sqrt{2}, \quad \int_{1-\sqrt{2}}^{1} \sqrt{2}, \quad \int_{-2-\sqrt{2}}^{-2} \sqrt{2}$. Here, $\sqrt{2}$ denotes the positive square root of 2. (In these cases, the rectangles cannot be broken up into fractions of unit squares as the rectangles in Fig. 6. But the results of the constructions can be checked by the numerical method.)

(c) The area lines $\int_{\sqrt{2}} \sqrt{2}$ and $\int_{1-\sqrt{2}} \sqrt{2}$.

(d) The areas under portions of horizontal lines without measuring the

altitudes of these lines or the lengths of the portions. Only one mensuration is indispensable in each case: that of the rise of an area line.

(e) $\int_1^3 O$ and several area lines of O. How are the lines $\int_1 O$ and $\int_2 O$ related? Is there an area line of O having the altitude 3 at 2?

12. (a) Prove $\int_a \mathrm{c} = \mathrm{c}(j - a)$.

(b) Show that, if l is any area line of the horizontal line c, there is a number b such that $l = \mathrm{c}j + b$.

(c) If $\mathrm{c} \neq 0$, where has the area line of c whose altitude at a is h the altitude 0?

(d) Find the particular area line of $.5$ having the altitude 3 at 2; and those of -2 having the altitudes $-.5$ at 1, and 3 at 1.

(e) Show that $\int_0^a 3\mathrm{c} = 3\int_0^a \mathrm{c}$, for every horizontal line c and every point a.

★13. Construct an area line l of the horizontal line 1.5 and determine its rise from 2 to 4.5.

(a) Without changing the point 0 on O, replace the original unit of length by a new unit half as long; that is to say, choose the former point .5 as 1. The horizontal line 1.5 now has the altitude 3; the former points 2 and 4.5 are now 4 and 9. Using the new unit, construct an area line l' of the horizontal line 3. How is the rise of l' from 4 to 9 in the new linear unit related to the rise of l from 2 to 4.5 in the old? How is the area in the new square unit related to the area in the old?

(b) Without changing the point 0, replace the original linear unit by one twice as long and repeat the constructions.

(c) How does, in general, a change of the linear unit affect (1) the rise of an area line of a given horizontal line between two points? (2) the area of the domain under the horizontal line between the two points?

★14. Let l be the line for which $l\,2 = 1$ and $l\,2.5 = 4$. Construct the slope line $\mathbf{D}\,l$. Then change the unit as in the preceding exercise and construct the slope line for the same line l. What is, in general, the effect of a change of the unit on the slope and the slope line of a nonvertical line?

★15. To determine the area under the horizontal line 400 from 1 to 3 graphically, one will want to measure vertical segments in a shorter linear unit than horizontal segments. Areas, then, are measured in unit rectangles, whose horizontal and vertical sides have the length of one respective unit. Choose a horizontal unit of 1" and a vertical unit of .01", construct an area line of the line *400*, and show that

(*a*) the area $\int_1^3 400$ is 400(3 − 1) unit rectangles;

(*b*) the rise of the area line from 1 to 3 is 800 vertical units.

Then construct the line 200 *j* whose rise from 3 to 5 is 400. Construct its slope line and show that the altitude of **D** (200 *j*) is 200 vertical units.

★16. Construct graphically $\int_5^2 .5$, $\int_3^1 0$, $\int_5^2 (-.5)$, and check the results by the numerical method.

★17. What is the meaning of those of the following symbols that have been defined? (The others will remain undefined throughout this book.) $\mathbf{D}(2j)$, $\mathbf{D}(\frac{1}{2}j)3$, $\mathbf{D}\,2$, $\mathbf{D}\,2$, $\mathbf{D}\,j2$, $\mathbf{D}(j2)$, $\mathbf{D}(\mathbf{D}j)$, $\mathbf{D}(\mathbf{D}j)3$,

$\mathbf{D}(\mathbf{D}j3)$, $\int_0 3$, $\int_1 5$, $\mathbf{D}\int_3^4 j$, $\mathbf{D}\int_3 j$.

3. THE IDEA OF NEWTON AND LEIBNIZ

a. The Relation between Slope and Area. Slope and area seem to be totally unrelated. Newton's teacher, Barrow, probably was the first to see clearly that, in fact, these concepts are very closely related – somewhat like square and square root – and this discovery is the very core of Calculus.

If n is a number different from 0, there is exactly one positive number that is called the square of n. For instance, the square of 3 is 9. If m is a positive number there are two numbers that are called the square roots of m, either root the negative of the other. For instance, the square roots of 9 are 3 and −3.

(i) *The square of a square root of a positive number* m *is* m.

(ii) *A square root of the square of a number* n ($\neq 0$) *is either* n *or* −n.

If *l* is a nonvertical line, there is one horizontal line that is called the slope line or derivative of *l*. For instance, **D** (3 *j* + *3*) = *3*. If c is any horizontal line, there are infinitely many lines that are called area lines or integrals of c, each of them parallel to any other. For instance, the horizontal line *3* has the area lines 3 *j* + 2, 3 *j*, 3 *j* − *6*; in fact, for any number d, the line 3 *j* + *d* is an area line of *3*.

Theorem I. *The slope line of any area line of a horizontal line* c *is* c; or, *the derivative of any integral of* c *is* c; in a formula,

(I) If c is any horizontal line, then $\mathbf{D}\int_a c = c$, for any a.

Theorem II. *Any area line of the slope line of a nonvertical line* l *is parallel to or identical with* l; or, *any integral of the derivative of* l *is parallel to* l *or identical with* l.

Indeed, if l is the line $cj + d$, then $\mathbf{D}\,l$ is the horizontal c; and if l' is any area line of c, then there is a number d′ such that $l' = cj + d'$. The line l' is parallel to l, if $d \neq d'$, and identical with l, if $d' = d$.

For instance, $\frac{1}{2}j$ has the altitude 2.5 at 5. Its slope line, $\mathbf{D}(\frac{1}{2}j)$, is the horizontal line $\frac{1}{2}$. The 5-area line of the latter is that line $\frac{1}{2}j + d'$ which has the altitude 0 at 5, that is, the line $\frac{1}{2}j - 2.5$. It is 2.5 units below $\frac{1}{2}j$;

in a formula, $\int_5 \mathbf{D}(\frac{1}{2}j) = \frac{1}{2}j - 2.5$.

More generally, $\int_a \mathbf{D}\,l$, the a-area line of $\mathbf{D}\,l$, has the altitude 0 at a.

The line l has the altitude $l\,a$ at a. Hence $\int_a \mathbf{D}\,l$ is $l\,a$ units below l;

in a formula,

(II) If l is any nonvertical line, then $\int_a \mathbf{D}\,l = l - l\,a$ for any a.

The relation between square and square root can also be expressed as follows:

(iii) *If* m *is a root of* n, *then* n *is the square of* m.

(iv) *If* n *is the square of* m, *then* m *is a root of* n.

Correspondingly, the following theorems can easily be verified:

Theorem III. *If* l *is an area line (or an integral) of* c, *then* c *is the slope line (or derivative) of* l.

Theorem IV. *If* c *is the slope line (or derivative) of* l, *then* l *is an area line (or integral) of* c.

An inspection of Fig. 7 will not reveal whether c was given and the

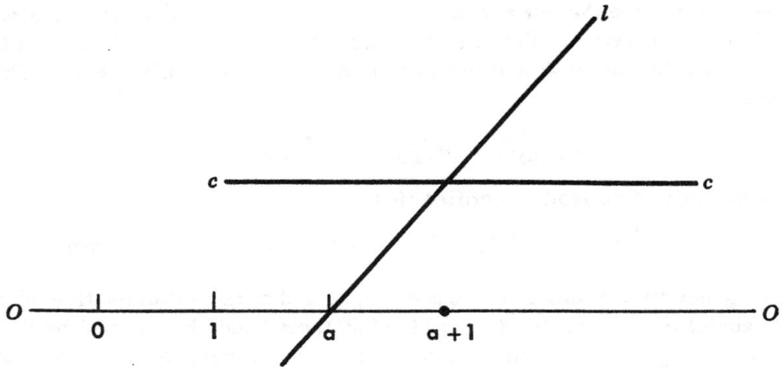

Fig. 7

a-area line of c was constructed or l was given and the slope line of l was constructed. The figure illustrates both

$$\int_a c = l \qquad \text{and} \qquad Dl = c;$$

and this duplicity of the meaning of Fig. 7 is a striking expression of the reciprocity of slope and area.

EXERCISES

18. Verify that similar relations of reciprocity exist between

(a) the tangents (in the sense of trigonometry) of angles that are not odd multiples of right angles, on the one hand, and the arctangents of numbers, on the other;

(b) the women who are mothers of women, on the one hand, and daughters, on the other.

b. The Calculus Solution of the Area Problem. There are two numbers having the square 9. They are called the square roots of 9. One of them is > 0 and is denoted by $\sqrt{9}$, the other one is < 0 and is denoted by $-\sqrt{9}$. Clearly $\sqrt{9} = 3$ and $-\sqrt{9} = -3$.

Some statements are valid for either root; e.g., the fourth power of either root equals 81; the absolute value of either root lies between 2 and 4; etc. For some purposes it is therefore convenient to have a symbol such as $\pm\sqrt{9}$ that stands for either root. In terms of the latter symbol

$$(\pm\sqrt{9})^4 = 81 \qquad \text{and} \qquad 2 < |\pm\sqrt{9}| < 4.$$

On the other hand, not being a specific number, $\pm\sqrt{9}$ is neither > 2 nor $\ngtr 2$. Obviously, only $\sqrt{9} > 2$, whereas $-\sqrt{9} \ngtr 2$.

There are infinitely many straight lines having as their common slope line or derivative the horizontal line *3*. They are called the *antiderivatives* of *3*. Each of them has the altitude 0 somewhere: one, and only one, at 0; one, and only one, at 2; and, for any number a, one and only one at a. These lines will be denoted by

$$\mathbf{D}^{-1}_{(0,0)}\,3, \quad \mathbf{D}^{-1}_{(2,0)}\,3, \quad \mathbf{D}^{-1}_{(a,0)}\,3,$$

respectively. One readily verifies that

$$\mathbf{D}^{-1}_{(0,0)}\,3 = 3j, \quad \mathbf{D}^{-1}_{(2,0)}\,3 = 3j - 6, \quad \mathbf{D}^{-1}_{(a,0)}\,3 = 3j - 3a.$$

More generally, for any three numbers a, b, and c, the antiderivative of the horizontal line c (of altitude c) which has the altitude b at a, will be denoted by $\mathbf{D}^{-1}_{(a,b)}$ c. (Of course, many of these lines coincide; see Exercise 20b.)

Since, by definition of antiderivatives, *3* is the slope line of the line $\mathbf{D}^{-1}_{(a,0)}\,3$, it follows from Theorem IV that $\mathbf{D}^{-1}_{(a,0)}\,3$ is an area line of *3*. Having the altitude 0 at a, $\mathbf{D}^{-1}_{(a,0)}\,3$ must be the a-area line of *3;* in a formula,

$$\mathbf{D}^{-1}_{(a,0)}\,3 = \int_a 3 \text{ and, more generally, } \mathbf{D}^{-1}_{(a,0)}\,c = \int_a c.$$

Theorems III and IV can be summarized in the following

Theorem V. *Any area line of* c *is an antiderivative of* c, *and any antiderivative of* c *is an area line of* c.

In view of Theorem V, the graphical solution of the area problem, which calls for the construction of area lines, can be restated; and this restatement is the

Calculus Solution of the Area Problem for Horizontal Lines. *To find the area under the horizontal line* c *from* a *to* b, *find any antiderivative of* c *and determine its rise from* a *to* b.

Some statements are valid for any antiderivative of *3;* for instance, the derivative of any antiderivative of *3* is *3*. For some purposes it is therefore convenient to have a symbol that stands for any antiderivative of *3*. A convenient symbol is $\mathbf{D}^{-1}\,3$. For instance,

$$\mathbf{D}\,(\mathbf{D}^{-1}\,3) = 3 \text{ and, for any c, } \mathbf{D}\,(\mathbf{D}^{-1}\,c) = c.$$

On the other hand, not being a specific line, $\mathbf{D}^{-1}\,3$ has at 2 neither the altitude 0 nor an altitude $\neq 0$. Only $\mathbf{D}^{-1}_{(2,0)}\,3$ has the altitude 0 at 2, whereas $\mathbf{D}^{-1}_{(1,0)}\,3$ has an altitude $\neq$ 0 at 2; in formulas,

$$\mathbf{D}^{-1}_{(2,0)}3\;2 = 0 \quad \text{and} \quad \mathbf{D}^{-1}_{(1,0)}3\;2 \neq 0.$$

$\mathbf{D}^{-1}\,3\;2$, and more generally, $\mathbf{D}^{-1}$ ca, for any c and any a, will remain undefined and therefore meaningless symbols.

The Calculus Solution of the area problem for horizontal lines, proved in this section, can be condensed in the formula

$$\int_a^b c = [\mathbf{D}^{-1}\, c]_a^b \text{ for any horizontal line c and any two numbers a and b.}$$

This simple formula contains the essence of Newton's and Leibniz' idea, the application of which to more complicated curves has marked an epoch in the history of mathematics and science.

EXERCISES

19. Verify the Calculus Solution of the area problem for

(a) a horizontal line below *O*;

★*(b)* the lines *.5* and *−.5* from 5 to 2 (see Exercise 16). Note that the result of the construction for the area always agrees with the sign rule: the product of the altitude and the base is positive if the factors are both positive or both negative, and negative if one factor is positive and one negative.

(c) Verify Theorem II for a line *l* with negative slope.

20. Show that

(a) $\mathbf{D}^{-1}_{(0,b)}\, c \neq \mathbf{D}^{-1}_{(0,b')}\, c$ for any two numbers b and b′ (≠ b);

(b) $\mathbf{D}^{-1}_{(a,b)}\, c = \mathbf{D}^{-1}_{(a',b')}\, c$, if and only if ac − b = a′c − b′;

(c) $\mathbf{D}^{-1}_{(a,b)}\, c = b + \int_a c$ for any two numbers a and b and for any horizontal line c;

(d) For any two numbers a and a′ (≠ a), is $\mathbf{D}^{-1}_{(a,0)}\, c = \mathbf{D}^{-1}_{(a',0)}\, c$? (Hint. Reconsider Exercise 11(e).)

(e) Analyze the meaning of the formulas

$$\mathbf{DD}^{-1}_{(a,b)}\, c = c \quad \text{and} \quad \mathbf{DD}^{-1}\, c = c, \text{ on the one hand,}$$

$$\mathbf{D} \int_a c = c, \text{ on the other.}$$

One of the main prerequisites for the understanding of calculus is the realization that the first formulas are immediate consequences of basic definitions, whereas the last expresses a profound discovery.

21. If $\int^{b} c$ denotes the line having the altitude $\int_{a}^{b} c$ at any point a on *O*, show that $\mathbf{D}\int^{b} c = -c$.

★22. Show that the Calculus Solution of the area problem has the following counterpart: *The slope of a nonvertical line l can be found by determining the altitude of a horizontal line having l as an area line.* In contrast to the calculus solution of the area problem, however, the above solution of the slope problem is of little importance.

CHAPTER II

GRAPHICAL SOLUTIONS OF THE TWO BASIC PROBLEMS

1. THE DERIVATIVE OF A POLYGON AND THE INTEGRALS OF A STEP LINE

Let l be a nonvertical line. If a and b are points on O, then the portion of l whose projection begins at a and ends at b is called a *segment* of l, more specifically, the segment *from* a *to* b. The point above a is said to be the *initial* point of the segment, the point above b its *terminal* point.

On the line l in Fig. 8 (a), the segment from a to b is drawn heavily, and so is the segment from a to b of the slope line **D** l. The latter segment will be called *the slope segment* of the former. Similarly, in Fig. 8 (b), the heavy segments on the area lines of the horizontal c are called *area segments* or *integrals* of the corresponding horizontal segment. Any two such integrals are parallel. Since they have the same projection on O, any integral of a horizontal segment can be obtained by moving any other of its integrals vertically.

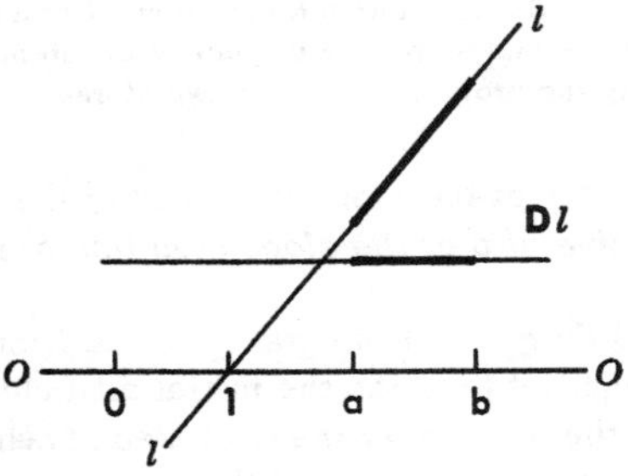

Fig. 8 (a)

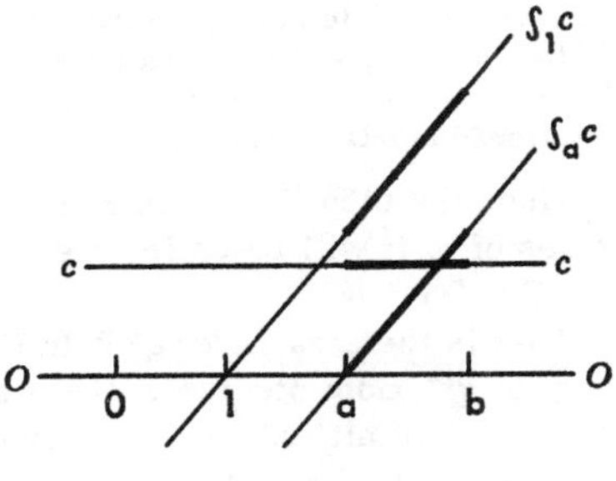

Fig. 8 (b)

By a *polygon* is meant in this book *a simple curve* (such as p in Fig. 9) *consisting of a finite number of segments, provided that the initial point of each segment (after the first) coincides with the terminal point of the preceding segment.* Each of the segments is called a *side* of p. Being a *simple* curve, a polygon intersects any vertical line in not more than one point. In particular, each side of a polygon is nonvertical.

In Fig. 9, for the sides of the polygon p the horizontal slope segments have been drawn. The projection of each of them (after the first) begins where the projection of the preceding segment ends. A finite chain of horizontal segments with this property is called a *step line;* each segment

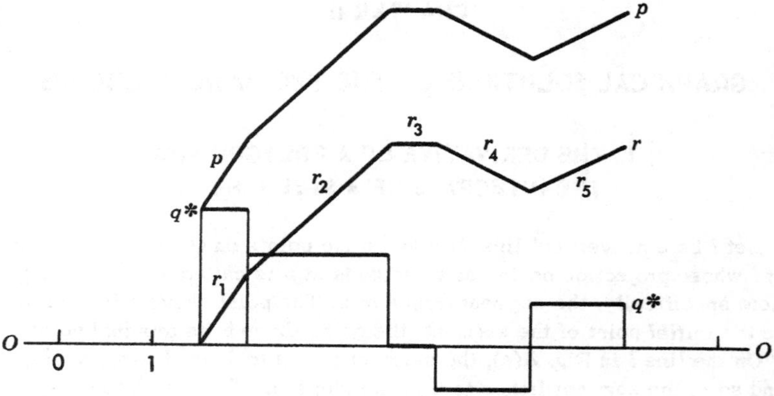

Fig. 9

in the chain a *step*. Step lines will be denoted by letters with asterisks; the one in Fig. 9 is denoted by q^*.

The asterisk is to indicate that the step line q^* (unless all its steps have the same altitude and form one horizontal segment) is not a simple curve. Some vertical lines intersect q^* in two points, e.g., the line through the terminal point of the first step and the initial point of the second step. Nor is (unless all steps have the same altitude) q^* a polygon. (Only the projections of the two aforementioned points coincide.)

Since the step line q^* in Fig. 9 consists of the slope segments of the sides of p, it will be called *the derivative of p* or *the slope step line of p*, briefly, **D** p.

What is the area under q^*? In Fig. 9 there is an integral r_1 of the first step of q^*; more precisely, the area segment that has the initial altitude 0 The terminal altitude of r_1 is equal to the area under the first step. Then there is in the figure the integral of the second step of q^* that begins where r_1 ends. The rise of this segment r_2 is equal to the area under the second step. Thus the terminal altitude of r_2 is equal to the sum of the areas under the first two steps. r_3 is the integral of third step that begins where r_2 ends, and so on. The terminal altitude of r_5 is equal to the area under the entire step line q^*.

Obviously, $r_1, r_2, \ldots, r_5$ form a polygon r which is an area polygon or integral of q^* in the following sense:

$$\int_a^b q^* = [r]_a^b ,$$

for any two points a and b belonging to the projection of q^*. More specifically, r is that integral of q^* which has the altitude 0 above the initial point of the projection of q^*. Therefore, r is called *the area polygon* or *integral of q^* with the initial altitude* 0.

The integrals of the steps of q^* could be selected so as to form a polygon (each area segment beginning where the preceding one ended) because on p. 10 also the lines parallel to k have been introduced as area lines or integrals of c; in other words, because all that is required of an area line of c is that its *rise* from a to b (and not necessarily its altitude at b) be equal to $\int_a^b c$.

Clearly, r is parallel to the polygon p whose derivative is q^*; and p is another integral of q^*. Thus $\int_a^b q^* = [p]_a^b$ for any two points a and b in the projection of q^*. It can easily be verified that Theorems I-IV of Chapter I hold verbatim for the derivative of a polygon and the integrals of a step line:

If q^* is a step line, then the derivative of any integral of q^* is q^*. If p is a polygon, then an integral of the derivative of p is identical with p or parallel to p and can be obtained by moving p vertically. If the polygon p is an integral of the step line q^*, then q^* is the derivative of p, and conversely.

A polygon is called the (a,a′)-*antiderivative* of the step line q^* and denoted by $\mathbf{D}^{-1}_{(a,a')}\, q^*$, if it has the altitude a′ at a and if q^* is the derivative of the polygon. If $\mathbf{D}^{-1} q^*$ is a symbol that stands for any antiderivative of q^*, the preceding theorem yields the following

Calculus Solution of the Area Problem for Step Lines. *To find the area under the step line q^* from* a *to* b, *determine the rise from* a *to* b *of any antiderivative of q^**. In a formula,

$$\int_a^b q^* = [\mathbf{D}^{-1}\, q^*]_a^b \quad \text{for any a and b in the projection of } q^*.$$

In Fig. 10 an integral has been constructed for a step line with eight

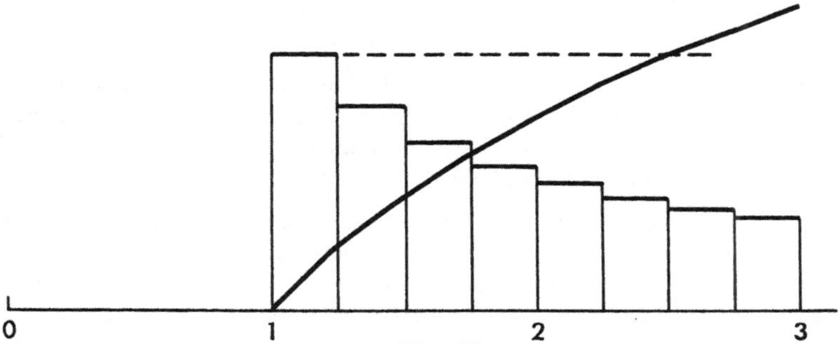

Fig. 10

steps of width 1/4 and altitudes 4/4, 4/5, . . . , 4/10, 4/11. For reasons explained later, that integral is a good approximation to the so-called *natural logarithmic curve,* whose altitude at x is the logarithm of x to the base e = 2.71828 . . . That curve attains the altitude 1 at e. The polygon in Fig. 10 attains the altitude 1 at 2.513 and thus provides a graphical approximation to the important number e. (For improvements of the construction, see Exercise 1.)

EXERCISES

Exercises 1-4 should be carried out on cross-section paper. A unit of 2" should be chosen.

1. Draw the step line beginning at 1 with eight steps of width 1/4 and the altitudes 4/5, 4/6, . . . , 4/11, 4/12. The 1st, 2nd, . . . step has the altitude of the 2nd, 3rd, . . . step in Fig. 10. Construct the integral with the altitude 0 at 1. It resembles that in Fig. 10 but, of course, lies below the latter. It attains the altitude 1 at a number that is larger than e. A good approximation to e can be obtained from the integral of an eight-step line with altitudes 8/9, 8/11, . . . , 8/23 (which lie between those just considered and those in Fig. 10). The reader should also determine the altitudes of that intermediate area polygon at 2 and at 3. He should compare his results with the natural logarithms of 2 and 3 or the ratio of his results with the ratio of the ordinary logarithms of 2 and 3.

2. Draw the following step lines, each consisting of eight steps of base 1/4, and construct the indicated integrals:

	Projection of the step line on *O*	Altitudes of the steps	Integral of initial altitude
(a)	from 0 to 2	1/4, 2/4, 3/4, . . . , 8/4	0
(b)	from −2 to 0	0/4, 1/4, 2/4, . . . , 7/4	0
(c)	from 2 to 4	0/8, 1/8, 2/8, . . . , 7/8	−1
(d)	from −1 to 1	−4/8, −3/8, −2/8, . . . , 3/8	2

The purpose of Exercises 2 and 4 is to familiarize the student with the integrals of step lines and the derivatives of polygons that follow certain simple patterns. In case of time limitations one or two constructions may be selected from each of these problems.

3. Draw integrals of a few step lines without measuring the altitudes and the bases of the steps. In each case, only one mensuration is indispensable: that of the rise of the area polygon, which yields the number of square units under the given step line. Also draw the derivatives of a few polygons without measuring the altitudes of the vertices.

4. Construct the derivative of a polygon with eight sides whose nine vertices have the projections 0/4, 1/4, 2/4, . . . , 7/4, 8/4 and the altitudes:

(a) 0/16, 1/16, 4/16, 9/16, 25/16, 36/16, 49/16, 64/16;

(b) 1/16, 4/16, 9/16, . . . , 64/16, 81/16;

(c) 0/32, 1/32, 8/32, 27/32, 64/32, 125/32, 216/32, 243/32, 512/32;

(d) $5\log_{10} 1$, $5\log_{10} 2$, $5\log_{10} 3, \ldots, 5\log_{10} 9$.

Two-place logarithms, which are sufficient for graphical constructions, can be obtained as follows. By definition, $\log_{10} n$ is *the exponent of that power of* 10 *which equals* n. From $10^0 = 1$ and $10^1 = 10$ it follows that $\log_{10} 1 = 0$ and $\log_{10} 10 = 1$.

The numbers $2^{10} = 1024$ and $10^3 = 1000$ differ little. Thus the 10-th roots of 2^{10} and 10^3 are approximately equal; consequently, if "$\sim$" denotes "approximately equal"

$$2 \sim 10^{3/10} \text{ and thus } \log_{10} 2 \sim .30.$$

From $4 = 2^2 \sim (10^{.30})^2 = 10^{.60}$ it follows that $\log_{10} 4 \sim .60$.

Similarly, $8 = 2^3$ implies that $\log_{10} 8 \sim .90$.

From $5 = 10/2 \sim 10^1/10^{.30} = 10^{1-.30} = 10^{.70}$ it follows that $\log_{10} 5 \sim .70$.

Furthermore, $3^4 = 81 \sim 80 = 8 \cdot 10 = 10^{.90} \cdot 10^1 = 10^{1.90}$. Taking fourth roots, one obtains $3 \sim (10^{1.90})^{1/4} = 10^{.475}$. Since $3^4 > 80$, we write $\log_{10} 3 \sim .48$.

Since $6 = 2 \cdot 3$ and $9 = 3 \cdot 3$, one obtains in a similar manner the logarithms of 6 and 9, while $\log_{10} 7$ can (rather inaccurately) be determined from the fact that $7^2 = 49 \sim 5 \cdot 10$.

Altogether, the following 2-place logarithmic table results:

n	1	2	3	4	5	6	7	8	9	10
$\log_{10} n$	.00	.30	.48	.60	.70	.78	.84	.90	.96	1.00

The table may be continued, since from $1.2 = \frac{3 \cdot 4}{10}$, $1.5 = \frac{3 \cdot 5}{10}$, etc. it follows that

$$\log_{10} 1.2 \sim .48 + .60 - 1 = .08, \quad \log_{10} 1.5 \sim .48 + .70 - 1 = .18,$$

$$\log_{10} 2.5 \sim .70 + .70 - 1 = .40, \text{ etc.}$$

Moreover, one can then set up a table of differences and apply interpolation.

5. Draw a polygon p. On O, denote by a and b the endpoints of the projection of p, and by $\hat{x}$ a point between a and b above which p has a peak or maximum. Show that the derivative $\mathbf{D}\, p$ has positive altitudes (i.e., is above O) to the left of $\hat{x}$ and negative altitudes (i.e., is below O) to the right of $\hat{x}$. If the top level is assumed on a plateau, then the slope of p at $\hat{x}$ and near $\hat{x}$ is 0; but $\mathbf{D}\, p$ is positive farther to the left, and negative farther to the right. Then study a point $\check{x}$ between a and b above which p has a minimum. Finally show that if p has a maximum (a minimum) at a, then the slope at a is negative (positive) and formulate the corresponding rule for b.

6. A segment joining two nonconsecutive vertices of a polygon p is called a *chord* of p. The polygon is said to be *cup-shaped* (*cap-shaped*) if every chord of p is above p (below p). Show that p is cup-shaped if, and only if, the slope of any side of p exceeds that of its left neighbor. On the left half of a cup, the slopes of the sides are negative but get closer to 0; on the right side of a cup, the sides have positive slopes and get steeper. Discuss a cap-shaped polygon.

★7. Suppose that 36 observations have been made with the following results:

3.1	4.7	4.7	2.8	4.3	5.4
2.6	4.2	4.4	3.2	4.2	3.8
3.8	4.2	5.3	4.5	4.1	3.4
3.0	5.7	5.1	3.6	3.9	3.5
5.4	4.8	4.8	3.6	4.9	3.7
4.0	4.4	4.0	5.1	4.6	4.6

Classify the observations in seven intervals of length .5: from 2.5 to 3 (2.5 included, 3 excluded); from 3 to 3.5 (3 included, 3.5 excluded); and so on; the last from 5.5 to 6. Call f_1 the number of observations falling into the first interval, f_2 the number falling into the second interval, and so on. Using $\frac{1}{4}$ inch as horizontal and as vertical unit, construct the (bell-shaped) step line h^* with seven steps (each one unit wide) whose altitudes are $f_1, f_2, \ldots, f_7$ units. The domain under h^* is called the *histogram* of the classified observations. Construct the (S-shaped) area polygon or integral $k = \int_{2.5} h^*$ of altitude 0 at 2.5. It is called the *cumulative frequency polygon* of the observations. What is the significance *(a)* of the altitude of k at 4? *(b)* of the rise of k from 3 to 4.5? *(c)* of those points on O (called the *lower quartile*, the *median*, the *upper quartile* of the observations) for which the altitude of k is 9, 18, 27 (that is, one half, one quarter, three quarters of its terminal height of 36)? Then construct the histogram using $\frac{1}{2}$" as horizontal and $\frac{1}{4}$" as vertical unit (see Exercise 15, p. 12).

★8. If l is a nonvertical straight line and a and b are two points on O, let $l\,[a,b]$ denote the segment on l whose projection on O is from a to b. The two segments

$$j\,[0,1] \quad \text{and} \quad (\tfrac{1}{2}j + \tfrac{1}{2})\,[1,2]$$

form a polygon p with a corner above 1. A line m passing through this corner is called a *line of support* of p above 1 if both segments lie on the same side of m. Show that $.75j + .25$ is, and $.25j + .75$ is not, a line of support of p at 1. Join the two horizontal segments

$$1\,[0,1] \quad \text{and} \quad \tfrac{1}{2}[1,2],$$

which constitute the derivative **D** p, by the vertical segment between them. The result is a non-simple broken line. Show that the added vertical segment contains just those points whose altitudes equal the slopes of lines of support of p at 1. State a general theorem about any corner of any polygon.

THE APPROXIMATE AREA UNDER A CURVE AND APPROXIMATE INTEGRALS

To find the area under a simple curve f approximately, replace f by a step line and determine the area under it.

If f is a *straight segment,* choose one step passing through the midpoint of f. The rectangle has precisely the same area as the trapezoid under f because the excess triangle and the deficit triangle balance one another.

If f is a *polygon,* choose a step line of which each step passes through the midpoint of the corresponding side of f. Again the area under the step line is precisely equal to the area under f.

If f is *curved,* choose steps for each of which the excess and the deficit seem to balance each other as accurately as possible. In Fig. 11, f is a quarter of a circle of radius 2. The approximate step line has six steps. The integral p (of initial altitude 0) of this step line rises to the altitude

$p2 = 3.2$ – a fair approximation to $\int_0^2 f$, which is $\pi = 3.14159\ldots$. The steps of the approximate step line in Fig. 11 have various widths. On the right side, where the curve is steeper, thinner steps have been chosen. As a result it was easier (than it would have been for six steps of equal width) to draw the heights of the steps in such a way that for each step the excess and the deficit balance one another. For the third step, excess and deficit have been shaded. More generally, in approximating a curve f by a step line follow the

Practical Rule. *Choose narrower steps where f is both curved and steep or sharply curved.*

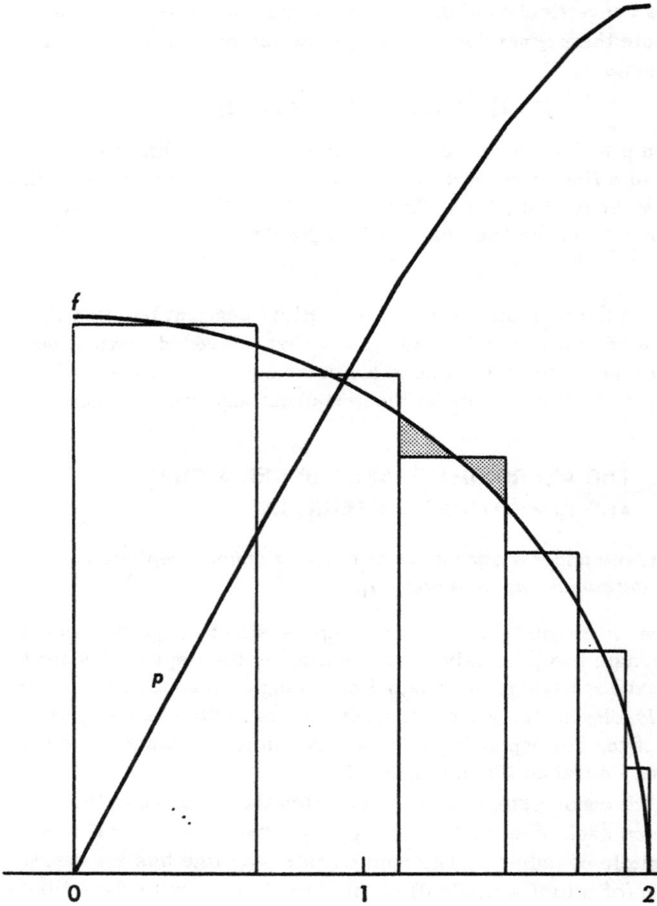

Fig. 11

In Fig. 11, not only is $p\,2$ approximately equal to $\int_0^2 f$, but for every x between 0 and 2 is the altitude of p approximately equal to $\int_0^x f$. This fact will be expressed by saying that p is an approximate integral of f. A step line approximating f with more and thinner steps has an area line p' that is an even better approximate integral of f; that is to say, the altitude

p' x is even closer to $\int_0^x f$ than is p x. This situation suggests the idea that the integrals of step lines approximating f whose steps are thinner and thinner come closer and closer to a precise integral of f – a curve $\int_a f$ whose altitude at any x is precisely equal to $\int_a^x f$.

EXERCISES

9. Construct an approximate integral of a semicircle of radius 1 with center 0.

10. Draw the segment from 0 to 3 of the line j for which $j\,\mathrm{x} = \mathrm{x}$ for any x. Clearly, $\int_0^3 j = \int_0^3 1.5$. But in what relation are the areas $\int_0^x j$ and $\int_0^x 1.5$ if $0 < \mathrm{x} < 3$? Obviously the area line $\int_0 1.5$ is not a good approximation to $\int_0 j$. To obtain one, replace the given segment of j by a step line q^* consisting of 12 steps of width 1/4, and construct $\int_0 q^*$. To obtain an even better approximate integral, construct $\int_0 q_1^*$, where q_1^* has 24 steps of width 1/8.

11. Construct approximations to the integrals $\int_1 j$ and $\int_{-2} j$. Replace the segment of j from 1 to 3 by a step line with 8 steps; the segment from -2 to 3 by a step line with 20 steps.

12. Construct approximations to *(a)* $\int_{-2} (.75j)$, *(b)* $\int_{-2} (.75j + 1)$, *(c)* $\int_{-2} (-j)$, *(d)* $\int_{-2} (-j + 1)$.

13. Let f be the quarter of a circle of radius 2 in Fig. 11. Divide the segment from 0 to 2 on O into five equal segments. Above each of them,

draw a rectangle passing through the highest point of f above the segment. The resulting step line is said to be *above* f. Construct its area polygon of initial altitude 0. Then draw the corresponding step line *below* f and construct its integral of initial altitude 0. The approximate integral of f in Fig. 11 lies between these two area lines.

3. POWER, EXPONENTIAL, AND SINE CURVES

Curves belonging to the following three important families are simple objects for graphical constructions.

a. Power Curves. Let c be any number. By the c-th power curve, denoted by j^c, we mean the curve which has altitude x^c above the point x on *O*, where x is any number for which there is a number x^c. For instance, there is a number $3^{1/2}$, but not a number $(-3)^{1/2}$. In Fig. 12 there are the positive half of the line *O* and the portions above it of the straight line j $(=j^1)$ and of the curves j^2, j^3 and $j^{1/2}$, $j^{1/3}$. The curves j^2 and $j^{1/2}$ are symmetric parabolas. More generally, j^n and $j^{1/n}$ are symmetric with respect to the line j. Every power curve has the altitude 1 at 1. For positive c, the curve j^c has the altitude 0 at 0.

If $x > 0$ and c is between b and d, then x^c is between x^b and x^d. Thus, for instance, above the positive half of *O*

$j^{2/3}$ lies between $j^{1/2}$ and j;

$j^{7/9}$ lies between $j^{2/3}$ and j;

$j^{3/2}$ lies between j and j^2;

$j^{7/3}$ lies between j^2 and j^3.

The curves $j^{1.4}$, $j^{1.41}$, $j^{1.414}$, ... (not shown in the figure) come closer and closer to a curve which is called $j^{\sqrt{2}}$. Similarly, the curves $j^{3.1}$, $j^{3.14}$, $j^{3.141}$, .. come closer and closer to a curve called j^{π}.

In Fig. 12 there are j^0 (which is the horizontal line *1*), j^{-1} (which is a hyperbola), j^{-2}, and $j^{-1/2}$.

Fig. 13 illustrates the behavior above the entire line *O* of those power curves whose exponent is the quotient of two integers, p and q, which may be assumed to have no common integer factor. Obviously then not both p and q can be even, which leaves the following three possibilities:

(1) *p is even, q is odd.* Then $j^{p/q}$ is symmetric with respect to the vertical line through 0. (Examples: $j^2 = j^{2/1}$, j^{-2}, $j^{2/3}$.)

(2) *p is odd, q is odd.* Then $j^{p/q}$ is symmetric with regard to the point 0 on *O*. (Examples: $j^3 = j^{3/1}$, $j^{1/3}$, $j^{7/3}$, j, j^{-1}.)

(3) *p is odd, q is even.* Then $j^{p/q}$ has no points with negative projections since the computation of $x^{p/q}$ for negative x would require the extraction of

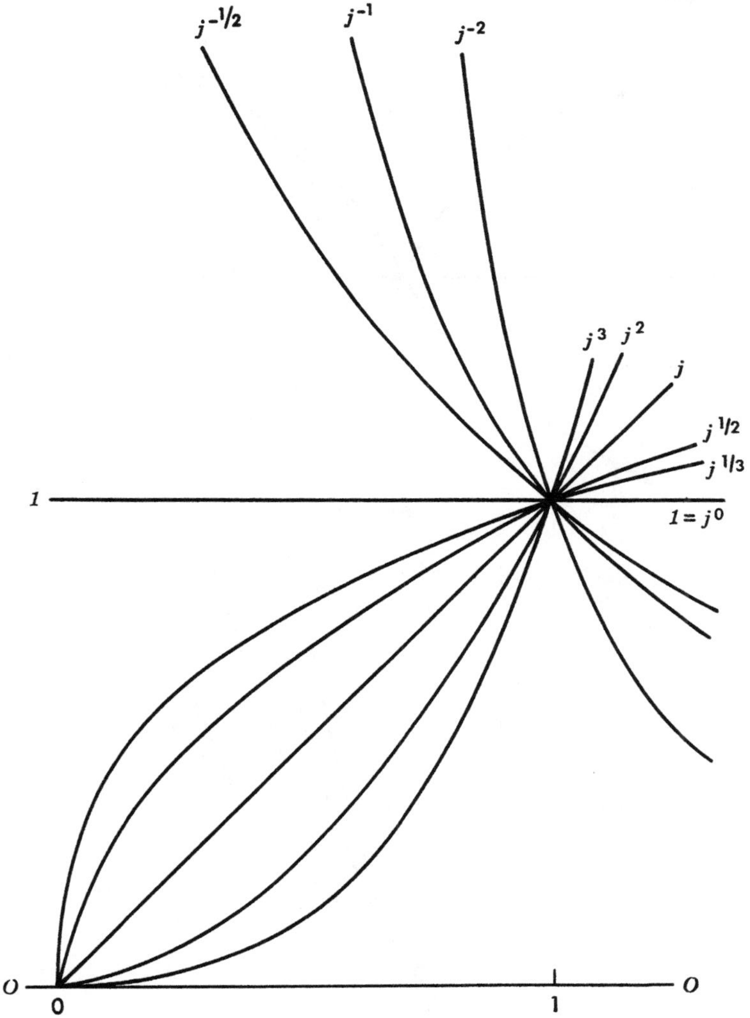

Fig. 12

a square root of a negative number. (Examples: $j^{1/2}$, $j^{3/2}$, $j^{-1/2}$.) Neither do $j^{\sqrt{2}}$ and j^{π} have points with negative projections. (See Exercise 19.)

$3j$ denotes the line having, at any x, the altitude 3x. Similarly, $3j^2$ will denote the curve whose altitude at x is $3x^2$; and $2j^3$ the curve whose altitude at x is $2x^3$; etc.

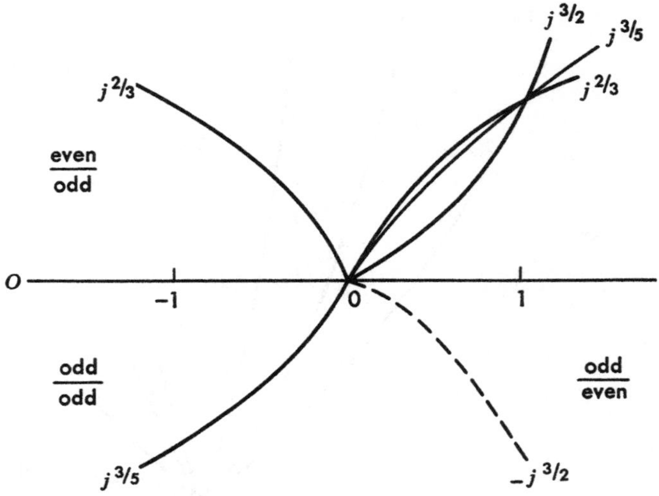
$j^{3/2}$
$j^{3/5}$
$j^{2/3}$
$j^{2/3}$
even/odd
O
−1
0
1
O
odd/odd
odd/even
$j^{3/5}$
$-j^{3/2}$

Fig. 13

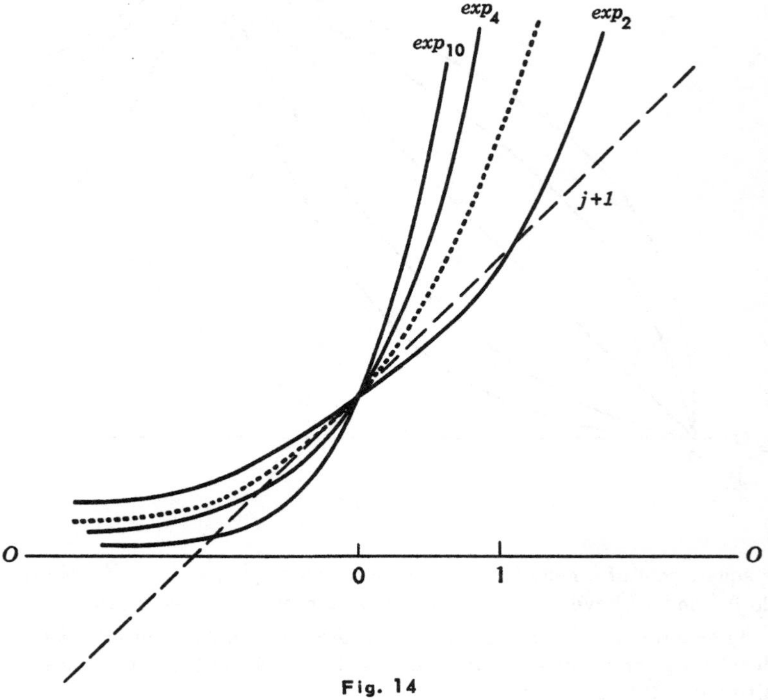
exp_4
exp_2
exp_{10}
$j+1$
O
0
1
O

Fig. 14

b. Exponential Curves. If b is a positive number, by the *exponential curve to the base* b one means the curve exp_b whose altitude at x is b^x. In Fig. 14, there are the curves exp_2, exp_4, exp_{10}. All exponential curves have the altitude 1 at 0. So does the straight line $j + 1$ whose altitude at x is x + 1. This line (which is, of course, not an exponential curve) is broken in Fig. 14. It has a second point in common with each of the exponential curves shown in the figure:

with exp_2 the point with projection 1 (since $2^1 = 1 + 1$);

with exp_4 the point with the projection $-\frac{1}{2}$ (since $4^{-1/2} = -\frac{1}{2} + 1$);

with exp_{10} a point still farther to the left.

c. Sine Curves. In Fig. 15 the arc beginning at 0 and ending at $\pi = 3.14159\ldots$ of the sine curve is drawn and denoted by *sin*. The altitude of this curve at x is equal to the sine of an angle subtending an arc of length x on the unit circle; that is, an angle of $\frac{180}{\pi}$ x degrees. The arc of the sine curve might be enclosed in a rectangle of altitude 1 and base π. In Fig. 15 there are, further, arcs of the curves whose altitudes at x are 2 *sin* x, *sin* 3x, *sin* $(x + \frac{\pi}{2})$ which equals *cos* x, and a *sin* b(x − c). The

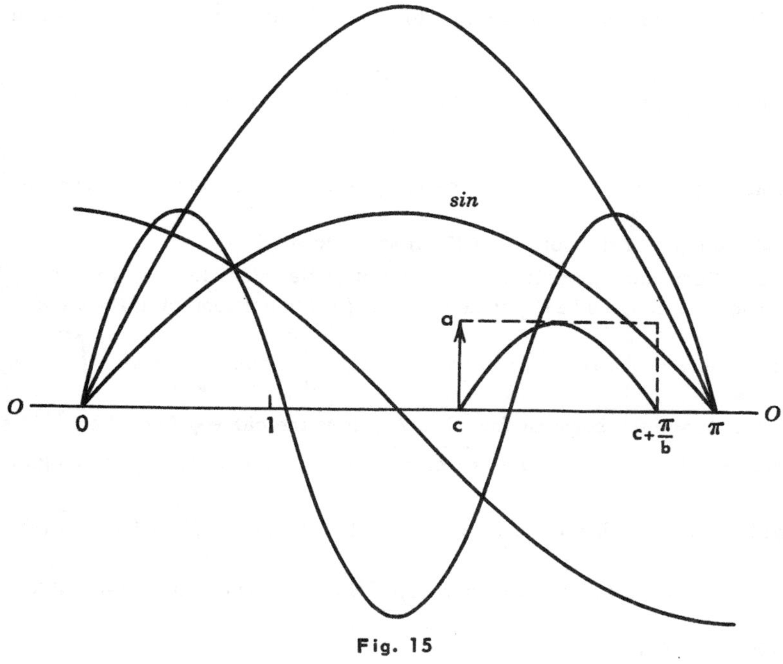

Fig. 15

last arc is enclosed in a rectangle of height a and base π/b which has a corner at the point c on O.

EXERCISES

In the following exercises use 1 inch as a unit.

14. Construct an approximation to $\int_{-2} j^2$, that is, a polygon p such that $p(-2) = 0$ and such that the rise of p between any two points a and b is approximately $\int_a^b j^2$. By means of this polygon, find $\int_{-2}^1 j^2$ approximately and also $\int_{-1}^2 j^2$. Then construct approximations to $\int_{-1} j^3$ and $\int_0 j^{1/2}$.

15. Construct an approximation to $\int_1 j^{-1}$ and by means of it determine approximately: *(a)* $\int_1^2 j^{-1}$, *(b)* $\int_1^3 j^{-1}$, *(c)* the point x on O for which $\int_1^x j^{-1} = 1$ (see Exercise 1, p. 22). Then construct an approximate integral of the portion of j^{-1} from -4 to $-.25$.

16. Construct approximate integrals of initial altitude 0 of *(a)* the portions of exp_2 and exp_4 from -2 to 2; *(b)* the portions of *sin* and *cos* from 0 to 2π. *(c)* Determine approximately $\int_0^\pi sin$, $\int_0^{\frac{\pi}{2}} cos$, $\int_0^\pi cos$.

17. Above the segment from 0 to 2π, draw the curve $\sqrt{1 - .64\, sin^2}$ whose altitude at x is $\sqrt{1 - .64\, sin^2 x}$. Construct an approximate integral with the initial altitude 0. By means of it, determine $\int_0^2 \sqrt{1 - .64\, sin^2}$.

18. Draw the curves $exp_{\frac{1}{2}}$ and $exp_{\frac{1}{4}}$. How are these curves related to exp_2 and exp_4?

*19. Determine fractions p/q of each of the three types (even/odd, odd/odd, odd/even) which are close to $\sqrt{2} = 1.41\ldots$ and draw the points with the projection -3 and the altitudes $(-3)^{p/q}$ (as far as there are such numbers). The construction makes it clear why $(-3)^{\sqrt{2}}$ is left undefined. The curve $j^{\sqrt{2}}$ has no points with negative projection. The same is true for j^{π}. Note further that $(-2)^{1/3}$ is negative; yet near 1/3 there are fractions p/q (of the type even/odd) for which $(-2)^{p/q}$ is positive. Consequently, the negative halves of the corresponding curves $j^{p/q}$ are *not* close to the negative half of $j^{1/3}$. Denote by exp_{-2} the totality of all points of altitude $(-2)^x$ above those points x for which there is a number $(-2)^x$. Describe exp_{-2} and plot some of its points.

4. APPROXIMATE DERIVATIVES OF A CURVE

First Method. *To find an approximate derivative of a curve f, replace f by a polygon and determine the derivative of the latter.*

In Fig. 16, the sine curve is drawn and on it nineteen points whose projections divide the segment on O from 0 to π into eighteen equal parts (each corresponding to 10 degrees). The curve is replaced by the polygon with nineteen points as vertices. The derivative of this polygon is a step line reminiscent of a cosine curve.

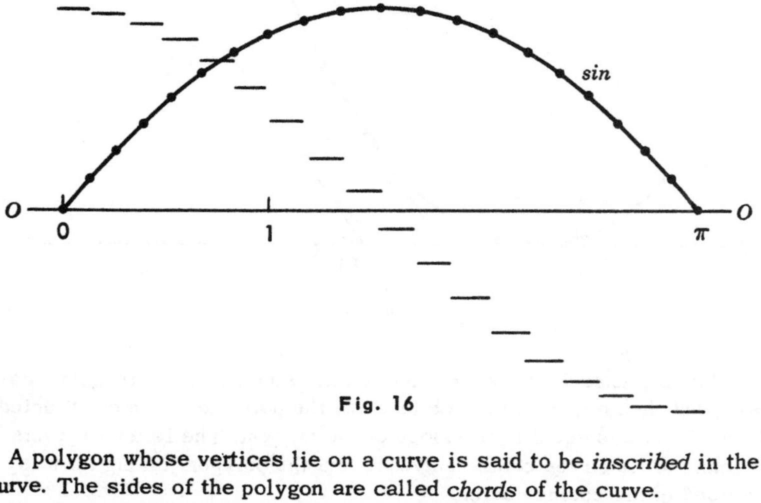

Fig. 16

A polygon whose vertices lie on a curve is said to be *inscribed* in the curve. The sides of the polygon are called *chords* of the curve.

To obtain a better approximate slope polygon of the sine curve, one would have to inscribe into the curve a polygon with shorter sides. Its derivative would have thinner steps and would resemble the cosine curve even more. This situation suggests the idea that the cosine curve is the precise slope curve of the sine curve.

Second Method. *To find an approximate derivative of a curve f, draw several tangents to f. For each point ξ of f at which the tangent has been drawn, construct the point ξ' with the same projection whose altitude is equal to the slope of that tangent. Join the points thus obtained by a polygon.*

In Fig. 17, above the point .75, the tangent, *t*, to the parabola j^2 has been drawn and the slope line of *t* was constructed in the usual way. But only the point of **D** *t* above .75 is marked heavily, while the rest of the slope line is dotted. The altitude of the heavy point above .75 equals the slope of the tangent to j^2 at .75.

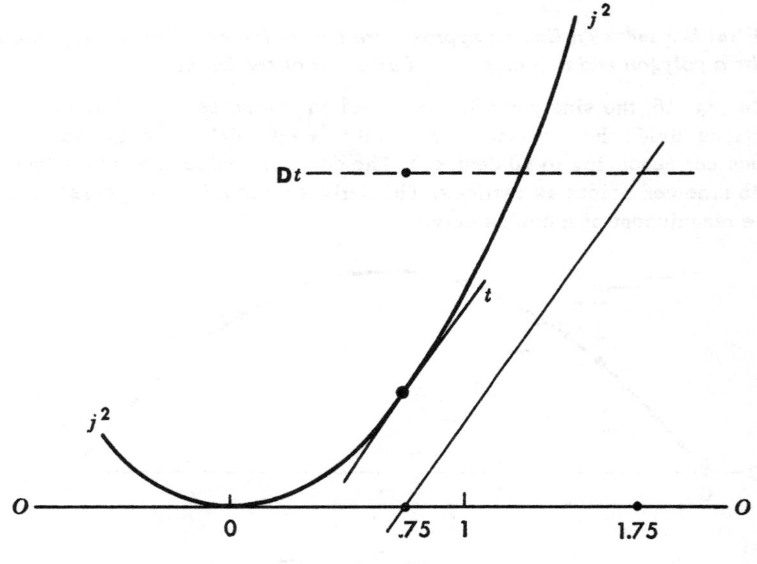

Fig. 17

In Fig. 18, at each of the 19 points on the sine curve the tangent was drawn, but then erased. For each tangent, the point has been constructed whose altitude is equal to the slope of the tangent. The latter 19 points are found to lie on the cosine curve. Hence the polygon joining them is inscribed in the cosine curve.

Further tangents to the sine curve would yield further points on the cosine curve and a polygon with shorter sides inscribed in the cosine

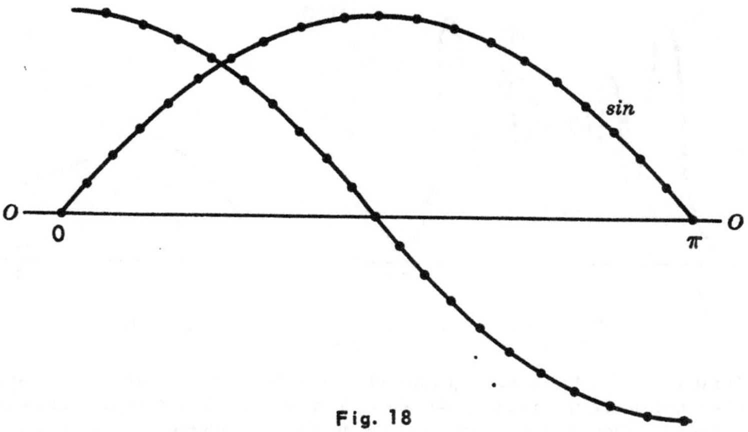

Fig. 18

curve. Again one is led to expect that the cosine curve is a precise slope curve of the sine curve.

Once a tangent to a curve is drawn, its slope can be determined in the same way as the slope of any other straight line. The question is how to draw the tangent to a curve at a point. For instance, how was t drawn in Fig. 17? First, the point of j^2 above .75 was joined by a straightedge to the point of j^2 above 1. Then the straightedge was moved in such a way that it joined the point above .75 to the point of j^2 above .875 and that motion was continued. As on j^2 points closer to that above .75 were chosen, the motion of the straightedge slowed down and finally came, for practical purposes, to a standstill. The line drawn in this rest position is t. Had the point of j^2 above .75 first been joined to that above .5 and had the straightedge then been moved, the result would have been the same.

Another way to construct t is as follows. Place a mirror, perpendicularly to the paper, through the point of j^2 above .75. At first, the arc of j^2 in front of the mirror and the reflection of this arc will form a corner. Move the mirror until the arc and its reflection continue one another smoothly. In this position the mirror is perpendicular to t. The line drawn by using the mirror in this position as a straightedge is called the *normal* to j^2 at .75. The line perpendicular to it through the point of j^2 is t.

In certain cases of infinitely many ups and downs of a curve near a point, the straightedge joining that point to neighbor points on the curve would keep going up and down without approaching a rest position. At such a point the curve has no tangent. Fig. 19(a) illustrates this case, above 0.

The curve in Fig. 19(b) has at 0 a right-side tangent and a left-side tangent, but no tangent, since the two half-tangents are different. At a point of this type a curve is said to have a *corner*.

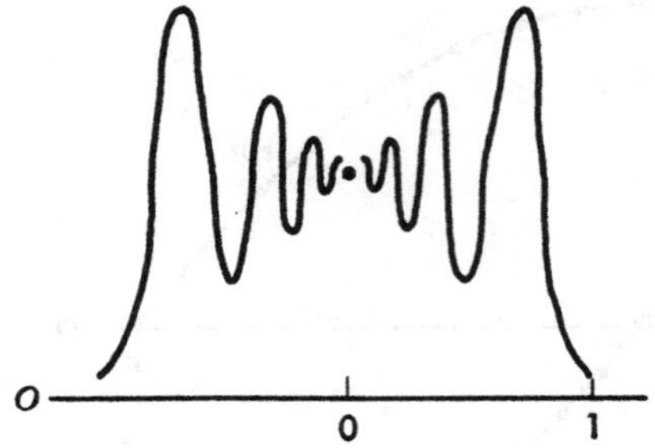

Fig. 19 (a)

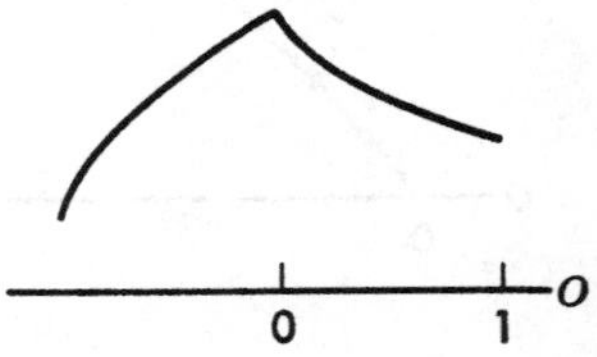

Fig. 19 (b)

The definition of tangents frequently given for circles and ellipses ("a straight line having exactly one point in common with the curve") is not applicable to more complicated curves. In Fig. 20 (a), the line parallel to the axis of the parabola has exactly one point in common with the curve without being a tangent. In Fig. 20 (b), the straight line *t* has two points in common with the curve and yet *t* is a tangent to the curve at one of the two points. In Fig. 20 (c), the line *O* is a tangent to the curve at two points.

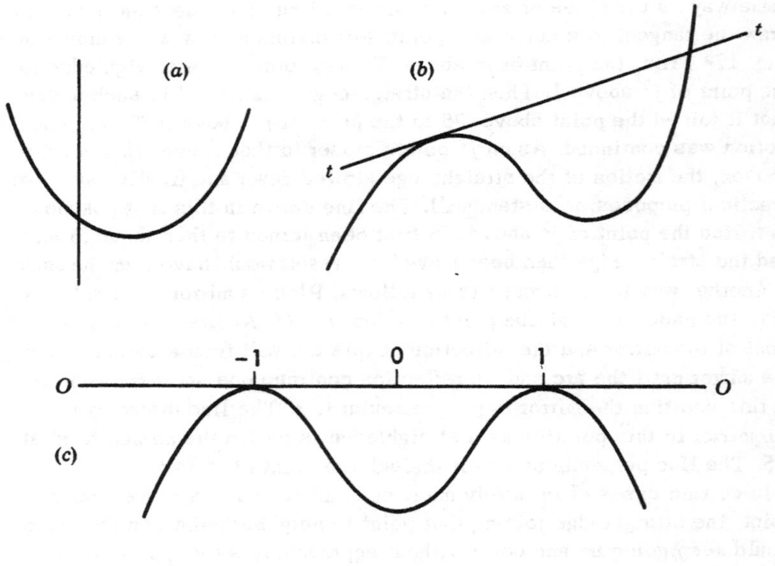

Fig. 20

EXERCISES

20. On the upper semicircle of radius 2 with center at 0, consider the arc whose projection is from -1.8 to 1.8. Inscribe a polygon and determine its derivative. Why can one not construct a good approximate derivative of the entire semicircle? Then consider the arc from 2.2 to 5.8 on the lower semicircle of radius 2 with center at 4. Draw tangents to this arc and determine an approximate derivative of the semicircle in this way.

21. By either method construct approximate derivatives of

(a) the parabolas j^2 and $3j^2$;

(b) the curves j^3 and $\frac{1}{3}j^3$;

(c) the arcs from .25 to 4 of the parabola $j^{1/2}$ and the hyperbola j^{-1};

(d) the arc from -4 to $-.25$ of the hyperbola j^{-1};

(e) the sine curve from $-\pi$ to 0 ;

(f) the cosine curve from $-\pi$ to π.

Draw the tangents at 0 to the curves exp_2 and exp_4 and determine their slopes, that is, $\mathbf{D}\, exp_2\, 0$ and $\mathbf{D}\, exp_4\, 0$.

22. Choose a horizontal unit of 1 inch, and a vertical unit of $\frac{1}{8}$ inch, and construct an approximate derivative of the curve $16j^2$.

5. COMPARISON OF INTEGRALS AND DERIVATIVES

In Fig. 21 (a), there is a straight line l, its (horizontal) slope line $\mathbf{D}\, l$, and an approximate integral of l. In Fig. 21 (b), a short segment of l has been replaced by a thin tooth. The approximate integral of the resulting polygon p and that of l differ little. But the derivative of p shows a big upward swing followed by a big downward swing, corresponding to the rising and falling part of the thin tooth.

While small changes of a curve have little effect on the integrals, they may completely change the derivatives. This instability of the derivative often impairs its value. If a curve has been even slightly misdrawn or miscopied, its derivative may be completely different from that of the correct curve. On the other hand, an approximate integral of the miscopied curve could still be used as an approximate integral of the correct curve.

A curve f is called *smooth* if for each small piece of f any two chords have almost equal slopes (and thus are almost parallel). The sine curve, the exponential curves, and the power curves j^c for positive c are examples of smooth curves. The curves in Fig. 19 are not smooth.

If f is a smooth curve and an inscribed polygon with short sides is replaced by one with still shorter sides, the derivative of the latter is only

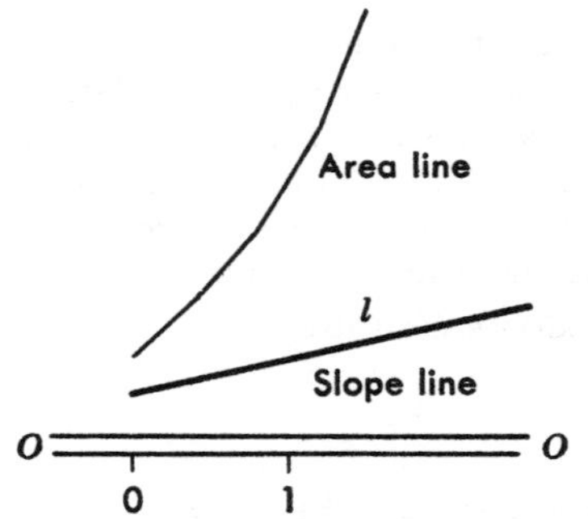

Fig. 21 (a)

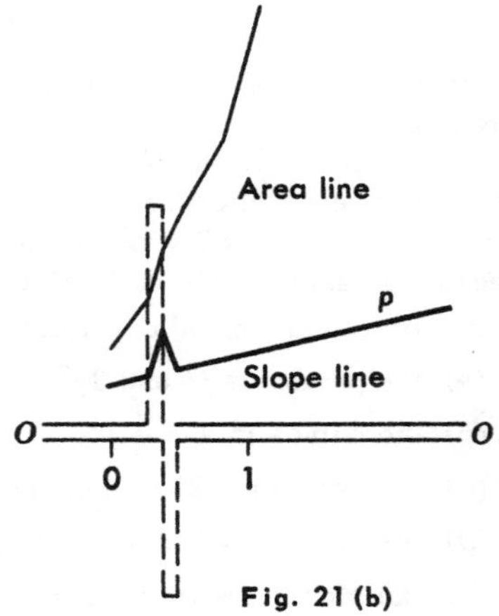

Fig. 21 (b)

slightly different from that of the former (and the derivatives of finer and finer inscribed polygons come closer and closer to **D** f). If f is not smooth, then the transition to a finer polygon may result in a complete change of the derivative (and f may not have any slope line **D** f).

6. THE FUNDAMENTAL IDEA OF CALCULUS SUGGESTED BY THE GRAPHICAL SOLUTION

In Fig. 11, the polygon p is an approximate integral of the circular arc f. Its derivative, **D** f, is the step line by which f has been replaced. If p' is a better approximate integral of f, the step line **D** p' is even closer to f.

These facts suggest that a precise integral of the circular arc is a curve whose slope line is f itself; in other words, that a precise area line of f is an antiderivative of f. This guess will later be substantiated.

In fact, it will be shown *that an integral of any curve f is an antiderivative of f and vice versa,* and that *the area under the portion of f from* a *to* b *is equal to the rise of an antiderivative of f from* a *to* b.

If, as Figs. 16 and 18 suggest, **D** *sin* = *cos* (the derivative of the sine curve is the cosine curve), then *sin* is an antiderivative of *cos*. This would imply that $\int_0^{\pi/2} cos$ is equal to $[sin]_0^{\pi/2}$ (the rise of the sine curve from 0 to $\pi/2$). This rise is $1 - 0 = 1$. It will be seen later that the area $\int_0^{\pi/2} cos$ indeed equals 1.

CHAPTER III

NUMERICAL SOLUTIONS OF THE TWO BASIC PROBLEMS

1. THE AREA UNDER A STEP LINE

Let q^* be a step line whose projection on O is from a to b, and let n be the number of steps of q^* (Fig. 22). To find $\int_a^b q^*$ numerically, measure the bases of the n steps,

$x_1 - a,\ x_2 - x_1,\ x_3 - x_2, \ldots, x_{n-1} - x_{n-2},\ b - x_{n-1},$

and their altitudes. If the latter are

$c_1,\ c_2,\ c_3, \ldots, c_{n-1},\ c_n,$

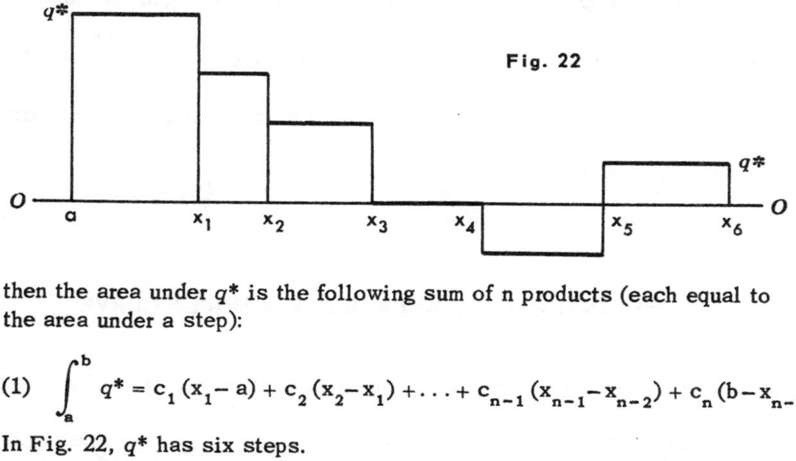

Fig. 22

then the area under q^* is the following sum of n products (each equal to the area under a step):

$$(1) \quad \int_a^b q^* = c_1(x_1 - a) + c_2(x_2 - x_1) + \ldots + c_{n-1}(x_{n-1} - x_{n-2}) + c_n(b - x_{n-1}).$$

In Fig. 22, q^* has six steps.

The step line r^* is called *regular* if the bases of all its steps are equal. If r^* has n steps, each has the base $\frac{b-a}{n}$ and the product sum in (1) can be simplified by factoring out $\frac{b-a}{n}$:

$$(2) \quad \int_a^b r^* = (c_1 + c_2 + \ldots + c_{n-1} + c_n) \cdot \frac{b-a}{n}.$$

$\int_a^b r^*$ can be found by an addition and one multiplication, whereas n multiplications and an addition were necessary to find $\int_a^b q^*$.

Another way of writing (2) is

$$(3)\qquad \int_a^b r^* = \frac{c_1 + c_2 + \ldots + c_{n-1} + c_n}{n} \cdot (b - a)\,.$$

The first factor in (3) is the average of the altitudes of the n steps. Thus *the area under a regular step line r* is equal to that of a rectangle whose projection on O coincides with that of r* and whose height is the average of the altitudes of the steps of r*.*

EXERCISES

1. Show that for any step line q^*, the area $\int_a^b q^*$ lies between $c_{min} \cdot (b - a)$ and $c_{max} \cdot (b - a)$, where c_{min} (and c_{max}) is the altitude of the lowest (the highest) step of q^*. What does that mean geometrically?

*2. If a step line q^* is symmetric about the point 0, then $\int_{-a}^{a} q^* = 0$ for any point a belonging to the projection of q^* on *O*. If q^* is symmetric about the vertical line through 0, then $\int_{-a}^{a} q^* = 2 \int_0^a q^*$.

2. THE APPROXIMATE AREA UNDER A SIMPLE CURVE

To find $\int_a^b f$ approximately by numerical methods, replace *f* by a step line as in Chapter II. But whereas for graphical purposes step lines with various bases were chosen, for numerical purposes it is advisable to replace *f* by a *regular* step line. To standardize the procedure, choose regular step lines for which the midpoint of each horizontal segment lies on the curve *f*. Such a step line will be said to be *adapted* to *f*.

Fig. 23 illustrates the procedure for the quarter circle about 0 of radius 2, which has the altitude $\sqrt{4 - x^2}$ above the point x. This curve, denoted by *f*, has been replaced by the step line r_5^*, adapted to *f* and consisting of 5 steps. The bases of the steps are of length 2/5. Their altitudes are equal to those of *f* above the points 1/5, 3/5, 5/5, 7/5 and 9/5, which are

$$\frac{\sqrt{99}}{5} = \frac{9.95}{5},\ \frac{\sqrt{91}}{5} = \frac{9.54}{5},\ \frac{\sqrt{75}}{5} = \frac{8.66}{5},\ \frac{\sqrt{51}}{5} = \frac{7.14}{5},\ \frac{\sqrt{19}}{5} = \frac{4.36}{5}\,.$$

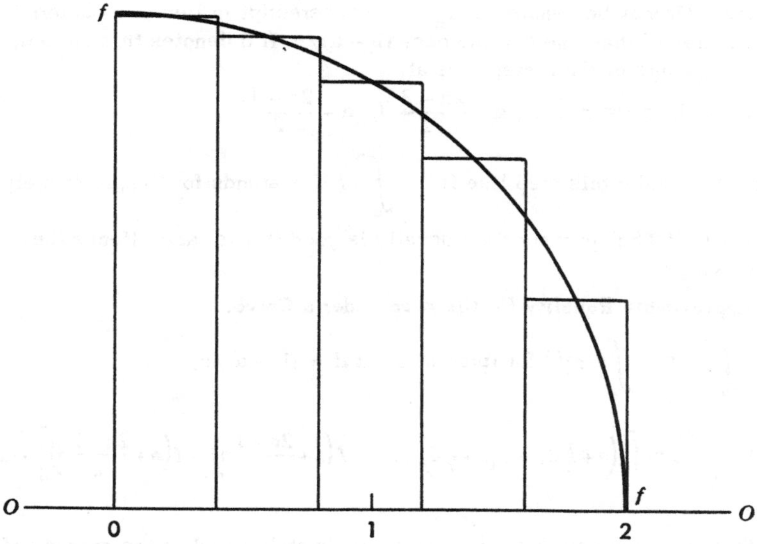

Fig. 23

According to (3)

$$\int_0^2 r_5^* = \frac{9.95 + 9.54 + 8.66 + 7.14 + 4.36}{5} \cdot \frac{2}{5} = 3.17.$$ Thus $\int_0^2 r_5^*$ differs from $\int_0^2 f$ (which equals $\pi = 3.14159\ldots$) by less than 1% of π.

For the line r_{10}^* that is adapted to f and has 10 steps

$$\int_0^2 r_{10}^* =$$

$$\frac{\sqrt{399}+\sqrt{391}+\sqrt{375}+\sqrt{351}+\sqrt{319}+\sqrt{279}+\sqrt{231}+\sqrt{175}+\sqrt{111}+\sqrt{39}}{10} \cdot \frac{2}{10}$$

$$= 3.15.$$

This number differs from π by less than 1/3% of π.

Now let f be any simple curve above the segment from a to b. For any integer n, there is one and only one step line that is adapted to f and has n steps. It may be denoted by $r_n^{(f)}$ (the superscript in lieu of an asterisk). Each step of that line has the base $(b - a)/n$. If d denotes this number, the midpoints of the n steps are at

$$a + \tfrac{1}{2}\,d,\quad a + \tfrac{3}{2}\,d,\ \ldots,\ a + \frac{2n-3}{2}\,d,\quad a + \frac{2n-1}{2}\,d.$$

The area under this step line is $\sim \int_a^b f$ if $\sim$ stands for "approximately equal to." The approximation usually is good if n is large. Hence the following

Approximate Equality for the Area under a Curve.

$$\int_a^b f \sim \int_a^b r_n^{(f)} \text{ for large n, or, if } d = (b-a)/n,$$

$$(4)\quad \int_a^b f \sim \left[f\left(a+\tfrac{1}{2}d\right) + f\left(a+\tfrac{3}{2}d\right) + \ldots + f\left(a+\frac{2n-3}{2}d\right) + f\left(a+\frac{2n-1}{2}d\right)\right]\cdot d.$$

One may also say that $\int_a^b f$ is approximately equal to the *average of the* n *mentioned values of f multiplied by* $b - a$. As an example, consider the area under the parabola j^2 from 0 to b. Here $d = b/n$. Hence

$$\int_0^b j^2 \sim \left[\left(\frac{1}{2}\cdot\frac{b}{n}\right)^2 + \left(\frac{3}{2}\cdot\frac{b}{n}\right)^2 + \ldots + \left(\frac{2n-3}{2}\cdot\frac{b}{n}\right)^2 + \left(\frac{2n-1}{2}\cdot\frac{b}{n}\right)^2\right]\cdot\frac{b}{n}$$

and, after a simple transformation,

$$\int_0^b j^2 \sim \left[1^2 + 3^2 + \ldots + (2n-3)^2 + (2n-1)^2\right]\cdot\frac{b^3}{4n^3}.$$

For $n = 10$,

$$\int_0^b j^2 \sim \left[1^2 + 3^2 + \ldots + 17^2 + 19^2\right]\cdot\frac{b^3}{4000} = \frac{1330}{4000}\,b^3 = .3325\,b^3;$$

e.g., $\int_0^2 j^2 \sim 2.66$. These are good approximations since it can (and later will) be shown that

$$\int_0^b j^2 = \frac{1}{3}\,b^3 \quad\text{and}\quad \int_0^2 j^2 = 2.\dot{6}\ldots$$

EXERCISES

3. Apply the Approximate Equality to the segment from a to b of the straight line cj. Compare the result with the expression, known from plane geometry, for the area of a trapezoid. Why is the result of (4) in this example accurate?

4. Find approximately

(a) $\int_0^2 j^3$ using 8 steps; (b) $\int_1^3 j^{-1}$ using 8 steps (see Ex. 1, Ch. II);

(c) $\int_1^2 j^{-1}$ using 10 steps; (d) $\int_{.5}^{10.5} log_{10}$ using 10 steps (see Ex. 4, p.23). By what modification of the procedure could $\int_1^{10} log_{10}$ be found from the logarithms of the integers 1, 2, . . . , 10?

*5. Find approximately $\int_0^1 j^2$ and $\int_0^1 j^{1/2}$. What is the sum of the two results? By geometric considerations, for each number, a, determine the number b for which $\int_0^a j^{1/2} + \int_0^b j^2 = a \cdot b$.

6. To obtain a very rough approximation of $\int_0^1 f$, one may compute $\int_0^1 r_1^{(f)} = f(1/2)$, the area under a single step. In general, one obtains a better approximation to $\int_0^1 f$ by computing $\frac{1}{2}\left[f(1/4) + f(3/4)\right]$. But in this computation one does not utilize the value $f(1/2)$ that had been determined in finding the area under $r_1^{(f)}$. With about the same effort that is required for the computation of the area under $r_2^{(f)}$, that is, by computing two values of f besides $f(1/2)$, one can determine

$$\int_0^1 r_3^{(f)} = \frac{1}{3}\left[f(1/6) + f(3/6) + f(5/6)\right],$$

which, in general, is even closer to $\int_0^1 f$ than is the area under $r_2^{(f)}$.

More generally, if $\int_0^1 r_n^{(f)}$ has been determined and a better approximation is desired, which should one rather compute *(a)* the area under $r_{2n}^{(f)}$ or under $r_{3n}^{(f)}$? *(b)* the area under $r_{4n}^{(f)}$ or under $r_{5n}^{(f)}$?

*7. Not for *every* curve f does $m < n$ imply that $\int_a^b r_n^{(f)}$ is a better approximation to $\int_a^b f$ than is $\int_a^b r_m^{(f)}$. Let f denote the polygon joining

the points of altitude:	7/4,	0 ,	7/4,	1 ,	7/4,	0 ,	7/4
above the points:	0 ,	1/4,	3/8,	1/2,	5/8,	3/4,	1 .

By plane geometry determine the precise number $\int_0^1 f$. Then find $\int_0^1 r_1^{(f)}$ and $\int_0^1 r_2^{(f)}$.

3. HOW ACCURATE IS THE APPROXIMATE EQUALITY? A FUNDAMENTAL INEQUALITY

The Approximate Equality for $\int_a^b f$ is unsatisfactory for two reasons: it is not accompanied by a stipulation as to *how large* n should be to yield a good approximation and it does not say *how good* the approximation is in any case. *Any* number is "approximately equal" to $\int_a^b f$ (and, in fact, to any other number) unless bounds for the inaccuracy of the approximation are stipulated.

Without such bounds, 2 feet may be said to approximately equal the height of an adult, 200 years approximately equal to his age.

Fortunately, bounds for the inaccuracy of the approximate area are available. Divide the segment from a to b into n equal parts and denote the altitudes of the *lowest* points of f above the 1st, the 2nd, . . . , the nth part, by $\check{c}_1^{(n)}, \check{c}_2^{(n)}, \ldots, \check{c}_n^{(n)}$. The n steps with those altitudes form a step line *below* f (see Ex. 13, p. 27). Corresponding to the division of the segment into n parts, the curve f is divided into n portions, $f_1, f_2, \ldots, f_n$. By the lowest point of f above the kth part is meant the altitude of that point in the portion f_k whose altitude above O is less than that of any other point of f_k. (If f_k should have several pits of minimum altitude, then $\check{c}_k^{(n)}$ denotes the altitude of any one of these lowest points of f_k.)

Similarly, if $\hat{c}_1^{(n)}, \hat{c}_2^{(n)}, \ldots, \hat{c}_n^{(n)}$ denote the altitudes of the highest points of f above the respective parts, the steps with these altitudes form a step line *above* f. Hence the

Fundamental Inequality. For any n,

$$(5)\quad \left[\check{c}_1^{(n)}+\check{c}_2^{(n)}+\ldots+\check{c}_n^{(n)}\right]\cdot\frac{b-a}{n} \leqq \int_a^b f \leqq \left[\hat{c}_1^{(n)}+\hat{c}_2^{(n)}+\ldots+\hat{c}_n^{(n)}\right]\cdot\frac{b-a}{n}.$$

Clearly, also the area under the step line $r_n^{(f)}$ (that is adapted to f and has n steps) lies between the same two bounds. Hence the difference between

$\int_a^b f$ and $\int_a^b r_n^{(f)}$ cannot exceed the difference between the two bounds,

that is,

$$(6)\quad \left[(\hat{c}_1^{(n)}-\check{c}_1^{(n)})+(\hat{c}_2^{(n)}-\check{c}_2^{(n)})+\ldots+(\hat{c}_n^{(n)}-\check{c}_n^{(n)})\right]\cdot\frac{b-a}{n}.$$

For any n, the number (6) is therefore a bound for the inaccuracy of the Approximate Equality (4).

In the example of $\int_0^b j^2$, for any portion of the parabola the initial point is the lowest, the terminal point the highest point. Thus the Fundamental Inequality yields

$$\left[\left(0\cdot\frac{b}{n}\right)^2+\left(1\cdot\frac{b}{n}\right)^2+\ldots+\left((n-1)\frac{b}{n}\right)^2\right]\cdot\frac{b}{n}\leqq\int_0^b j^2$$

$$\leqq\left[\left(1\cdot\frac{b}{n}\right)^2+\ldots+\left(n\cdot\frac{b}{n}\right)^2\right]\cdot\frac{b}{n},$$

or

$$(7)\quad \left[0^2+1^2+\ldots+(n-1)^2\right]\cdot\frac{b^3}{n^3}\leqq\int_0^b j^2\leqq\left[1^2+2^2+\ldots+n^2\right]\cdot\frac{b^3}{n^3}.$$

The difference between this upper and this lower bound is $n^2\cdot\frac{b^3}{n^3}$ or b^3/n. The smaller that difference, the shorter the interval to which $\int_0^b j^2$

has been confined. In the case of b = 2, the difference is .008 if n = 1000, and .0008 if n = 10,000. In other words, by choosing n = 10,000, one can determine two bounds (i.e., numbers with $\int_0^2 j^2$ between them) that differ by less than .001.

The fact that π lies between the numbers 3.1425 and 3.1415 (that differ by not more than .001) is expressed by saying that π equals 3.142 to three decimals. Thus by choosing n = 10,000, one can determine $\int_0^2 j^2$ to three decimals.

The actual determination of those bounds presupposes the computation of $1^2 + 2^2 + \ldots + 10{,}000^2$. Fortunately, these sums can be found without going through the tedious process of actual addition, because of the following law:

$$(E_n) \quad 1^2 + 2^2 + \ldots + (n-1)^2 + n^2 = \frac{1}{3}n^3 + \frac{1}{2}n^2 + \frac{1}{6}n \text{ for any n.}$$

Arithmetical laws of this kind are frequently discovered (as are physical laws) by a guess based on the inspection of special cases. For instance, if one observes that

$$1 = 1, \quad 1 + 3 = 4, \quad 1 + 3 + 5 = 9, \quad 1 + 3 + 5 + 7 = 16,$$

he may guess that, for any integer n, the sum of the first n odd numbers is equal to n^2, that is,

$$1 + 3 + \ldots + (2n-3) + (2n-1) = n^2.$$

A few further trials confirm the guess. Beyond such further verifications in special cases, arithmetical laws (in contrast to physical laws) are capable of general proofs. For instance, one proof of the law concerning odd numbers reads as follows:

Assume the law to be valid for the first n − 1 odd numbers; that is to say, assume that

$$1 + 3 + \ldots + (2n-3) = (n-1)^2.$$

By adding 2n − 1 on both sides, one concludes

$$1 + 3 + \ldots + (2n-3) + (2n-1) = (n-1)^2 + (2n-1).$$

Since the right side equals n^2, the law has been proved for n under the assumption of its validity for n − 1. Now the law is valid for 4 (and smaller numbers); thus, according to what has been proved, it is valid for 5. Since it is valid for 5, it is valid for 6; and so on. For this reason mathematicians are satisfied that the law is valid for any n. A reasoning of this kind is called a *proof by induction*.

It is easy to guess the law about the sum of odd numbers. It requires greater mathematical inventiveness to guess the law about the sum of the squares expressed in (E_n). But once one has somehow conjectured this law, he can fairly easily substantiate the guess by a proof by induction. First one verifies the formula

$$(E_1) \quad 1^2 = \frac{1}{3} + \frac{1}{2} + \frac{1}{6}.$$

Then one proves (E_n) under the assumption that the law is valid for n−1; that is to say, that

$$(E_{n-1}) \quad 1^2 + 2^2 + \ldots + (n-1)^2 = \frac{1}{3}(n-1)^3 + \frac{1}{2}(n-1)^2 + \frac{1}{6}(n-1).$$

The sum on the right side equals $\frac{1}{3}n^3 - \frac{1}{2}n^2 + \frac{1}{6}n$. Adding n^2 to both sides, one obtains (E_n). Thus the law holds for 2, for 3, . . . and for any n.

Applying (E_n) to (7) one finds:

$$(7') \quad \left[\frac{1}{3} - \frac{1}{2n} + \frac{1}{6n^2}\right] b^3 \leqq \int_0^b j^2 \leqq \left[\frac{1}{3} + \frac{1}{2n} + \frac{1}{6n^2}\right] b^3, \text{ for any n.}$$

For instance,

$$\left[\frac{1}{3} - \frac{1}{2 \cdot 10^3} + \frac{1}{6 \cdot 10^6}\right] \cdot 8 \leqq \int_0^2 j^2 \leqq \left[\frac{1}{3} + \frac{1}{2 \cdot 10^3} + \frac{1}{6 \cdot 10^6}\right] \cdot 8, \text{ and}$$

$$\left[\frac{1}{3} - \frac{1}{2 \cdot 10^4} + \frac{1}{2 \cdot 10^8}\right] \cdot 8 \leqq \int_0^2 j^2 \leqq \left[\frac{1}{3} + \frac{1}{2 \cdot 10^4} + \frac{1}{2 \cdot 10^8}\right] \cdot 8.$$

The larger n in (7′), the smaller the terms within the brackets, except $\frac{1}{3}$. Therefore, the above inequalities suggest that $\int_0^b j^2$ cannot be anything but $\frac{1}{3}b^3$; in particular, that $\int_0^2 j^2 = \frac{8}{3} = 2.\dot{6}$.

If f is the quarter of the circle of radius 2, the Fundamental Inequality reads:

$$\left[\sqrt{4 - \left(\frac{2}{n}\right)^2} + \sqrt{4 - \left(\frac{4}{n}\right)^2} + \ldots + \sqrt{4 - \left(\frac{2n}{n}\right)^2}\right] \cdot \frac{2}{n} \leqq \int_0^2 f$$

$$\leqq \left[\sqrt{4 - \left(\frac{0}{n}\right)^2} + \sqrt{4 - \left(\frac{2}{n}\right)^2} + \ldots + \sqrt{4 - \left(\frac{2n-2}{n}\right)^2}\right] \cdot \frac{2}{n}.$$

The difference between this upper and this lower bound is $\sqrt{4 - \left(\frac{0}{n}\right)^2} \cdot \frac{2}{n} = \frac{4}{n}$, which is small if n is large. However, to find the bounds themselves one has to add n square roots and no fortunate circumstances make this tedious task unnecessary. For instance, if one chooses n = 1000, he has to compute the following sum of 1000 numbers:

$\sqrt{4 - \left(\frac{2}{1000}\right)^2} + \sqrt{4 - \left(\frac{4}{1000}\right)^2} + \ldots + \sqrt{4 - \left(\frac{2000}{1000}\right)^2}$; and if the bounds are to differ by at most .004, then n must be chosen at least as large as 1000.

In general, both accuracy and speed are desirable. In most cases — and the following remarks apply to approximate solutions of problems in general — the tendencies for speed and accuracy conflict with one another. To arrive at an answer quickly one will choose a small number of steps; but then the result may

be too inaccurate. To attain high accuracy may take too much time. Frequently one has no choice: only a limited amount of time is available, or a definite accuracy must be attained. But if one has to solve a problem in which neither the number of steps nor the accuracy is prescribed, then he has to reach a compromise between the conflicting tendencies and somehow to choose n.

EXERCISES

8. How large must n be chosen if $\int_0^2 j^2$ is to be enclosed between two bounds differing by less than .000001, and what are these bounds? Answer the same question for $\int_0^1 j^2$, $\int_0^3 j^2$, $\int_{-1}^0 j^2$, and $\int_{-1}^2 j^2$.

9. Assuming that $\int_0^b j^2 = \frac{1}{3} b^3$, find *(a)* $\int_a^b j^2$, *(b)* $\int_{-b}^b j^2$ and determine which is closer to $\int_a^b j^2$: the average of $\check{c} \cdot (b - a)$ and $\hat{c} \cdot (b - a)$ (where $\hat{c}$ and $\check{c}$ are the maximum and minimum altitude of j^2 between a and b) or $\int_a^b r_1^{(j^2)}$?

10. By mathematical induction, prove $1^2 + 3^2 + 5^2 + \ldots + (2n-1)^2 = \frac{4}{3} n^3 - \frac{1}{3} n$. Use this formula in the Approximate Equality for $\int_0^b j^2$. Show that, for each n, the number $\int_0^b r_n^*$ in (4) lies between the bounds in (7′). Compare the difference between $\int_0^b r_n^{(j^2)}$ and $\int_0^b j^2$ with the difference, b^3/n, between those two bounds.

11. Let c be any number. Apply the Fundamental Inequality

(a) to the trapezoidal area, $\int_a^b (cj)$;

(b) to the area under the parabola whose altitude above x is cx^2, that is, to $\int_0^b (cj^2)$;

(c) to $\int_{-a}^{0} (cj^2)$.

$\star$12. Apply the Fundamental Inequality first to $\int_0^b j^3$ (Cavalieri, 1647) and then, in particular, to $\int_0^1 j^3$ (choosing $n = 100$) and to $\int_0^2 j^3$ (choosing $n = 200$). How large must n be chosen if the bounds for $\int_0^1 j^3$ are to differ by less than .001? Before one can answer these questions, he must find $1^3 + 2^3 + \ldots + (n-1)^3 + n^3$. In this case, the reader may be able to guess the law by inspecting the four numbers

$$1^3, \quad 1^3 + 2^3, \quad 1^3 + 2^3 + 3^3, \quad 1^3 + 2^3 + 3^3 + 4^3 .$$

Once the right conjecture is formed, it will be easy to substantiate it by a proof by induction.

13. Apply the Fundamental Inequality to $\int_0^b j^8$. For which n do the upper and lower bounds differ by less than .01? The actual computation of the bounds is too difficult, since there is no simple expression for $1^8 + 2^8 + \ldots + n^8$.

14. Apply the Fundamental Inequality to the semicircle of radius 2 with center at 0; and to the quarter of the ellipse whose projection on *O* is from 0 to 3, and whose altitude above x is $\frac{1}{2}\sqrt{9 - x^2}$. Choose the number of steps.

15. Non-regular step lines were used by Fermat (one of the great precursors of Newton and Leibniz) as early as 1644 in his determination of the area under any power curve $j^{p/q}$ where p and q are integers and $p/q > 0$. To find $\int_0^b j^{p/q}$ approximately, choose a positive number $t < 1$ and divide the segment from 0 to b into infinitely many segments in the following way (illustrated in Fig. 24 in the case $p/q = 1/2, b = 1.3, t = 1/2$): From right to left, the first step is from tb to b, the second from t^2b to tb, ... ,

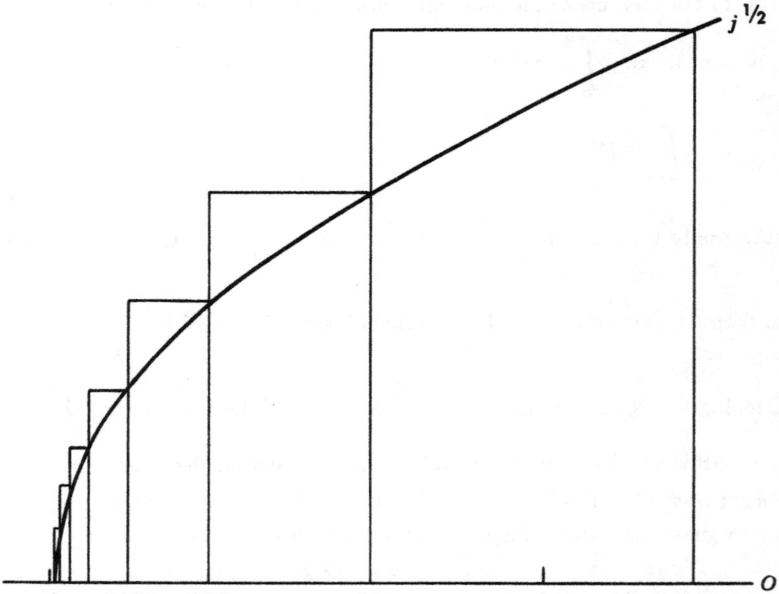

Fig. 24

the kth step from $t^k b$ to $t^{k-1}b$, for any integer k. Thus the width of the kth step is $bt^{k-1}(1-t)$. If $p/q > 0$, then the maximum altitude of $j^{p/q}$ in any interval is at its terminal point. Thus the step of altitude $(t^{k-1}b)^{p/q}$ will be *above* $j^{p/q}$, the step of altitude $(t^k b)^{p/q}$ *below* $j^{p/q}$, in the kth interval. The area under the step line (with infinitely many steps) above the curve is the sum of a geometric progression which is easily seen to be $b^{(p/q)+1} \cdot \dfrac{1-t}{1-t^{(p/q)+1}}$. The area under the step line below the curve is the last product multiplied by $t^{p/q}$. Hence

$$b^{(p+q)/q} \cdot \frac{1-t}{1-t^{(p+q)/q}} \cdot t^{p/q} \leqq \int_0^b j^{p/q} \leqq b^{(p+q)/q} \cdot \frac{1-t}{1-t^{(p+q)/q}} \cdot$$

(If $-1 < p/q < 0$, the situation is reversed.) To obtain a good approximation one has to choose t close to 1. For instance, if $p/q = 2$, then

$$b^3 \cdot \frac{1-t}{1-t^3} t^2 \leqq \int_0^b j^2 \leqq b^3 \cdot \frac{1-t}{1-t^3} \cdot$$

If $t = .999$, the ratio of the bounds is $.999^2$. Compute the bounds.

Show that Fermat's method is applicable even if $-1 < p/q < 0$, although in this case $\int_0^b j^{p/q}$ is the area of a domain that cannot be enclosed in a rectangle of finite altitude. In contrast to the cases where $p/q > 0$, for $p/q < 0$ the curve $j^{p/q}$ has its highest point above the initial point of each interval. Why is the method inapplicable to $\int_0^b j^{p/q}$ for $p/q < -1$? (Hint. What is the law about geometric progressions?)

Apply Fermat's method to $\int_a^\infty j^{p/q}$ for $p/q < -1$, where $\int_a^\infty$ means the area between the curve and the half of the line O to the right of a (> 0). In this case, too, the highest point of the curve is above the initial point of each interval. But the half line from a on must be divided into infinitely many segments getting longer and longer; namely, for a number $t > 1$, into the intervals from a to ta, from ta to t^2a, from t^2a to t^3a, etc.

16. Show that, for any positive integer n,

$$\frac{1}{n+1} + \frac{1}{n+2} + - \cdots + \frac{1}{2n} < \int_1^2 j^{-1} < \frac{1}{n} + \frac{1}{n+1} + - \cdots + \frac{1}{2n-1}.$$

What is the difference between the two bounds? For $n = 1, 2, 3$, the lower bounds are

$$\frac{1}{1+1} = 1 - \frac{1}{2},$$

$$\frac{1}{2+1} + \frac{1}{3+1} = 1 - \frac{1}{2} + \frac{1}{3} - \frac{1}{4},$$

$$\frac{1}{3+1} + \frac{1}{4+1} + \frac{2}{5+1} = 1 - \frac{1}{2} + \frac{1}{3} - \frac{1}{4} + \frac{1}{5} - \frac{1}{6}.$$

By induction, prove

$$\frac{1}{n+1} + \frac{1}{n+2} + \cdots + \frac{1}{2n} = 1 - \frac{1}{2} + \frac{1}{3} - \frac{1}{4} + \cdots + \frac{1}{2n-1} - \frac{1}{2n},$$

and hence

$$1 - \frac{1}{2} + \frac{1}{3} - \frac{1}{4} + - \cdots + \frac{1}{2n-1} - \frac{1}{2n} < \int_1^2 j^{-1} < 1 - \frac{1}{2} + \frac{1}{3} - \frac{1}{4} + - \cdots + \frac{1}{2n-1}.$$

(Brouncker, 1668.)

4. THE APPROXIMATE SLOPE OF A CURVE AT A POINT

An approximate numerical computation of the slope of j^2 at .75 follows the construction of the tangent to j^2 at the point above .75 described on page 34. One computes the slopes of the chords of j^2 between .75 and 1 and between .75 and .875. They are

$$\frac{1^2 - .75^2}{1 - .75} = 1.75 \quad \text{and} \quad \frac{.875^2 - .75^2}{.875 - .75} = 1.625.$$

For any $b \neq .75$, the chord of j^2 between .75 and b has the slope

$$\frac{b^2 - .75^2}{b - .75} = b + .75 = 1.5 + (b - .75).$$

If b is close to .75, that chord is close to the tangent of j^2 at .75. Hence for the slope of the latter, which is $\mathbf{D}\, j^2\, .75$, one finds

$$\mathbf{D}\, j^2\, .75 \sim 1.5 + (b - .75) \sim 1.5 \quad \text{if} \quad b \sim .75.$$

More generally,

Example 1. $\mathbf{D}\, j^2\, a \sim \dfrac{b^2 - a^2}{b - a} = b + a \sim 2a$ if $b \sim a$.

Example 2. If n is any positive integer (1 or 2 or 3 or...), then, if $b \sim a$,

$$\mathbf{D}\, j^n a \sim \frac{b^n - a^n}{b - a} = b^{n-1} + ab^{n-2} + \ldots + a^{n-2}b + a^{n-1}.$$

For instance, $\mathbf{D} j^3\, a \sim \dfrac{b^3 - a^3}{b - a} = b^2 + ab + a^2$, where $b \sim a$. If $b = a + .01$, then $b^2 + ab + a^2 = 3a^2 + .03a + .0001$. Hence $\mathbf{D}\, j^3\, a \sim 3a^2 + .03a + .0001$; for instance $\mathbf{D}\, j^3 2 \sim 12.0601$. Similarly, by choosing $b = a - .0001$ it is seen that $\mathbf{D}\, j^3 2 \sim 11.99939999$.. For $n = 9$,

$$\mathbf{D}\, j^9\, a \sim \frac{b^9 - a^9}{b - a} = b^8 + ab^7 + a^2b^6 + \ldots + a^7b + a^8, \text{ where } b \sim a.$$

Example 3. For the curve $\frac{1}{3} j^3$ having the altitude $\frac{1}{3} x^3$ at x,

$$\mathbf{D}\left(\frac{1}{3} j^3\right) a \sim \frac{(1/3)b^3 - (1/3)a^3}{b - a} = \frac{1}{3}(b^2 + ab + a^2) \text{ where } b \sim a.$$

Similarly, for any number k,

$$\mathbf{D}\, (k\, j^2)a \sim \frac{kb^2 - ka^2}{b - a} = k(b + a) \text{ where } b \sim a.$$

Example 4. $\mathbf{D}\, \sin\, a \sim \dfrac{\sin b - \sin a}{b - a} = \dfrac{2}{b - a} \sin \dfrac{b-a}{2} \cdot \cos \dfrac{b+a}{2}$,

where $b \sim a$. For $a = 0$, $\mathbf{D}\, \sin 0 \sim (\sin b)/b$ where $b \sim 0$.

More generally, for any simple curve *f*, the slope of the chord between a and b is $\dfrac{f b - f a}{b - a}$, called the *difference quotient* between a and b. Consequently, if f has a tangent at a, then its slope (which is denoted by $\mathbf{D} f \mathrm{a}$) satisfies the following

Approximate Equality for the Slope of f at a.

(8) $$\mathbf{D} f \mathrm{a} \sim \frac{f\mathrm{b} - f\mathrm{a}}{\mathrm{b} - \mathrm{a}} \quad \text{for b} \sim \text{a (but} \neq \text{a)}.$$

Of the chords of j^2 between a and the various points between a − 1/n and a + 1/n, the steepest is the one joining a and a + 1/n, the flattest is the chord between a and a − 1/n. The slopes of these two chords are, as one readily sees, 2a + 1/n and 2a − 1/n, respectively. The slope of the tangent to j^2 at a lies between these two numbers; that is to say,

(9) $$2\mathrm{a} - 1/\mathrm{n} \leqq \mathbf{D} j^2 \mathrm{a} \leqq 2\mathrm{a} + 1/\mathrm{n}, \quad \text{for any n}.$$

For instance, by choosing n = 10 and 100, one finds that $\mathbf{D} j^2 .75$ lies between 1.4 and 1.6, and between 1.49 and 1.51. More precisely, the slope of j^2 at .75 can be neither > 1.5 nor < 1.5; in other words it cannot be anything but 1.5. More generally, $\mathbf{D} j^2 \mathrm{a}$ cannot be anything but 2a.

However, this procedure is not always applicable. Suppose, for instance, one tried to determine $\mathbf{D} j^3 0$ in this way. The steepest chord of j^3 between 0 and any point in the interval between $-\frac{1}{\mathrm{n}}$ and $\frac{1}{\mathrm{n}}$ is the chord between 0 and $\frac{1}{\mathrm{n}}$ (which happens to be identical with the chord between 0 and $-\frac{1}{\mathrm{n}}$). Its slope is $1/\mathrm{n}^2$. But there is no flattest chord since each chord has a positive slope and there is always a flatter chord with a smaller slope.

In some cases where no flattest and/or steepest chord exists, bounds for the slope of a tangent can be found in other ways. For instance, in the case of $\mathbf{D} j^3 0$ it is clear that no chord can have a negative slope, wherefore $\mathbf{D} j^3 0 \geqq 0$. Consequently, $\mathbf{D} j^3 0$ is enclosed between 0 and $1/\mathrm{n}^2$, which implies that $\mathbf{D} j^3 0$ cannot be different from 0.

EXERCISES

17. Find approximately $\mathbf{D} j^{-1} \mathrm{a}$, $\mathbf{D} j^{-1} 1$, $\mathbf{D} j^{-1} 2$, $\mathbf{D}(16 j^2)\mathrm{a}$, $\mathbf{D}(\frac{1}{9} j^9)\mathrm{a}$, $\mathbf{D}(\frac{1}{\mathrm{n}} j^{\mathrm{n}})\mathrm{a}$, $\mathbf{D} j^{-2} \mathrm{a}$, $\mathbf{D}\, sin\, 0$, $\mathbf{D}\, sin\, \frac{\mathrm{a}}{2}$, $\mathbf{D}\, cos\, \mathrm{a}$, $\mathbf{D}\, cos\, 0$, $\mathbf{D}\, exp_2 \mathrm{x}$, $\mathbf{D}\, exp_4 \mathrm{x}$.

18. Set up inequalities for $\mathbf{D} j^{-1} 1$, $\mathbf{D} j^{-1} \mathrm{x}$, $\mathbf{D}(\frac{1}{3} j^3)\mathrm{a}$, $\mathbf{D}\, sin\, 0$, $\mathbf{D}\, sin\, \pi$, $\mathbf{D}\, cos \frac{\pi}{2}$, $\mathbf{D}\, exp_2 0$, $\mathbf{D}\, exp_4 0$. Choose n = 100 and n = 1000. Obtain the numbers $2^{.001}$, *sin* .01, etc. from tables. (*sin* .01 is the sine of the angle of .01 radians = $.01 \cdot \frac{180}{\pi}$ degrees). What are the precise values of $\mathbf{D}(\frac{1}{3} j^3)\mathrm{a}$, $\mathbf{D} j^{-1} 1$, $\mathbf{D} j^{-1} \mathrm{a}$ suggested by the inequalities?

19. Set up inequalities for $\mathbf{D}\, cos\, 0$ and $\mathbf{D}\, sin \frac{\pi}{2}$.

5. THE FUNDAMENTAL IDEA OF CALCULUS SUGGESTED BY THE NUMERICAL METHOD

By applying the Fundamental Inequality to j^2 one finds (see p. 47), that $\int_0^b j^2$ cannot be anything but $\frac{1}{3}b^3$. Hence the area curve $\int_0 j^2$ of j^2 that has the altitude 0 at 0 should, at any b, have the altitude $\frac{1}{3}b^3$. In other words, $\int_0 j^2$ should be the curve $\frac{1}{3}j^3$.

From Exercise 18 it is known that $\mathbf{D}(\frac{1}{3}j^3)a$ cannot be anything but a^2. Thus the slope curve (or derivative) of the curve $\frac{1}{3}j^3$ should be the parabola j^2. This implies that $\frac{1}{3}j^3$ is an antiderivative of j^2.

Comparison of the two results reveals that the area curve $\int_0 j^2$ is an antiderivative of j^2, and that the area $\int_0^b j^2$ is equal to the rise from 0 to b of that antiderivative, $\frac{1}{3}j^3$. This bears out the general theorem suggested by the graphical methods.

Another confirmation is obtained as follows: $\int_0^b (cj)$ is the area of a right triangle of base b and height $c \cdot b$. Thus, by plane geometry, $\int_0^b (cj) = \frac{c}{2}b^2$. Consequently, the area curve $\int_0 (cj)$ has, at any b, the altitude $\frac{c}{2}b^2$; that is to say, it is the parabola $\frac{c}{2}j^2$. If this area curve is also an antiderivative of cj, then cj must be the derivative of $\frac{c}{2}j^2$. From Exercise 11a this is known to be actually the case. (Incidentally, this is one of the cases mentioned in Exercise 22 on page 18 where a derivative can be determined from the solution, by elementary geometry, of an area problem.)

If (and only if) one has gone through the difficulties of Exercises 12, 13, and 14, he will fully appreciate the tremendous importance of the fundamental idea of calculus.

To find the number $\int_0^b j^3$ approximately by the numerical method, the following steps have to be taken: one has (1) to guess that $1^3 + 2^3 + \dots + n^3 = (1 + 2 + \dots + n)^2 = \left[\frac{n(n+1)}{2}\right]^2$, (2) to prove this guess by induction, (3) to derive the inequality for $\int_0^b j^3$. All this is feasible but cumbersome. Using, on the other hand, the fundamental idea of calculus, all one has to do is to notice that

$$\mathbf{D}j^4 = 4j^3, \quad \mathbf{D}(cj^4) = 4cj^3, \quad \mathbf{D}(\tfrac{1}{4}j^4) = j^3.$$

Since $\frac{1}{4}j^4$ is an antiderivative of j^3, Newton and Leibniz conclude that $\int_0^b j^3$ is the rise of $\frac{1}{4}j^4$ from 0 to b, that is, $\frac{1}{4}b^4$.

To determine approximately the number $\int_0^b j^8$ by the numerical method, as the first step one would either have to find a simple expression for $1^8 + 2^8 + \dots + n^8$ (which is not an easy task) or to add a large number of 8th powers (which is very tedious and time consuming). But if one notices (see Exercise 17) that $\mathbf{D}(\frac{1}{9}j^9) = j^8$ and hence $\frac{1}{9}j^9$ is an antiderivative of j^8, then all he has to do is to determine the rise of $\frac{1}{9}j^9$ from 0 to b, and he obtains

$$\int_0^b j^8 = \frac{1}{9}b^9; \quad \text{for instance,} \quad \int_0^1 j^8 = \frac{1}{9} \quad \text{and} \quad \int_0^2 j^8 = \frac{512}{9},$$

and this result is precise, not approximate.

To find the approximate area under the circle or the ellipse numerically, one has to add many square roots and, since he could hardly find a simple expression for the sum, he would have to go through the terrific trouble of the addition. The fundamental idea of calculus saves this enormous labor. All one has to do is to determine an antiderivative of the circle or the ellipse. This problem will be solved in Chapter VIII. The rise of the antiderivative is the desired number.

Whenever an antiderivative of a curve f can be found, the calculus solution of the area problem for f is precise and utterly simple.

Why then, one might ask, study at all the graphical and numerical methods, which are quite cumbersome and yield only approximate results? The answer is that Newton's and Leibniz' idea has a serious limitation. While the method of calculus is simple and precise where it is applicable, *there are important curves for which an antiderivative cannot be found and to which therefore the calculus solution of the area problem cannot be applied.*

An example is the curve f whose altitude at x is $\sqrt{1 - .64 \sin^2 x}$. The area problem for curves of this type is important because the period of a pendulum and the length (not the area) of an ellipse can be found if, and only if, $\int_a^b f$ can be determined. (For the latter reason, area lines or integrals of curves such as f are called *elliptic integrals.*) Since an antiderivative cannot be found, graphical or numerical methods must be applied to determine $\int_a^b f$ approximately.

The same is true for the bell-shaped curve having the altitude 2^{-x^2} at x. The area problem for curves of this type is of paramount importance for statistics.

EXERCISES

20. Denote, for the purpose of this exercise, the curve having the altitude 2^{-x^2} at x by g. From the Approximate Equality, find $\int_0^1 g$, and apply the Fundamental Inequality to $\int_0^3 g$. Use $n = 10$. Obtain numbers such as $2^{-.01}$ from logarithm tables.

21. Find approximately the area from 0 to π under the curve of Exercise 17 in Chapter II (an elliptic integral) and compare the result with that obtained by the graphic method.

CHAPTER IV

THE IDEA AND THE USE OF FUNCTIONS

1. REMARKS ON NUMBERS

a. Types of Numbers. In calculus, as here presented, only numbers that are real and finite are considered. For the sake of simplicity, therefore, throughout this book *number* is used as an abbreviation for *finite real number*. The number concept will not here be analyzed. It will be taken for granted that the reader knows how to perform simple operations on numbers, can handle equalities and inequalities, and the like.

Important types of numbers are:

(1) *Integers,* such as 0, 1, −3, $5 \cdot 10^6$.

(2) *Rational* numbers, which are equal to the quotients of two integers, such as

$$\frac{1}{4} = .25,\ -\frac{4}{3} = -1.\dot{3}\ldots,\ \frac{9}{33} = .\dot{2}\dot{7}\ldots,\ 4.8 \cdot 10^{-10},\ 3.1,\ 3.14,\ 3.141;$$

(dots above digits indicate the period of a decimal fraction); any integer is equal to a quotient with the denominator 1, and thus is a rational number.

(3) *Irrational* numbers, that is, numbers which are not rational, such as

$$\log_{10} 2,\ \sqrt{2},\ \sqrt[3]{5},\ (1+\sqrt{5})/2,\ \pi = 3.141\ldots,\ e = 2.71828\ldots.$$

$\sin^2 40° + \log_{10}\sqrt{10} + \cos^2 40°$ has not the *form* of a quotient of two integers. Yet that number can easily be shown to be equal to 3/2, and thus is rational. In contrast, $\log_{10} 2$ can be proved to be unequal to any rational number. If p and q stand for any two positive integers, then 10^p is an integer ending in 0 (one, followed by p zeros), whereas 2^q is an integer ending in 2, 4, 6, or 8 and not in 0. Hence $10^p \neq 2^q$ and, consequently, $10^{p/q} \neq 2$. Thus 10 raised to a power that is a quotient of two positive integers is not equal to 2. However, by the definition of the logarithm of a number, 10 raised to the power $\log_{10} 2$ is equal to 2. Hence $\log_{10} 2$ is not equal to the quotient of two positive integers. (Nor, of course, is $\log_{10} 2$ equal to the quotient of two negative integers, since, if it were, it would also be the quotient of two positive integers.) Thus $\log_{10} 2$ is not a rational number. (The number $\sqrt{2}$ was proved to be irrational by Pythagoras, the number π by Euler.)

Excluded from calculus are:

(1) $\sqrt{-1}$, $2 - \sqrt{-3}$, and the like — called *imaginary* and *complex* numbers in algebra, where they are introduced to make the equations $x^2 + 1 = 0$ and $x^2 - 4x + 7 = 0$ soluble.

(2) Objects of more advanced mathematical studies, referred to as *infinitely large* and *infinitely small* numbers.

b. The Designation of Specific Numbers. The symbols for numbers ($0, 1, -2, \frac{1}{4}, \ldots, e$) are not in this book italicized. The letters e and π denote specific numbers between 2.71828 and 2.71829 and between 3.14159 and 3.14160, respectively.

Another usage must be mentioned. If an investigation requires numerous references to $e^{\pi} + \pi^{e} + \sqrt[3]{5}$, (as would, e.g., a proof that this number is irrational), then instead of repeating every time the cumbersome symbol $e^{\pi} + \pi^{e} + \sqrt[3]{5}$, one would say at the outset:

"Let a denote the number $e^{\pi} + \pi^{e} + \sqrt[3]{5}$" or "Set $a = e^{\pi} + \pi^{e} + \sqrt[3]{5}$." Thereafter (until further notice) one would refer to the number by the letter a and, conversely, would reserve a for the sole purpose of designating that specific number. Such a practice will in this book be called an *ad hoc use* of the letter a – where "ad hoc" means "for this particular purpose." In another study, the letter a may serve as an abbreviation for another long symbol. Of course, other letters also (b, c, . . . , k, . . .) are put to *ad hoc* uses.

c. Numbers and Points. If on a line two points, 0 and 1, have been chosen (as on *O* in Chapter I), then with any number x one can associate the point (or "image") that is x units to the right of the point 0. In this way, any point on *O* is associated with exactly one number (or "mark") indicating how many units the point is to the right of 0. It is convenient (and usually harmless) to use the words "number" and "point on *O*" synonymously; e.g., "4" will designate both the number 4 and the point 4.

d. The Reach of Operations. The reader has learned in his school years that $2 \cdot 3 + 4$ stands for the sum of the product $2 \cdot 3$ and 4, and not for the product of 2 and the sum $3 + 4$. The latter is denoted by $2 \cdot (3+4)$, whereas $2 \cdot 3 + 4$ is an abbreviation of $(2 \cdot 3) + 4$. This fact is sometimes expressed by saying that, on either side of a dot indicating multiplication, only the number next to the dot is within its reach. In contrast, on either side of the + sign indicating addition, the product next to the sign is within its reach. (Such a product may, of course, consist of more than two factors or of only one factor, as in $5 \cdot 2 \cdot 3 + 4$.)

The purpose of this convention is to save parentheses, for example, in expressions such as $(2 \cdot 3) + 4$ and $(2 \cdot 3) + (4 \cdot 7)$, which are written, briefly, $2 \cdot 3 + 4$ and $2 \cdot 3 + 4 \cdot 7$. This convention has become second nature.

e. Differences and Quotients. Likewise, in grade school one learns that there is only one number which, when added to 2, yields the sum 6, and that this number is denoted by $6 - 2$. It equals 4. Similarly, there is a number $0 - 2$, a number $2 - 0$, and a number $0 - 0$. Moreover, there is exactly one number which, when multiplied by 2, yields the product 6. It is denoted by $6/2$ and equals 3. Similarly, there is a number $0/2$, which equals 0. But there is no number whose product with 0 is 2. Thus 2/0 cannot be defined in the same way as other quotients. Neither does the symbol 0/0 denote any specific number, since the product of 0 and *any* number is 0. Thus, division of any number by 0 is undefined and results in a symbol that in this book will remain meaningless.

2/0 and 0/0 are comparable to combinations of words such as "the Pacific coast of the State of New York" and "that city in Illinois which lies in the U.S." Expressions of this kind are highly misleading because one suggests the existence of something that does not exist, the other, the uniqueness of something that is not unique.

f. Negations of Statements. Occasionally it will be necessary to negate statements such as $2 = 3$ and $-1 > 0$. For this purpose it is convenient to use symbols $\neq$, $\ngtr$, and the like, and to express the negations in statements such as $2 \neq 3$, $-1 \ngtr 0$ (instead of which one may, of course, also say $-1 \leqq 0$).

Negating a conjunction, e.g., the conjunction of inequalities,

$$4^2 - 1 > 4 \text{ and } \pi^2 - 1 > \pi \text{ and } (\sqrt{2})^2 - 1 > \sqrt{2} \text{ and } 0^2 - 1 > 0$$

does not mean negating *each* of the inequalities. It means negating at least one of them; that is to say, asserting at least one of the statements

$$4^2 - 1 \leqq 4,\quad \pi^2 - 1 \leqq \pi,\quad (\sqrt{2})^2 - 1 \leqq \sqrt{2},\quad 0^2 - 1 \leqq 0.$$

g. Classes of Numbers. The class concept will not in this book be analyzed. The class consisting of the four numbers 0, 2, 3, 4 will be denoted by $\{0, 2, 3, 4\}$. Here, the arrangement of the elements of the class is irrelevant; e.g., $\{0, 2, 3, 4\} = \{0, 3, 4, 2\} = \dots$. Similarly, the class consisting of the two numbers 1 and -1 is denoted by $\{-1, 1\}$ or $\{1, -1\}$; the class consisting of the single number 0 by $\{0\}$. The following classes of numbers are often referred to in calculus:

(1) the *(universal)* class of *all* numbers or all points on *O*, denoted by U;

(2) the class of all positive (all negative) numbers or of all points on the right (the left) side of the point 0, denoted by P (by N);

(3) the *(vacuous)* class, containing *no* number (no point), denoted by V.

Throughout this book, capitals in roman type will be used to denote classes of numbers (except where all elements of the class are enumerated and enclosed in braces, as in $\{1, -1\}$). If A and B are two classes of numbers, one frequently has to consider

(1) A + B, the class of all numbers belonging to at least one of the classes A, B;

(2) $A \cdot B$, the class of all numbers belonging to both classes A and B. For instance, $P + N$ is the class of all numbers except 0, whereas $P \cdot N = V$. Furthermore, $\{0, 2, 3, 4\} + \{1, 2\} = \{0, 1, 2, 3, 4\}$ and $\{0, 2, 3, 4\} \cdot \{1, 2\} = \{2\}$.

If every number belonging to the class C belongs also to the class A, then C is called a *subclass* of A.

EXERCISES

1. State for each of the following numbers whether it is rational (and in particular an integer) or irrational, and substantiate the statements:

(a) $\sqrt{2 \sin 45°}$;

(b) $\log_{10} (\sqrt{20}) + \log_{10} (\sqrt{50})$;

(c) $10^2 \cdot 10^{1/2}$

(d) $\log_{10} 50$;

(e) $\log_{10} \log_{10} 1000$;

(f) $\sqrt[3]{2}$.

2. What is within reach of the – sign for subtraction *(a)* on the right side; *(b)* on the left side?

3. Denote by $U - \{0, 2, 3, 4\}$ the class of all numbers not in $\{0, 2, 3, 4\}$ and more generally, for any class A, by $U - A$ the class of all numbers not in A. Show *(a)* that P is a subclass of $U - N$; more precisely, that $U - N = P + \{0\}$; *(b)* that A is a subclass of $A + B$; *(c)* that, if C is a subclass of A, then $U - A$ is a subclass of $U - C$.

★4. What rational numbers are equal to two different decimal fractions? Can a number be equal to more than two different decimal fractions? (Hint: $3 \cdot \frac{1}{3} = 1 = 1.\dot{0}$ and $3 \cdot (.\dot{3}) = .\dot{9}$. Hence $1.\dot{0} = .\dot{9}$.)

2. NUMERICAL VARIABLES

a. Examples. The conjunction of the three statements

$$2^4 - 9 \cdot 2^3 + 26 \cdot 2^2 - 24 \cdot 2 = 0,$$

$$3^4 - 9 \cdot 3^3 + 26 \cdot 3^2 - 24 \cdot 3 = 0,$$

$$4^4 - 9 \cdot 4^3 + 26 \cdot 4^2 - 24 \cdot 4 = 0,$$

can be expressed in one single formula containing a *letter* and accompanied by what may be called a *legend;* that is, by directions concerning the use of the letter. For instance, the conjunction may be expressed as follows:

(1) $x^4 - 9 \cdot x^3 + 26 \cdot x^2 - 24 \cdot x = 0$, where x may be replaced by 2 and by 3 and by 4.

In algebra, it is customary to omit dots indicating multiplication – a usage followed in formulas (2)–(7) below. In (1) the dots have been retained so that, at least in one example, statements concerning specific numbers can be obtained by the *mere replacement* of the letter x by numbers.

The conjunction may also be expressed in

(2) $t^4 - 9t^3 + 26t^2 - 24t = 0$, where t may be replaced by 2 and by 3 and by 4.

The letters x in (1) and t in (2) are what are called *numerical variables* or *number variables* – letters accompanied by indications concerning their replacement by numbers.

To be quite precise, one should say that the letter x is to be replaced by *numerals*, that is, *designations of numbers*. However, the distinction between numerals and numbers (and, in general, between symbols and what they symbolize) will not in this book be as carefully maintained as it is in modern books on logic, which distinguish the number 4 and the numeral "4." In calculus, the statement that x is to be replaced by certain numbers can hardly lead to confusion.

The class of all numbers by each of which, according to the legend, a letter may be replaced in a formula is called the *scope* of the variable in the formula. The scope of x in (1) and of t in (2) is the class $\{2, 3, 4\}$.

This scope might be extended to $\{0, 2, 3, 4\}$, since a valid statement also results if x in (1) is replaced by 0. Replacements by 5, –½, and π result in false formulas; that is to say,

(3) $x^4 - 9x^3 + 26x^2 - 24x \neq 0$, if x is replaced by 5, $-\frac{1}{2}$, and π.

In algebra one proves that

(4) $x^4 - 9x^3 + 26x^2 - 24x = x(x - 2)(x - 3)(x - 4)$, where x may be replaced by any number.

The scope of x in (4) is the class U of all numbers. A product equals 0 if and only if at least one of the factors equals 0, and a factor on the right side of (4) equals 0 if x is replaced by 0 or 2 or 3 or 4, and only then. Consequently,

(5) $x^4 - 9x^3 + 26x^2 - 24x = 0$ for any number x in $\{0, 2, 3, 4\}$, and

(6) $x^4 - 9x^3 + 26x^2 - 24x \neq 0$ for any number x not in $\{0, 2, 3, 4\}$,

where the wording of the legends has been simplified. Formulas (5) and (6) are often united in

(7) $x^4 - 9x^3 + 26x^2 - 24x = 0$ if and only if x is in $\{0, 2, 3, 4\}$.

In the statement

(8) $\frac{2}{a^2-1} = \frac{1}{a-1} - \frac{1}{a+1}$ for any number a not in $\{1, -1\}$,

the letter a is a numerical variable whose scope is the class U – $\{1, -1\}$ of all numbers not in $\{1, -1\}$ (see Exercise 3); that is to say, replacing the letter a in (8) by any number other than 1 or –1 results in a valid statement. Replacing a in (8) by 1 or –1 results in a statement containing the meaningless symbol $\frac{1}{1-1}$.

The scope of y in

(9) $\mathit{log}_{10}(100y) = 2 + \mathit{log}_{10}\, y$ for any number $y > 0$

is P, the class of all positive numbers.

The legend in (9) is sometimes given the form "where y stands for any positive number." Accordingly, a numerical variable is also defined as a letter that, in a formula, stands for any element of a certain class of numbers (Weierstrass, 1880). Or the formula in (4) may be preceded by the remark "if x is any number, then" as an abbreviation of the legend "where x may be replaced by any number."

Late Greek arithmeticians used an abbreviation of the word arithmos (= number) as a numerical variable. The wide use of numerical variables began in the Renaissance, especially after the publication of a book by the French algebraist Vieta (1591).

b. A General Law and a Typographical Convention. In a statement containing one numerical variable, that letter may, without any change of meaning, be replaced by any other letter that is not reserved for the designation of specific numbers as are e and π and, occasionally *ad hoc*, a, b, ... (p. 58). An example is the transition from (1) to (2). In (9), y might be replaced, without any change of the meaning, by x, by a, by c,... .

Throughout this book, numerical variables are denoted by lower-case letters in roman type; e.g., *by* x, t, a, c, – *never by capitals and never by italics. Conversely, lower-case letters in roman type (other than* e*) are reserved for numerical variables or ad hoc designations of numbers.* This convention makes the word "number" in legends such as those in (4), (8), and (9) superfluous.

From page 1 on, points on the line *O* and numbers (such as altitudes, areas, and slopes) have been denoted in roman type; lines (such as *O*, *3*, and *j*) in italics. (See p. 5).

c. Miniature Scopes. Two cases deserve special mention.

If the scope consists of exactly *one* number, then the numerical variable really designates one specific number (as do e and π and *ad hoc* designations of numbers).

Quite important is the case of scopes consisting of *two* numbers, especially, of the two square roots of a positive number. That the fourth power of either root of 9 equals 81, and either root increased by 2 belongs to the interval between −8 and 13, might be expressed in terms of a numerical variable with the scope $\{\sqrt{9}, -\sqrt{9}\} = \{3, -3\}$, for instance, in the form

$r^4 = 81$ and $-8 < r + 2 < 13$,
where r may be replaced by either $\sqrt{9}$ or $-\sqrt{9}$.
However, it is customary to use a *self-explanatory* symbol, i.e., a symbol that does not require any legend, such as $\pm\sqrt{9}$ in

$$(\pm\sqrt{9})^4 = 81, \quad -8 < (\pm\sqrt{9} + 2) < 13, \quad |\pm\sqrt{9}| = 3.$$

The inequalities $\pm\sqrt{9} < -5$, $-1 < (\pm\sqrt{9} + 2) < 1$, and $|\pm\sqrt{9}| = 10$ are false.

d. Several Numerical Variables. In algebra one proves that

(10) $x^4 - 9x^3y + 26x^2y^2 - 24xy^3 = x(x-2y)(x-3y)(x-4y)$, where x and/or y may be replaced by any two numbers – briefly, for any x and any y.

If only y is replaced by 1, the result (if powers of 1 are omitted) is (4). If both x and y are replaced by 1, the result (if powers of 1 and factors 1 are omitted) is

$$1 - 9 + 26 - 24 = (1-2)(1-3)(1-4).$$

The meaning of (10) is not changed if x and y are replaced by any two non-identical letters, e.g., by a and b or by x and a, or even if x and y are interchanged, as in

(10′) $y^4 - 9y^3x + 26y^2x^2 - 24yx^3 = y(y-2x)(y-3x)(y-4x)$ for any x and any y.

e. Conjunctive Use of Variables. In all preceding examples, replacement of a numerical variable by a number belonging to its scope results in a *statement*. The numerical variable is a device by means of which many (even infinitely many) statements may be condensed into one. Besides this, as it were *indicative*, use of variables, there are numerous other usages; for instance, what might be called a *conjunctive* use of variables in definitions of classes of numbers, as in

P is the class of all x such that $x > 0$; in a formula, $P = \{x \mid x > 0\}$; and similarly, $N = \{x \mid x < 0\}$; $P + N = \{x \mid x \neq 0\}$. (see p. 59). Thus $\{x \mid \ldots\}$ is read "the class of all x such that … ."

If, in the definition of P, x is replaced by 3, the result is the valid inequality $3 > 0$; but if x is replaced by -2, the resulting inequality $-2 > 0$ is not valid. The formula in such a definition becomes valid if and only if x is replaced by a number belonging to the class that is being defined. In other words, that class consists just of those numbers which make the formula valid. For instance, according to (7),

$$\{ x \mid x^4 - 9x^3 + 26x^2 - 24x = 0 \} = \{ 0, 2, 3, 4 \} .$$

Further examples are

$\{ t \mid t > 1 \}$ is the class of all numbers >1 (the ray to the right of 1);

$\{ a \mid 1 \leq a \leq 2 \}$ is the "closed interval", $\{ a \mid 1 < a < b \}$, the "open interval" between 1 and 2.

f. Imperative Use of Variables. In still another, as it were *imperative*, way variables are used in so-called equations. Such an equation is a task or an order to the reader. He is to look for a number (or for all numbers) of a certain kind. For instance,

Find x such that $x^2 - 1 = 0$. Solution: $x = 1$ or -1.

Find t such that $t^4 - 9t^3 + 26t^2 - 24t = 0$. Solution: t belongs to $\{ 0, 2, 3, 4 \}$.

In the formulas

$$x^3 - x = x(x + 1)(x - 1) \text{ and } x^2 - 1 = (x + 1)(x - 1) \text{ for any } x,$$

x^3 and x^2 are symbols that stand for any number belonging to a certain class and thus, in this restricted sense, numerical variables with the scopes U and U – N, respectively. However, these formulas are not accompanied by articulate rules as to how x^3 and x^2 may be replaced by numbers. Obviously one cannot change them regardless of what he does to x, for instance to 8 while x is replaced by 1 or left unchanged. If the first formula above were accompanied by the legend "where x^3 may be replaced by any number, provided x is replaced by the 3rd root of that number," then x^3 would become an indicatively used numerical variable. Variables without articulate directions as to their replacement by specific elements of their scopes will not be considered in this book. In the literature, symbols such as x^3 and x^2 in the formulas first given are sometimes referred to as *dependent* variables.

EXERCISES

5. What are the numerical variables, if any, in the following formulas (most of which result from statements by the omission of legends);

$$\int_a^b 3 = 3 \cdot (b - a); \quad \mathbf{D} j\, 2 = 1; \quad \mathbf{D}\, l\, 2 = \frac{[l]_2^b}{b - 2}; \quad \mathbf{D}\,(2j) = 2;$$

$$\int_a \mathbf{D}\, l = l - l\, a; \quad \mathbf{D} \int_a 3 = 3\,; \quad (\pm\sqrt{2})^2 = 2; \quad (\sqrt{9})^2 = 9.$$

Formulate those of the statements that contain numerical variables without using letters – that is, in words.

★6. What are the numerical variables

(a) in the Approximate Equality (4) on p. 42 if n therein is replaced by 10;

(b) in the Fundamental Inequality on p. 45 if n therein is replaced by 3.

What is the scope of the numerical variable n in the equality (E_n) on p. 46? Show that the numerical variables in the formula $\mathbf{D}\, j\, a = \dfrac{[j]_a^b}{b - a}$ do not have independent scopes. (Hint: The denominator must not be equal to 0.) What in that formula may be replaced by any element of a certain class is the pair (a,b) rather than a and b separately. Describe that class.

7. What becomes of (10) if both x and y are replaced by the same letter, say t? Find statements that contain two variables and cease to be valid if both are replaced by the same letter. Are the statements obtained from (10) if x and y are replaced *(a)* by 2 and 3, *(b)* by 3 and 2 identical? If they are not, how can (10) and (10′) have the same meaning?

8. Negating the statement "$x^2 - 1 > 0$ for any x" amounts to asserting that in $x^2 - 1 > 0$ one cannot replace x by each number (though one may replace it by some numbers). The latter assertion is often expressed in the form of a so-called *existential statement* "there is at least one number x such that $x^2 - 1 \not> 0$" or "$x^2 - 1 \leqq 0$ for some x". Also one can say $\{ x \mid x^2 - 1 \leqq 0 \} \neq V$. Negate the following statements: *(a)* $x^2 - 1 = 0$ for any x; *(b)* $t^2 - 1 \neq 0$ for any t. Is "$t^2 - 1 \leqq 0$ for some t" equivalent to negating "$x^2 - 1 > 0$ for any x"?

9. *(a)* Describe in words (without the use of numerical variables) the following classes: $\{x \mid x \geqq 0\}$; $\{x \mid x = 0 \text{ or } x > 0\}$; $\{x \mid x \not< 0\}$; $\{x \mid x = 0\}$; $\{x \mid x = 0 \text{ or } x \geqq 1\}$.

(b) Verify $\{x \mid 1 \leqq x \leqq 2\} = \{x \mid 1 < x < 2\} + \{x \mid x = 1 \text{ or } x = 2\}$; $\{x \mid |x| > 1\} = \{x \mid x > 1\} + \{x \mid x < -1\}$; $\{1, -1\} = \{x \mid x = 1 \text{ or } x = -1\}$.

(c) Describe in a different way $\{x \mid x \leqq 2\} + \{x \mid x \geqq 1\}$; what are $\{x \mid x < 1\} \cdot \{x \mid x > 2\}$ and $\{x \mid x > 0\} + \{x \mid x < 1\}$?

(d) Describe in various formulas the class (i) of all numbers whose absolute values are > 2 and < 3; (ii) of all numbers whose squares are < 2; (iii) of all numbers which are either between 2 and 3 or between –3 and –2.

(e) Draw all classes discussed in this exercise on a line with a scale. It is customary to indicate closed intervals by brackets, and open intervals by parentheses; e.g., 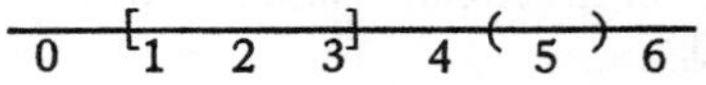 .

10. The class P may be called *the scope of* x *in the symbol* $\log_{10} x$, since if x is replaced by a positive number the result is a number, whereas if x is replaced by 0, by a negative number, by a sphere, etc. the result is not defined (and therefore meaningless) in calculus. Determine the scope in this sense of x in each of the following symbols:

(a) $\log_{10} \log_{10} x$; *(e)* $\sqrt{3 - x^2}$; *(i)* $\dfrac{x^2 - 1}{x - 1}$;

(b) $\log_{10} \sin x$; *(f)* $2^{1/x}$; *(j)* $\dfrac{(x-1)^2}{x - 1}$;

(c) $\sqrt{x^3 - x^2}$; *(g)* $10^{-1/x^2}$;

(d) $\sqrt{x^2 - x^3}$; *(h)* $\dfrac{x^2}{x}$; *(k)* $\dfrac{x - 1}{x^2 - 1}$.

3. THE CONCEPT OF FUNCTION

Example 1. If, with any number, one pairs 16 times the square of that number (e.g., with 0 the number 0, with 1 the number 16, with 1.5 the number 36), then mathematicians say that one has defined a function. The result of that process of pairing, the *function*, is a certain class of pairs of numbers – a class which includes, e.g., the pairs (0,0), (1,16), (1.5,36), (–1,16), $(\sqrt{2}, 32)$, since $16 \cdot 0^2 = 0, \ldots,$ $16(\sqrt{2})^2 = 32$; and which does not include, e.g., the pairs (16,1),

(3, 1/3), and (1, 5), since $16 \cdot 16^2 \neq 1, \ldots, 16 \cdot 1^2 \neq 5$. The process can be conveniently described in terms of a numerical variable: for any x, the number $16x^2$ is paired with x – briefly, with any x, $16x^2$ is paired, or $16t^2$ is paired with any t. The function may be defined as the class of all pairs $(x, 16x^2)$ for any x, or as the class of all pairs $(a, 16a^2)$ for any a, ... Here the scope of the numerical variables x, t, a, ... is the class of all numbers. In this and the following examples, replacement of a numerical variable by a specific number yields a specific pair of numbers belonging to a certain class of such pairs. (The variables are used conjunctively.)

Example 2. If $(4x + 1) \cdot (4x - 1) + 1$ is paired with any x, the operation that yields the number paired with x is not the same as in Example 1, but the result of the operation is. The resulting function, that is, the class of all pairs $(x, (4x + 1) \cdot (4x - 1) + 1)$ for any x, is identical with the class of all pairs $(x, 16x^2)$ for any x; that is to say, any pair belonging to either class belongs also to the other. Hence the two functions are said to be equal. They are also equal to the class of all pairs $(\frac{1}{4}t, t^2)$ for any number t. For t = 0 one obtains the pair (0, 0); for t = 4, the pair (1, 16); for t = 6, the pair (1.5, 36); etc. For no number t can one obtain the pairs (16, 1), (3, 1/3), etc. Also the class of all pairs $(a^3, 16a^6)$ for any a is equal to the function in Example 1.

Example 3. By pairing with any number a that is $\neq 0$ the reciprocal of a, one defines a function different from that in Examples 1 and 2: the class of all pairs (a, 1/a) for any $a \neq 0$. The pairs (3, 1/3), (1/3, 3), and $(\sqrt{2}, \sqrt{2}/2)$ do belong to this function; the pairs (1, 16), (0, 0), (0, 5) do not. Many words can be saved by using symbols for classes of pairs of numbers similar to those for classes of numbers mentioned at the end of Section 2. The function in Example 3 might be described as the class $\{(x, 1/x) \mid x \neq 0\} = \{(a, 1/a) \mid a \neq 0\} = \{(x, 1/x) \mid x \text{ in } P + N\}$ $= \ldots$. Similarly, all functions in Examples 1 and 2 are equal to $\{(x, 16x^2) \mid x \text{ in } U\} = \{(t, 16t^2) \mid t \text{ in } U\}$.

Example 4. The class $\{(t, \frac{1}{4}\sqrt{t}) \mid t \geq 0\}$ results from pairing with each non-negative number $\frac{1}{4}$ times its positive square root. The pairs (0, 0), $(4, \frac{1}{2})$, (16, 1) do belong to this function; the pairs (–16, 1), (16, –1), (–16, –1) do not. The class $\{(16x^2, x) \mid x \geq 0\}$ is equal to the class just defined.

Example 5. The class $\{(t, -\frac{1}{4}\sqrt{t}) \mid t \geq 0\}$ is a function different from that in Example 4. The pairs (0, 0), $(4, -\frac{1}{2})$, (16, −1) do belong to it; the pairs (16, 1) and (−16, 1) do not. The class is equal to $\{(16x^2, x) \mid x \leq 0\}$.

Example 6. The class $\{(16x^2, x) \mid x \text{ in } U\}$ of all pairs $(16x^2, x)$ for any x is not called a function because this term is restricted to the result of a process whereby with any number (or any number in a certain class of numbers) exactly one number is paired. The class in Example 6 contains the pairs (16, 1) and (16, −1) and thus results from a process of pairing with 16 two numbers, 1 and −1. (Incidentally, the class contains all the pairs belonging to either function of the Examples 4 and 5, and only those pairs). On the other hand, the function in Example 1 does contain the pairs (1, 16) and (−1, 16).

Example 7. The class of all pairs (x, *sin* x) for any x, where *sin* x is the sine of an angle of x radians, is a function. The number *sin* x may be computed (see Chapter X) or may be looked up in sine tables. Any number that is ≥ -1 and ≤ 1 is paired with infinitely many numbers; e.g., 0 with 0, π, $-\pi$, 2π, -2π, ····

Example 8. An extreme case in this direction is the class $\{(x,3) \mid x \text{ in } U\}$ of all pairs (x,3) for any x. It results from pairing the number 3 with each number.

Each of the preceding functions is a class of pairs of numbers obtained by pairing exactly one number with any number belonging to a certain class. The latter class of numbers is called the *domain* of the function. It is the class U (of all numbers) in Examples 1, 2, 7, and 8; the class P + N (of all numbers $\neq 0$) in Example 3; the class U − N in Examples 4 and 5. In each pair belonging to a function, the *first* member is a number (belonging to the domain) *with* which a number is paired; the *second* member is a number *which* is paired with the first; and it is called a *value* of the function. For instance, 16 in the pair (1,16) belonging to the function in Example 1 is called the value which that function assumes for 1 – briefly, its value for 1. The term "*the* value for 1" is appropriate since only one number is paired with 1. Determining this value for 1 is called *evaluating* the function for 1. In Example 3, evaluation of the function for 2 yields the value 1/2 or .5; for a, the value 1/a.

The class of all values of a function is called its *range*. In Examples 3 and 4, the range is identical with the domain. The range contains all nonpositive numbers in Example 5; all numbers that are ≥ -1 and ≤ 1 in

Example 7; it consists of the single number 3 in Example 8. A function (as in Example 8) that assumes the same value for any number of the domain is said to be *constant*. Any function whose range contains more than one number is called *non-constant*.

This is the only use of the words "constant" and "non-constant" that so far has been introduced in this book. In particular, the terms "constant number" and "non-constant number" have not been, *and will not be*, herein defined; nor will, for instance, the term "lucky number."

Each of the definitions (Examples 1-8) makes use of a numerical variable — a letter with a legend stipulating by what numbers the letter may be replaced to yield pairs of numbers belonging to the function. The class of those numbers, the scope of the numerical variable, is identical with the domain of the function. E.g., in $\{(x, 1/x) \mid x \neq 0\}$, the class P + N is both the scope of x and the domain of the function. The latter may also be defined (without the use of numerical variables) as the class of all pairs of numbers $\neq 0$ in which the second member is the reciprocal of the first. (P + N is still the domain of the function.)

With the preceding examples in mind, the reader can easily understand the following general definition. *A function is a class of pairs of numbers that does not contain two pairs with equal first members and unequal second members.* The domain (the range) of a function is the class of all numbers that are the first (the second) members in pairs belonging to the function. A function is obtained by pairing exactly *one* number belonging to its range with any number belonging to its domain. Hence $\{(x, 16x^2) \mid x \text{ in } U\}$ (which includes (1, 16) and (−1, 16)) is a function, whereas $\{(16x^2, x) \mid x \text{ in } U\}$ (which includes (16, 1) and (16, −1)) is not.

Example 9. A function need not contain infinitely many pairs of numbers. The class of all (x, y) such that $x^2 + y^2 = 0$ contains only one pair, namely (0, 0), and, like any class consisting of one pair of numbers, falls under the definition of a function. The class of all (x,y) such that $x^2 + y^2 + 1 = 0$ contains no pair of numbers at all. It will be referred to as the *vacuous* function and designated by $\emptyset$. Clearly, the domain of $\emptyset$ is the empty class of numbers, and so is its range.

In Example 9, use has been made of two numerical variables, x and y. Similarly, the function in Example 1 may be defined as the class of all pairs (x,y) such that $y = 16x^2$; that in Example 3, as

(11) the class of all (a,b) such that $a \neq 0$ and $b = 1/a$; etc.

Example 10. (11) might also be defined as the class of all (a, b) such that $a \cdot b = 1$ (in which case the remark $a \neq 0$ is superfluous, since $a \cdot b = 1$ rules out $a = 0$) or as

(11′) the class of all (x, y) such that $x \cdot y = 1$;
of all (x, a) such that $x \cdot a = 1$ (or $a \cdot x = 1$);
of all (y, x) such that $y \cdot x = 1$ (or $x \cdot y = 1$); and the like.

Example 10 contains what are called *implicit definitions* of a function. In contrast, the definitions of that same function in Example 3 or by (11) are called *explicit*. So is its definition as

(11″) the class of all (x,y) such that $x \neq 0$ and $y = 1/x$.

From (11″), the value y for x can be found by computing the reciprocal of x, thus *by an operation not involving the value* y *itself*. To evaluate the function for x on the basis of the implicit definition (11′), first in the condition $x \cdot y = 1$, in which y is involved, both sides must be divided by x.

Similarly, the definition of a function as the class of all (x,y) satisfying the condition $x^2 - \frac{1}{16} y = 0$ is implicit, since operations involving y (adding $\frac{1}{16}$ y to both sides, multiplying by 16, and reading from right to left) have to be performed to make y computable by operations on x such as those described in Example 1. It is slightly harder to realize that the same function may also be defined (implicitly) as the class of all (x, y) such that $16x^4 - yx^2 - y^3 + 16x^2y^2 = 0$. One has to notice that the left side can be written in the form $(16x^2 - y) \cdot (x^2 + y^2) = 0$.

The class of all (y,x) such that $16x^2 - y = 0$ (studied in Example 6) is not a function. In many cases it is very difficult to decide whether a class of pairs (x, y) such that x and y satisfy some condition is a function, and if so, to find its value for given x.

Now return to a plane as described on page 1, with the horizontal line *O*, and consider any of the functions in the preceding examples. For any number x belonging to its domain, mark that point whose altitude above the point x on *O* is equal to the value of the function for x. The class of all points marked is called the *graph* of the function. Since the function pairs not more than one number with any number, the graph has not more than one point in common with any vertical line and thus is what on page 1 has been called simple.

In more advanced branches of mathematics, functions are studied whose domains and ranges consist of complex numbers. The functions considered in calculus are then referred to as *real functions*.

Some texts refer to what in this section have been called functions as *one-valued functions* and introduce many-valued functions by pairing one *or more* numbers with each number of a domain. Examples are the pairing of $\sqrt{x}$ and $-\sqrt{x}$ with any $x \geq 0$, that is, the class of all pairs (x^2, x) for any $x \geq 0$, and the class of all pairs (sin x, x) for any x. The graphs of these "many-valued functions" are the reflections in the line j of the curves j^2 and *sin*, respectively. For such non-simple curves, neither the slope at x nor the area from a to b is defined. In fact, derivatives and integrals are defined only for so-called *branches* of those "many-valued functions," that is, for ("single-valued") functions that can be, as it were, extracted from them; e.g., for the functions $j^{1/2}$ and $j^{-1/2}$ or for the function (called *arcsin*) that consists of all pairs (*sin* x, x) for any x between $-\pi/2$ and $\pi/2$. Never are derivatives and integrals defined for "many-valued functions" themselves. Therefore, in calculus, the introduction of this concept does not serve any useful purpose. (Even in the theory of complex functions, only very special types of many-valued functions are actually studied.)

If A is a class of numbers, then the class consisting of all pairs (a, 1) for any a belonging to A, and all pairs (b, 0) for any number b not belonging to A, is often referred to as the *characteristic function* of A. Its value for any x indicates, as it were, the frequency of x in the class: 1, if x is an element of A; 0, if it is not. E.g., the characteristic function of the class $\{0, 3, 4\}$ consists of the pairs (0,1), (3,1), (4,1), and (x,0) for any $x \neq 0, 3, 4$. According to the use of the word class in mathematics, a class cannot contain an element several times or with a frequency or multiplicity greater than 1. However, the same element of a class may be somehow associated with several elements of another class; it then is said to have a frequency > 1 with respect to the other class.

The class of all numerals in any of the six boxes below

[0] [4] [3] [4] [0] [4]

is $\{0, 3, 4\}$. But of the boxes, two contain "0," one "3," and three "4." If with any x the number of boxes containing "x" (also called the *frequency* of "x" in the boxes) is paired, the resulting class of all pairs

$$(0,2),\ (3,1),\ (4,3),\ (x,0) \text{ for any } x \neq 0,\ 3,\ 4,$$

is called the *frequency function* of the class $\{0, 3, 4\}$ *with respect to the class of boxes.*

A class with a frequency function (of which the class is the domain) is a concept of paramount importance in algebra, statistics, etc. Oddly enough, this concept has no traditional name. It will hereinafter be referred to as a *weighted class.* For instance, the class of all roots of the equation $x^2(x-3)(x-4)^3 = 0$ is $\{0, 3, 4\}$. The weighted class of all roots (each weighted by its multiplicity is $\{0, 3, 4\}$ with the same frequency function that $\{0, 3, 4\}$ has with respect to the boxes.

EXERCISES

11. Find examples of two functions that are not equal although their domains are equal and their ranges are equal. In one example, let both domains be P, both ranges U; in another, both domains N, both ranges U; in a third, both domains as well as both ranges U.

12. Which of the following are pairs of equal functions?

(a) the class of all pairs (x, $\sin^2$ x) for any x and the class of all pairs (t, (1 + cos t) · (1 – cos t)) for any t;

(b) the classes of all (x, sin x) and (x, $\sqrt{1 - \cos^2 x}$) for any x;

(c) the classes of all (x, sin x) and (x, $\sqrt{1 - \cos^2 x}$) for any x that is $\geq -\pi/2$ and $\leq \pi/2$.

(d) the classes of all (x, sin x) and (x, $\sqrt{1 - \cos^2 x}$) for any x that is ≥ 0 and $\leq \pi$.

13. Find the numbers, if any, common (1) to the domains, (2) to the ranges, of the following pairs of functions:

(a) the class of all pairs (a, $a^2 - 1$) for any a, and the class of all pairs (x, (x – 1) · (x – 1)) for any x;

(b) the class of all (x, $\sqrt{x^2}$) for any $x \geq 0$, and the class of all (x, $\sqrt{1 - x^2}$) for any x that is ≥ -1 and ≤ 1;

(c) the functions assuming the values $\sqrt{x^2 - 3}$ and $\sqrt{1 - x^2}$ for any x belonging to the scope of the respective symbol (see Exercise 10).

14. Consider the classes of all pairs (x,y) such that

(a) $x^2 + 3y = 4$; *(b)* $x + 3y^2 = 4$; *(c)* $x^2 + 3y^2 = 4$;

(d) $x^3 + y^3 = 1$; *(e)* $y = x + 2^x$; *(f)* $x + y = 0$;

(g) $y = 2^x + 2^{-x}$; *(h)* $x = 2^y + 2^{-y}$; *(i)* $x = 2^y - 2^{-y}$;

(j) $y = \log_{10} \log_{10}(1 - x^2)$.

Which classes are functions? Which of these functions are explicitly defined and which implicitly? Find explicit definitions for the latter and determine the domains and ranges of all the functions.

15. Which of the following classes (each containing four pairs of numbers) are functions? (In Chapter VII, this exercise will assume great importance.)

(a) (3, 0), (4, 1), (5, 2), (6, 3);

(b) (3, 0), (4, 1), (1, 0), (2, 1);

(c) (3, 2), (4, 6), (6, 2), (7, 1);

(d) (2, 3), (6, 4), (2, 6), (7, 1);

(e) (2, 3), (6, 4), (2, 4–1), (3, 7);

(f) (2+1, 2), (6, 4), (3, 3), (4+2, 4);

(g) (2+1, 2), (2^2, 8 − 2), (log_{10} 1,000,000, log_{10} 100), (7, *cos* 0).

For each function, determine the domain and the range and draw the four-point graph. (Do not join the four points by any fancy curve or polygon.) Which of the functions are equal?

16. Draw the graph of the so-called *sign function* consisting of (0, 0), all pairs (x, 1) for any x > 0, and all pairs (x, −1) for any x < 0. It is customary to write

sgn 0 = 0; *sgn* x = 1 for any x > 0; *sgn* x = −1 for any x < 0.

Then draw the graphs of the functions assuming the values

(a) $sin|x|$; *(b)* $|sin\ x|$; *(c)* $|sin|x||$; *(d)* x^2/x; *(e)* $log_{10} x$;

(f) $log_{10}|x|$; *(g)* $|log_{10} x|$; *(h)* $|log_{10}|x||$, for any x belonging to the scope of the respective symbols (cf. Exercise 10).

17. Draw the graphs of

(a) the function assuming the value 0 for any integer, the value 1 for any number exceeding an even integer by less than 1, and the value −1 for all other numbers;

(b) the function pairing with any number x the difference: x minus the largest integer < x.

Define the functions (a) and (b) as classes of pairs of numbers. (Some beginners are inclined to believe that such functions are without any practical significance. They should be assured that these and similar functions play (under the names of rectangular and triangular waves) a great role in wave analysis.)

★18. Draw a few points of the graph of the function whose domain is the class of all rational numbers of the form p/q where p and q are integers, q odd, and which assumes for p/q the value $(-3)^{p/q}$. For theoretical purposes, mathematicians consider even the function assuming the value 1 for any rational number, and the value 0 for any irrational number. Draw a few points of its graph above the interval from 0 to 1.

4. NAMES AND SYMBOLS FOR FUNCTIONS

Some functions do have traditional names and symbols, others do not. Of the functions mentioned in Section 3 of this chapter, only those in Examples 3, 4, and 7 do. They are called the positive square root, the negative square root, and the sine, and are denoted by $\sqrt{\ }$, $-\sqrt{\ }$, and *sin*, respectively. For instance,

$$sin = \{(x, sin\ x) \mid x \text{ in } U\} = \{(a, sin\ a) \mid a \text{ in } U\} \ldots.$$

The value of the function *sin* for x (for 2) is designated by placing *sin* in front of x (of 2). Similarly, the value of $\sqrt{\ }$ for a (for 9) is designated by $\sqrt{a}$ (by $\sqrt{9}$). The following table contains these and other examples.

Name	Symbol	Value for x	Name of the Graph
sine	*sin*	*sin* x	sine curve
arctangent	*arctan*	*arctan* x	arctangent curve
logarithm to the base b	log_b	$log_b\ x$	logarithmic curve
exponential function to the base b	exp_b (rare)	b^x or, rarely, $exp_b\ x$	exponential curve
n-th root	$\sqrt[n]{\ }$	$\sqrt[n]{x}$	missing
n-th power	missing	x^n	n-th power curve or j^n
1st power or identity	missing	x	*j*
constant function of value 3	missing	3	the horizontal line *3*

Oddly enough, the n-th power function, that is, the class of all pairs (x, x^n) for any x, in spite of its paramount importance, does not possess a traditional symbol. There does not even exist a symbol for the first power, which, for any x, assumes a value identical with x, and thus has been called the *identity function*. In absence of such symbols, these functions are traditionally designated by their values for x as "the function x^n" and "the function x." Similarly, the constant function *3* is usually referred to by its value as "the function 3." For the sake of consistency, most mathematicians use also the terms "the function *sin* x" and "the function $log_{10}x$", even though in these cases symbols for the functions are available, namely, *sin* and log_{10}.

No reason is apparent why the identity, the n-th power, and the constant function of the value 3 should *not* have symbols, whereas there are numerous reasons why they *should*. Therefore, in this book, those functions will not remain anonymous. They will be denoted by the symbols *j*, j^n, and *3* that, in the first chapters, have been introduced for their graphs, just as the *log* function is denoted by the symbol used for its graph, the *log* curve. The values of the functions at 2 and at x will be denoted by

$$j\,2, \quad j^n 2, \quad 3\;2 \quad \text{and} \quad j\,x, \quad j^n x, \quad 3\;x,$$

just as are the altitudes of their graphs above 2 and x. These values and altitudes are 2, 2^n, 3 and x, x^n, 3, respectively. Thus the following table is to replace the last lines of the preceding one.

Name	Symbol	Value for x	Name of the Graph
n-th power	j^n	x^n or $j^n x$	n-th power curve or j^n
1st power or identity	j	x or j x	identity line or j
constant function of value 3	*3*	3 or *3* x	the (horizontal) line *3*.

The functions here tabulated (in conjunction with the other trigonometric and arc functions, *cos, tan,..., arcsin,...*) form a basis for the introduction of numerous functions, whose definitions (by means of addition, multiplication, and substitution) will be systematically treated in Sections 7 and 8. For instance, the class of all pairs (x, $16x^2$) for any x will be denoted by $16j^2$; the function whose graph is the upper semicircle of radius 2 about 0, by $\sqrt{4-j^2}$.

Throughout this book, the symbols designating functions and their graphs are printed in italic type: cos, log_b, j, 3, ... Conversely, until Chapter VII, lower-case italics are reserved for the designation of functions and what in Section 5 will be called function variables.

3 and 3, *log* and *log* 3, *j* and *j* x (= x) are related somewhat like a party of married couples and a spinster.

Just as a roman letter, a, b, ... , may be used *ad hoc* as an abbreviation for the cumbersome symbol of a specific number (p. 58), an italic letter (usually *f, g, ...*) is often used as an *ad hoc* designation of a specific function whose symbol one wishes to avoid repeating. For instance, one might write

"Let *f* be the function assuming for any x the value

$$sin\,(cos\;x + 2^x\,arctan\;x^4)/(4 + x^2)^{1/3}\text{"}$$

(if that function had to be studied). Thereafter, until further notice, one would reserve the symbols *f* and *f* x for the sole purpose of designating that specific function and its unwieldy value for x.

A letter, y, that may be replaced by any value of the function *cos* clearly is a numerical variable whose scope is Ran *cos*, the range of the function *cos*.

For instance, one can say

$-1 \leq y \leq 1$ for any y that is a value of *cos* or

$-1 \leq t \leq 1$ for any t that is a value of *cos*.

Instead, one might use a self-explanatory variable (in the sense of p. 63) for instance, val *cos*. Then one can write without any legend

$$-1 \leq \text{val } cos \leq 1, \quad \text{val } exp > 0, \quad \text{val } j^2 \geq 0, \quad \text{val } j^2 > -1.$$

Similarly, one might denote a numerical variable whose scope is Dom *cos* (the domain of *cos*) by arg *cos* (where arg stands for *argument*, a term sometimes used for the elements of the domain of a function). For instance,

$$\text{arg } log_{10} > 0 \quad \text{and} \quad \text{arg } j^{-1} \neq 0.$$

Such self-explanatory variables (which are less arbitrary than, and therefore in many ways preferable to, other variables) obviously follow rather the ways mathematicians *talk* than the ways they *write*. (The indefinite article in "a value," "an argument," "a square root" etc. is the linguistic analogue of the written variable).

*The reader might determine which of the following two definitions is correct

log_{10} is the class of all pairs (arg log_{10}, val log_{10});

cos is the class of all pairs (arg *cos*, *cos* (arg *cos*)).

EXERCISES

19. Find Dom j^n and Ran j^n (the domain and the range of j^n) for the various values of n. Hint. In either problem four possibilities occur: U, P + N, U – N, and P.

20. Tabulate the names, symbols, and values for x of the signum function (see Exercise 16) and of the absolute-value function, denoted by *abs*, defined as the class of all pairs (x, | x |) for any x. Draw the graphs of the two functions. How are they related? Express the laws $|a + b| \leqq |a| + |b|$ and $|x| = \sqrt{x^2}$ using the symbol *abs*. Show that, for any number, the value of one of the functions *abs*, *sgn*, and *j* is the product of the other two.

21. For each of the functions tabulated in this section (and in Exercise 20), determine domain and range. Wherever possible, record the result in a formula; e.g., Dom log_{10} = P, Ran log_{10} = U.

22. The symbol 3^2 (or b^c) designates values of two totally different functions; the value of j^2 for 3 (or of j^c for b); and the value of exp_3 for 2 (or of exp_b for c). Draw the points of altitude 3^2 on the curves j^2 and

exp_3. Similarly, the value of the function $c^2(j + j^3)$ for b is equal to the value of the function $(b + b^3)j^2$ for c. Give further examples of functions related in this way.

5. FUNCTION VARIABLES

a. The Concept and Its Use. Each of the formulas

$$D \int_0 3 = 3, \quad D \int_0 0 = 0, \quad D \int_0 (-4) = -4$$

deals with a specific horizontal line (or constant function). More important is the following general statement: The derivative of the 0-integral of *any* constant function is that constant function; in symbols,

(12) $D \int_0 c = c$, where c may be replaced by any constant function

—briefly, for any constant function *c*.

The letter c in the preceding formula is accompanied by a legend concerning its replacement by functions of a certain kind, namely, constant functions such as *3, 0, −4*. Each such replacement results in a statement about a specific function. Thus c might be called a *function variable* whose scope is the class of all constant functions. It is used *indicatively* (p. 63). The meaning of (12) is not changed if c is replaced by any other (lower-case italic) letter, as in

$$D \int_0 b = b \text{ and } D \int_0 f = f, \text{ for any constant functions } b \text{ and } f,$$

respectively.

The typographical convention maintained in this book makes the replacement of variables by specific elements of their scopes very simple: Number variables (or numerical variables), printed in roman type, may be replaced only by symbols for specific numbers, which (except for π) are also printed in roman type; function variables, printed in lower-case italics, only by symbols for specific functions, which (with few exceptions, such as the so-called gamma function, are also printed in lower-case italics.

One cannot in (12) replace c by either the number 3 or the numerical variable c. In either case a meaningless symbol would result, since a number has no integral or area curve.

Example 1. The definitions of domain and range can be restated as follows: Dom *f* (and Ran *f*) denotes the class of all numbers that are the first (the second) member in a pair belonging to *f*. If, in this definition, one replaces *f* by log_{10}, he obtains the definition of Dom log_{10}, which is P, (and of Ran log_{10}, which is U). Clearly, *f* might be replaced, without any change of the meaning, by say *g*.

Many statements contain function variables *and* numerical variables.

Example 2. $\mathbf{D}\int_a c = c$ for any constant c and any a. (Our typographical convention makes it superfluous to say "*number* a" and "constant *function* c".) This is a generalization of (12).

Example 3. The fact that the values of the functions *cos* for 3 and log_{10} for x are denoted by *cos* 3 and log_{10}x can be generalized as follows: The values of *f* for 3 and for x are denoted by *f* 3 and *f* x, for any function *f* and any x such that Dom *f* includes 3 and x, respectively. In other words, if a pair with the first member 3 or x belongs to *f*, its second member is *f* 3 and *f* x, respectively. Hence *f* may be defined as the class of all pairs (x, *f* x) such that x belongs to Dom *f*, for any *f*. Here *f* may be replaced by any specific function, e.g., by log_{10} which is the class of all pairs (x, log_{10}x) for any x in Dom log_{10}.

f 3 and *f* x are symbols for the values of *f* that were frequently used in the 18th century. Later, mathematicians began to fear that these symbols might be mistaken for products and replaced them by *f*(3) and *f*(x), read "*f* of 3" and "*f* of x," designations which prevail in the current literature. Yet, in replacing *f* by multi-letter standard symbols of specific functions such as *cos* and log_{10} these parentheses are omitted; no one writes *cos*(3) or log_{10}(x).

Example 4. Two functions, *f* and *g*, are said to be *equal,* in symbols, *f* = *g*, if any pair of numbers belonging to either function belongs also to the other. If *f* = *g*, then Dom *f* = Dom *g*, and *f* x = *g* x for any x in that common domain. Conversely, if *f* and *g* are two functions such that Dom *f* = Dom *g* and *f* x = *g* x for any x in the domain, then *f* and *g* are equal (that is, equal classes of pairs of numbers). E.g., if *f* x = 16 x^2 and *g* x = (4x + 1) · (4x − 1) + 1 for any x, then *f* = *g*.

If Dom *f* = Dom *g*, then negating *f* = *g* is asserting that in *f* x = *g* x one cannot replace x by every number; in other words, that *f* x ≠ *g* x for at least one number x.

Example 5. If *f* x ≠ *g* x for any x belonging to (Dom *f*) · (Dom *g*), (which implies, without in general being implied by, the negation of *f* = *g*), then *f* and *g* will be said to be *disjoint,* in symbols, *f* || *g*. Clearly, *f* || *g* if and only if the two classes of pairs of numbers, *f* and *g*, have no pair in common. The functions *O* and exp_2 are disjoint; in symbols, *O* || exp_2; the functions *O* and *sin* are unequal but not disjoint.

b. Graphs and Arithmons. In the plane introduced on page 1, any function *f* is represented by a simple class of points (called the *graph* of *f*). The projection of this class on the line *O* is Dom *f*. Conversely, any simple class of points in the plane is the graph of the function (which might be called the *arithmon* of the class). A pair of numbers belongs to this function if and only if its first member is the projection on the line *O* of a point in the class, and its second member is the altitude of that point above its projection.

In view of this correspondence, the words "function" and "simple class of points in the plane" will be used synonymously, just as the words "number" and "point on *O*."

The representation by simple curves and points on *O* shows that functions and numbers are coordinated mathematical objects. Just as a function is the result of pairing a number with each number belonging to a certain class of numbers (called the domain of the function), a number is the stimulus eliciting a number from each function belonging to a certain class of functions (namely, those that contain the number in their domains). The number *cos* 3 is not only the value of *cos* for 3; it is also the response to 3 of *cos*. This remark points the way to a postulational treatment of numbers and functions analogous to that of points and lines in plane geometry.

c. The Reach of Function Symbols. Symbols such as log_{10} 3 + 4 and *sin* a + b are ambiguous as is 2 · 3 + 4. The conventions tabulated below in the first three lines are generally, if tacitly, adopted. Only the number immediately following a function symbol is within its reach.

The symbol	does mean	and does not mean
2 · 3 + 4	(2 · 3) + 4	2 · (3 + 4)
log_{10} 3 + 4	(log_{10} 3) + 4	log_{10} (3 + 4)
sin a + b	(*sin* a) + b	*sin* (a + b)
f a + b	(*f* a) + b	*f* (a + b)

By these conventions parentheses are saved: those in the second column may be omitted (although they may, of course, also be retained).

With regard to *f* a · b no conventions are universally adhered to. We shall write (*f* a) · b and *f* (a · b) depending upon the meaning.

The symbol *sin* 3 x in classical mathematics is ambiguous. Traditionally, it stands for *sin*(3 · x). But now and then college teachers are exasperated by beginners who, in computing the derivative, treat *sin* 3x as though it meant (*sin* 3) · x. (The beginners may never have seen an unambiguous definition of the symbol *sin* 3 x .)

EXERCISE

23. In Exercise 5 (p. 65), name the specific functions and the function variables, and determine the scopes of the latter. Do the same for

$\mathbf{D}(3\ j) = 3;\quad \mathbf{D}^{-1}_{(0,0)}\ 3 = 3\ j;\quad \mathbf{D}(\mathbf{D}^{-1}_{(a,0)}\ 3) = 3;\quad \mathbf{D}(\mathbf{D}^{-1}\ 3) = 3.$

6. RESTRICTIONS AND EXTENSIONS. CLASSES OF NUMBERS

If each pair of numbers belonging to the function f belongs also to the function g, then g is called an *extension* of f, and f a *restriction* of g. The class $\{ (x, 1/x) \mid x > 0 \}$ is a restriction of j^{-1}. It has the domain P and will be denoted by j^{-1} (on P). Similarly,

$$j^2 \text{ (on N)} = \{ (x, x^2) \mid x < 0 \} \text{ and } \mathit{1} \text{ (on P + N)} = \{ (x, 1) \mid x \neq 0 \} .$$

Clearly, any restriction of a constant function is constant.

More generally, if f is any function and A is any subclass of Dom f, then

(13) f (on A) $= \{ (x, f\,x) \mid x \text{ in } A \}$.

Similarly, if B is a subclass of Ran f, we define

(14) f (into B) $= \{ (x, f\,x) \mid f\,x \text{ in } B \}$.

For instance, log_{10}(into P) is the class of all pairs (x, log_{10} x) such that log_{10} x > 0, that is, x > 1. Hence, if P′ *(ad hoc)* denotes the class of all numbers > 1, then log_{10} (into P) = log_{10} (on P′). The following formulas can easily be proved:

j^{-1} (into P) = j^{-1} (on P); $-j$ (into P) = $-j$ (on N);

f (on Dom f) = f ; f (into Ran f) = f for any f.

Let A denote the class $\{ x \mid -\pi/2 \leqq x \leqq \pi/2 \}$. The function *sin* has only one restriction relative to A, *sin* (on A). But that restriction has infinitely many extensions. One of them is *sin*, that is, *sin* (on U). Another one assumes the value 1 to the right of $\pi/2$, and the value -1 to the left of $-\pi/2$.

U, P, and N denote specific classes. In (13) and (14) A and B are what may be called *class variables*, symbols that may be replaced by the designation of any class of numbers. In the preceding example, A was used to designate *ad hoc* a specific class.

Functions yield descriptions of classes of numbers. For instance, although $j \neq$ *abs*, there are some numbers x such that j x = *abs* x (that is, x = | x |), namely, the numbers belonging to the class U − N. This class might therefore be described by the symbol $\{ j = abs \}$. Of course, just as 12 = 11 + 1 = 3 · 4 = 6 · 2 = ... , so

$$U - N = \{exp_2 \geqq \mathit{1}\} = \{ j^3 \geqq O \} = \{ j \geqq O \} = \{ sgn \neq -\mathit{1} \} = \cdots .$$

More generally, $\{ f \leqq g \}$ will denote the class of all x belonging to (Dom f) · (Dom g) for which $f\,x \leqq g\,x$; and $\{ f = g \}$ is defined similarly. For instance,

$\{ f = O \}$ is the class $\{ x \mid f\,x = 0 \}$ of what is called the *zeros of the function* f and the *roots of the equation*

$$f\ x = 0.$$

The class $\{ j^2 - 1 = 0 \}$ consists of the numbers 1 and −1; the class $\{ sin = 0 \}$ of 0, π, $-\pi$, 2π, -2π, ... For any a > 0, the class $\{ exp_{10} = a \}$ consists of exactly one number, called log_{10} a. Clearly, $\{ exp_{10} = a \}$ is the vacuous class, V, if a ≦ 0. Also $\{ j^2 = -1 \}$ = $\{ 0 = 1 \}$ = V. On the other hand, $\{ j^2 = j^2 \} = \{ 0 = 0 \} = U = \{ f = f \}$ for any f such that Dom f = U. More generally, $\{ f = f \}$ = Dom f.

EXERCISES

24. Draw the graphs of *1* (on P); j^2 (on P); j^3 (into N); *exp* (on $\{ 0 \leqq j \}$); *3* (on $\{ 0 \leqq j \leqq 1 \}$); *sin* (on $\{ -\pi \leqq j \leqq \pi \}$).

25. Find three different extensions of each of the following functions: *(a)* log_{10}; *(b)* *cos* (on P); *(c)* the functions in Exercise 13*(b)*. Find all extensions of j^{-1}.

26. Show that *(a)* Ran *cos* = $\{ -1 \leqq j \leqq 1 \}$; *(b)* Ran exp_2 = $\{ j > 0 \}$ = $\{ sgn = 1 \}$; *(c)* Ran *sgn* = $\{ j^3 = j \}$; and prove that $\{ tan = j \}$ contains infinitely many numbers.

27. Find three different descriptions of each of the following: the class consisting of *(a)* −1, 0, 1; *(b)* the single number 2; *(c)* all positive integers; *(d)* the number 0 and the numbers ≧ 1 (Hint: See Exercise 10*(c)*); *(e)* all numbers ≧ 1 and all numbers ≦ −1.

★28. Every function is both an extension and a restriction of itself. Show that two functions are equal if and only if either is an extension of the other. g is called a *proper* extension of f if g is an extension of f but not equal to f. Show that the only functions that have no proper extensions are those whose domain is U. Note that if A is any subclass of Dom f and B any subclass of Ran f, then

$$\text{Dom}\,[f\,(\text{on A})] = \text{A} \quad \text{and} \quad \text{Ran}\,[f\,(\text{into B})] = \text{B}.$$

7. ADDITION AND MULTIPLICATION OF FUNCTIONS

On page 5, the line having at any x the altitude $\frac{1}{2}$ x + 1 (which may be obtained by superposition of the lines $\frac{1}{2} j$ and *1*) was denoted by $\frac{1}{2} j + 1$. Similarly, $j + j^3$ will denote the curve obtained by superposition of j and j^3, as well as the function whose graph is that curve. Generally, for any f and g, we denote by $f + g$ the curve having the altitude $f\,x + g\,x$, or the function assuming the value $f\,x + g\,x$, for any x belonging to the domains of both f and g. For instance, $j + log_{10}$ assumes the value $j\,x + log_{10}\,x$ for any $x > 0$; (for $x \leqq 0$, $log_{10}\,x$ is not a number, nor will a value of $j + log_{10}$ be defined). If $(j + log_{10})\,x$ denotes the value of $j + log_{10}$ for x, the preceding definition can be restated in the formula

$$(j + log_{10})\,x = j\,x + log_{10}\,x \;=\; x + log_{10}\,x \text{ for any } x > 0.$$

Clearly, $\mathrm{Dom}(j + log_{10}) = (\mathrm{Dom}\ j) \cdot (\mathrm{Dom}\ log_{10}) = P$. More generally,

$$(f + g)\,x = f\,x + g\,x \text{ for any x in } (\mathrm{Dom}\ f) \cdot (\mathrm{Dom}\ g).$$

$(\mathrm{Dom}\ f) \cdot (\mathrm{Dom}\ g)$ denotes, of course, the class of all numbers belonging to both domains.

In Fig. 25, there are the curves *sin* and 2 *cos* as well as *sin* + 2 *cos*. The sum has the same altitude as has 2 *cos* at 0 and π, where the altitude of *sin* is 0; the same altitude as has *sin* at $\pi/2$, where the altitude of 2 *cos* is 0; the altitude 0, where 2 *cos* and *sin* have opposite altitudes, as at 2.03.

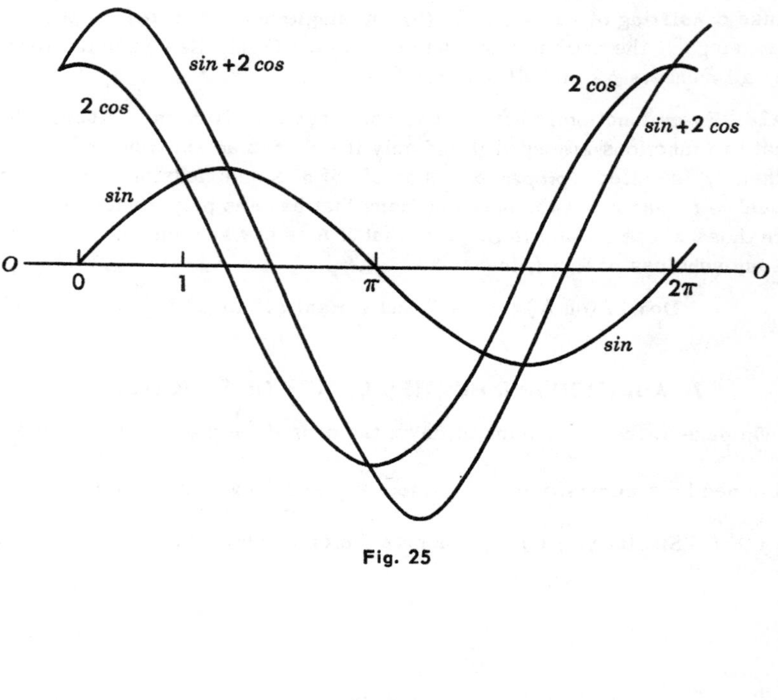

Fig. 25

Multiplication is defined similarly. The product $f \cdot g$ is the function for which

$$(f \cdot g)\,x = f\,x \cdot g\,x \quad \text{for any x in (Dom } f) \cdot (\text{Dom } g).$$

Note the dot in the symbol for the product of two functions. In this book (with one exception to be mentioned on p. 86) **the dot indicating multiplication of functions will never be omitted.**

In Fig. 26, there are the curves j^{-1} and *sin* as well as $j^{-1} \cdot sin$. The product has the same altitude as has *sin* at 1, where the altitude of j^{-1} is 1; the same altitude as has j^{-1} at $\pi/2$, where the altitude of *sin* is 1; the altitude 0 at π, where *sin* has the altitude 0; a positive altitude, between 0 and π, where both j^{-1} and *sin* are above *O*; a negative altitude between π and 2π, where j^{-1} is above, and *sin* below, *O*. The point 0 does not belong to the projection of the product, since it does not belong to the domain of j^{-1}. But, using sine tables, one finds that, near 0, the altitude of the product is close to 1.

In Fig. 26, the horizontal line *1* has been drawn. To find the point of $f \cdot g$ above the point x on *O*, proceed as follows (In the figure $f = j^{-1}$ and $g = sin$):

1. On *O*, choose an arbitrary point x′, different from x.
2. Join x′ to the point on the line *1* above x; that is, draw the line *l*.
3. Join x′ to the point of the curve *f* above x; that is, draw the line *m*.
4. Draw the line *l′* parallel to *l* through the point of the curve *g* above x.
5. Draw the line *m′* parallel to *m* through the point x″ where *l′* intersects *O*.

The point of intersection of *m′* with the vertical line through x belongs to the curve $f \cdot g$. Indeed, let p denote the altitude of this point above *O*. The similarity of the two pairs of right triangles with the vertices at x′ and at x″ implies that

$$p : g\,x = f\,x : 1 \quad \text{and hence} \quad p = f\,x \cdot g\,x.$$

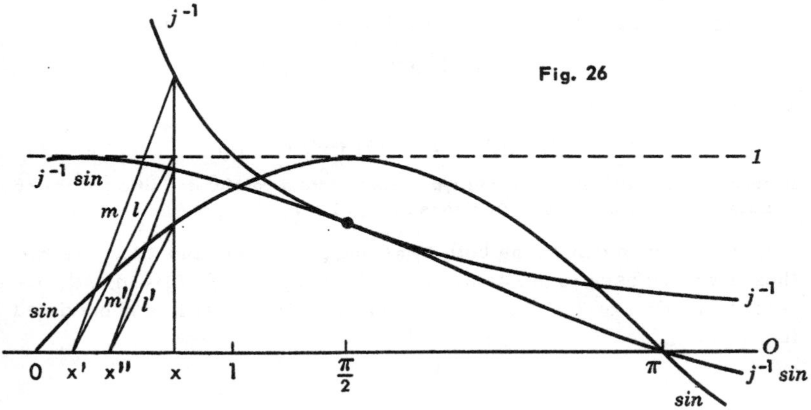

Fig. 26

Since, by definition of $f \cdot g$, $fx \cdot gx$ is the altitude of that curve above x, the point at the altitude p above x belongs to the curve $f \cdot g$.

The domains of the semi-parabolas $j^{1/2}$ and $(-j)^{1/2}$ (which must not be confused with $-j^{1/2}$) have only 0 in common. Thus the sum as well as the product of these two functions contains only the pair (0, 0).

The domains of the function log_{10} and the function assuming the value $log_{10}(-x)$ for any $x < 0$ are disjoint (i.e., have no number in common). Hence, the sum as well as the product of the two functions is the vacuous function.

Addition and multiplication in the realm of functions follow laws that are very similar to (but not quite identical with) the laws concerning the corresponding operations in the realm of numbers.

Commutative Laws: $f + g = g + f$ and $f \cdot g = g \cdot f$.

Associative Laws: $f + (g + h) = (f + g) + h$ and $f \cdot (g \cdot h) = (f \cdot g) \cdot h$.

The Distributive Law: $(f + g) \cdot h = f \cdot h + g \cdot h$.

for any three functions f, g, h.

The commutative and distributive laws jointly imply

$$h \cdot (f + g) = h \cdot f + h \cdot g.$$

According to the commutative laws, the order of the terms is immaterial. According to the associative laws, it is immaterial where, in denoting the sum or the product of three functions, parentheses are inserted. Hence these parentheses may be dispensed with altogether. $f + g + h$ and $f \cdot g \cdot h$ are unambiguous.

This rule can readily be extended to sums and products of any finite number of functions, $f_1 + f_2 + f_3 + f_4$, $g_1 \cdot g_2 \cdot g_3 \cdot g_4$, and so on. For infinitely many terms, the associative law is not even valid for numbers, since

$$1 + [(-1) + 1] + [(-1) + 1] + [(-1) + 1] + \cdots = 1$$
$$[1 + (-1)] + [1 + (-1)] + [1 + (-1)] + \cdots = 0,$$

so that the expression

$$1 + (-1) + 1 + (-1) + \cdots$$

is ambiguous. Difficulties of this type necessitate a careful definition of the sum of infinitely many numbers or functions (see p. 129).

In expressions containing both sums and products of functions, parentheses will be used in the same way as for numbers: $f \cdot (g + h)$ will denote the product of f and $g + h$, whereas $(f \cdot g) + h$ will be abbreviated to $f \cdot g + h$.

Neutral Elements. $O + f = f = f + O$ and $1 \cdot f = f = f \cdot 1$ for any f. In the realm of numbers, 0 and 1 are neutral, and they are the *only* neutral elements: $a + b \neq a$, if $b \neq 0$. Similarly, if Dom f = U, then $f + g \neq f$ if $g \neq O$, and the only function the addition of which leaves f unchanged is O. But if Dom $f \neq$ U, then f remains unchanged also upon addition of the restriction O (on Dom f) as well as of any extension of that restriction.

Opposite Elements. For any function f,

$-f$ denotes the class of all pairs (x, $-f$ x) for any x in dom f;

$1/f$, the class of all pairs $(x, \frac{1}{f\,x})$ for any x in dom f such that $f\,x \neq 0$.

These functions, called the *negative* and the *reciprocal* of f, have the following properties:

Dom $(-f)$ = Dom f and Dom $(1/f)$ = (Dom f) $\cdot \{ f \neq O \}$;

$f + (-f) = O$ (on Dom f) and $f \cdot (1/f) = 1$ (on Dom $(1/f)$). For instance, the product of j and $1/j$ (or j^{-1}) is 1 (on P + N) and not 1; and $\cos \cdot (1/\cos) = 1$ (on $\{\cos \neq 0\}$).

The negative as well as the reciprocal of any constant function is constant, except $1/O$ and the restrictions of $1/O$, all of which are the vacuous function, $\emptyset$.

Clearly, $g = -f$ implies $-g = f$. Hence

$$-(-f) = f \text{ and } -(-(-f)) = -f.$$

But all that $g = 1/f$ implies is $1/g = f$ (on $\{f \neq O\}$). For instance, $j^{-1} = 1/j$; yet $1/(1/j) \neq j$, since $1/j^{-1} = j$ (on P + N). However, if Ran g does not contain 0, then $1/(1/g) = g$. Now 0 does not belong to Ran $1/f$ for any f. Consequently,

$$1/(1/(1/f)) = 1/f.$$

It is customary to write $f - g$ for $f + (-g)$ and f/g for $f \cdot (1/g)$. Clearly, $1/g = 1/g$.

The addition of equal functions (as that of equal numbers) is often rendered as a multiplication. One writes

3 cos for cos + cos + cos; $2j^4$ for $j^4 + j^4$; $4f$ for $f + f + f + f$. Similarly, if $g + g + g = 4f$, one sets $g = \frac{4}{3} f$.

One writes even π *sin* for the function whose value for any x is the limit (in the sense defined in Chapter V) of the numbers

$$3.1 \sin x, \quad 3.14 \sin x, \quad 3.141 \sin x, \ldots .$$

In this sense, a function b f can be defined for any f and any number b. It can be proved that b f = $b \cdot f$, where b is the constant function of value b. For instance,

$$3 \cos = 3 \cdot cos = cos + cos + cos; \quad 4 f = 4 \cdot f = f + f + f + f \,.$$

Thus, if the *first* factor of a product is a *constant* function, we may (and usually will) omit the dot and replace the italic by the corresponding roman letter. (In longhand, one would omit the dot and the underscoring.) But one must preserve the dot and the italic type if the *second* factor in a product is constant – to avoid confusion between

the *functions* $f \cdot 3$ (= $3 \cdot f$ = 3 f), $log \cdot 2$ (= $2 \cdot log$ = 2 log), etc.

and the *numbers* f 3, log 2, etc., which are values of f and log, etc.

The commutative law holds, of course, also for products with one or two constant factors: $f \cdot c = c \cdot f$; and is not in any way violated by the fact that this product may be denoted by c f but not by f c.

The addition of a function and a number will not in this book be defined, just as the sum of a function and a sphere is left undefined. However, it is sometimes convenient to denote the sum of a function such as 3 j and a constant function, say 2, by 3 j + 2 (instead of 3 j + 2 or $3 \cdot j + 2$). Or one might write

"The graph of a j + b is a straight line for any a and b" instead of

"The graph of $a \cdot j + b$ is a straight line for any two constant functions a and b."

The sum of the constant function of the value 2π and another function, say *arcsin*, cannot in print be conveniently denoted by anything but 2π + *arcsin*. (In longhand, one might underscore 2π.)

This convention can be summarized as follows: *If in a sum, one term is an (italicized) designation of a function or function variable and the other is a roman type symbol* (2, 5, ... , b, c, ...), *then the latter designates a constant function (2, 5, ...) or stands (as would* b, c, ...*) for any constant function.*

EXERCISES

29. Using 1 inch as the unit, construct graphs of

(a) $j + j^2$, $j + exp_2$, $exp_2 + exp_{1/2}$ on $\{-2 \leqq j \leqq 2\}$.

(b) j, $j - \frac{1}{6} j^3$, $j - \frac{1}{6} j^3 + \frac{1}{120} j^5$ on $\{-\pi \leqq j \leqq \pi\}$.

(c) $j \cdot sin$ (on $\{-\pi \leqq j \leqq \pi\}$) and $j \cdot exp_{1/2}$ (on $\{-1 \leqq j \leqq 2\}$).

(d) $j + abs$, 2 j + abs, $\frac{1}{2} j - abs$; what is the shape of a j + b abs for any two numbers a and b?

30. Compute the following values of functions:

(a) $(1 + j + j^2)2$, $(1 + j + j^2)x$, $(a - j)^n b$;

(b) *to* five decimals, the values of the functions in Ex. 29 b) for $\pi/18$, $\pi/9$, $\pi/3$ (compare them with the corresponding values of *sin*);

(c) $(f - fa)x$, $(f - f \cdot a)\,a$;

(d) $\left(\frac{j^3 - 1}{j - 1}\right) x$, $\left(\frac{f - fa}{j - a}\right) b$ for any $b \neq a$.

31. Draw a simple curve f and construct the curves $-f$ and $1/f$. In particular, apply the latter construction to *cos* (thus obtaining *sec*), and to $\sqrt{1 - j^2}$. For two specific simple curves, *ad hoc* denoted by f and g, construct f/g.

32. What, if anything, can be said about Ran $(f + g)$ and Ran $(f \cdot g)$ if Ran $f = \{-2 \leqq j \leqq 3\}$ and

(a) Ran $g = \{-5 \leqq j \leqq 7\}$;

(b) Ran $g = \{5 \leqq j \leqq 7\}$;

(c) Ran $g = P$;

(d) Ran $g = U$?

What if

(e) Ran f = Ran $g = P$;

(f) Ran $f = P$ and Ran $g = U$;

(g) Ran f = Ran $g = U$?

★33. Show that $f + g$ is the class of all pairs of numbers (x, y) with the following property: if
(1) (x,s) belongs to f; (2) (x, t) belongs to g,
then $s + t = y$.

★34. For any number a let c_a denote the constant function of the value a (which in this book is usually denoted by italic a). Prove that $c_{a+b} = c_a + c_b$ and show that this law cannot be expressed if c_a is denoted by a.

8. SUBSTITUTION

A third operation connecting functions (which has no analogue for numbers) is substitution. The result of substituting g into f is a function denoted by fg.

If j^3 is to be substituted into j^2, the value which the resulting function j^2j^3 assumes for x is found as follows: first compute j^3x, the cube of x or x^3; then compute the value which j^2.assumes for the number x^3, that is, the square of x^3. This number, $j^2(j^3\ x) = (x^3)^2$, is the value $(j^2j^3)x$. Obviously, it equals x^6 for any x. Hence $j^2j^3 = j^6$ (whereas $j^2 \cdot j^3 = j^5$).

If j^2 is to be substituted into j^3, first compute $j^2\ x = x^2$, then the value of j^3 for x^2. This number $j^3(j^2\ x) = (x^2)^3$ is the value $(j^3j^2)x$. It is likewise x^6. Hence $j^3\ j^2 = j^6$.

To find the value for x of the function $(3\,j\ + 1)j^2$ obtained by substituting j^2 into $3j + 1$, first compute $j^2 x = x^2$, and then the value of $3\,j + 1$ for x^2, that is, $3\ x^2 + 1$. This is the value for x of $(3\,j + 1)j^2$. It is also the value for x of the function $3\ j^2 + 1$. Hence $(3\,j + 1)j^2 = 3\,j^2 + 1$ (whereas $(3\,j + 1)\cdot j^2 = 3\,j^3 + j^2$) .

To substitute $3\,j + 1$ into j^2, first compute the value $3\,j + 1$ for x, that is, $3\ x + 1$; then compute the value of j^2 for $3\ x + 1$, that is, $(3x + 1)^2 = 9x^2 + 6x + 1$. Hence $j^2(3j + 1) = 9j^2 + 6j + 1$.

The above examples show that

$$j^3j^2 = j^2j^3\ (= j^6) \text{ but } (3j + 1)\,j^2 \neq j^2(3j + 1), \text{ since}$$

$3\,j^2 + 1 \neq 9\,j^2 + 6\,j + 1$. Whereas $f + g = g + f$ and $f \cdot g = g \cdot f$, the function fg may or may not be equal to the function gf.

The general definition of fg is as follows: fg is the function whose value for x is equal to the value that f assumes for g x; in a formula,

$$(fg)x = f(g\ x). \tag{15}$$

Here, of course, x must belong to Dom g and g x to Dom f. Consequently,

$$\text{Dom } fg = \text{Dom } g \text{ (into Dom } f). \tag{16}$$

On the other hand, g x belongs to Ran g and $f(g\ x)$ to Ran f. Therefore,

$$\text{Ran } fg = \text{Ran } f \text{ (on Ran } g). \tag{17}$$

The reader will notice the symmetry of formulas (16) and (17) (of which the latter has no analogue for addition and multiplication; see Exercise 32) – a symptom of the fact that substitution is the functional operation par excellence.

If in the definition (15)

f is replaced by	and *g* by	then *fgx* is	for any x such that
sin	*log*	$sin\ log\ x = sin\,(log\ x)$	x is in Dom *log* = P.
log	*sin*	$log\ sin\ x = log\,(sin\ x)$	*sin* x is in Dom *log* = P.
j^{-1}	$j^{1/2}$	$j^{-1}j^{1/2}\ x = 1/\sqrt{x}$	x is in Dom $j^{1/2}$ = P and $\sqrt{x}$ is in Dom j^{-1} = P + N.
f	$j^{1/2}$	$f\,j^{1/2}\,x = f(\sqrt{x})$	x is in P and $\sqrt{x}$ is in Dom *f*.
j^{-1}	*g*	$j^{-1}g\ x = 1/g\ x$	x is in Dom *g* and $g\,x \neq 0$.
3	*g*	$3\ g\ x = 3$	x is in Dom *g*.
f	*3*	$f\ 3\ x = f\ 3$	
exp_2	j^{-1}	$exp_2\,j^{-1}\,x = 2^{1/x}$	$x \neq 0$.

Fig. 27 illustrates the last example. There are $f = exp_2$ and $g = j^{-1}$ and *fg*. The point of *f g* above the point x on *O* has been constructed with the use of the (dashed) line *j*.

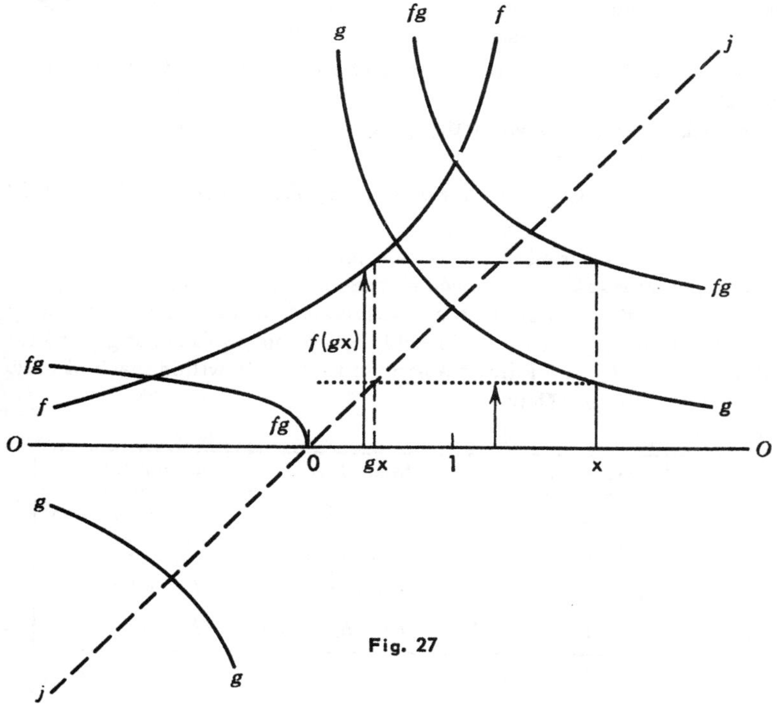

Fig. 27

To obtain that point, proceed as follows:

(1) *Through the point on g above x, draw the (dotted) horizontal line* (its altitude is g x).

(2) *Through the point where that dotted line intersects j, draw the vertical line;* (the point of intersection, like any point on j, has a projection on O equal to its altitude; the latter being g x, the vertical line passes through the point g x on O).

(3) *Through the point where that vertical line intersects f, draw the (dashed) horizontal line;* (the point of intersection has the altitude $f(g\ \mathrm{x})$; so does, therefore, the dashed horizontal line).

(4) *The point of the dashed horizontal line that is above* x *is the point of the curve fg that was to be constructed;* (for, above x, it has the altitude $f(g\ \mathrm{x})$, which is fg x).

Substitution into a number has not been, and will not be defined in this book. For instance, *cos* cannot be substituted into 3. This is why 3 *cos* can be used without ambiguity to denote *3* · *cos* or *cos* + *cos* + *cos*.

Substituting into f the constant function *3* must not be confused with evaluating f for 3. The two operations can be performed under the same condition, namely, if and only if the number 3 belongs to Dom f. But whereas the value f 3 is a number, the result of the substitution is the constant function f *3* whose value, for any x, is f 3. For instance, $j^2 3 = 9$ and $j^2 \mathit{3} = \mathit{9}$. More generally, f c is the constant function of the value f c.

The substitution of any function g into any constant function c yields a restriction of c:

$$cg = c \text{ (on Dom } g\text{); for instance, } 3\ log = 3 \text{ (on Dom } log) = 3 \text{ (on P).}$$

Clearly, c d = c, for any two constant functions c and d.

Many parentheses can be saved by the following convention: Only immediately adjacent functions are within another's reach with regard to substitution; e.g., if f is followed by $g + h$, only g (and not $g + h$) is within reach of f; if h follows a product $f \cdot g$, it is within reach of only g (and not of $f \cdot g$). Thus

The expression	stands for	and not for
$fg + h$	$(fg) + h$	$f(g + h)$
$fg \cdot h$	$(fg) \cdot h$	$f(g \cdot h)$
$f + gh$	$f + (gh)$	$(f + g)h$
$f \cdot gh$	$f \cdot (gh)$	$(f \cdot g)h$

The following general laws can be derived from the definition of substitution:

Associative Law: $f(gh) = (fg)h$.

Distributive Laws: $(f + g)h = fh + gh$ and $(f \cdot g)h = fh \cdot gh$.

The similar formulas

$$h(f + g) = hf + hg \quad \text{and} \quad h(f \cdot g) = hf \cdot hg$$

are valid for some functions and not valid for others (as is the formula $fg = gf$). The reader should verify, for instance, that

$j(f + g) = jf + jg$ $(= f + g)$ for any f and g, whereas

$j^2(f + g) \neq j^2f + j^2g$ $(= f^2 + g^2)$ unless either $f = O$ or $g = O$.

According to the associative law, the results of substituting gh into f, and of substituting h into fg, are equal. In denoting the result of a chain of substitutions it is immaterial where one inserts parentheses. Hence, in such a chain, parentheses may be dispensed with altogether; fgh is unambiguous just as are $f \cdot g \cdot h$ and $f + g + h$.

Neutral Element. In substitution, j plays a neutral role as do the functions O in addition and 1 in multiplication:

(18) $\quad jf = fj = f$, for any function f.

Opposite Elements. A number x will be said to be *essential* for the function f if, and only if,

(1) x belongs to Dom f;

(2) $f\,x = f\,x'$ implies $x = x'$ for any number x' in Dom f.

In other words, x is essential for f if x is the only number for which f assumes the value $f\,x$.

Clearly, every number is essential for j, for any odd power, and for *exp;* every positive number, for *log;* every number $\neq 0$, for j^{-1}. No number is essential for *cos*, for *sin* $(2j)$, and for $3\,tan + 4$. For instance, $cos\ x = cos\ (x + 2\pi)$.

For any function f, the class of all numbers that are essential for f is a subclass of Dom f which will be called the *essential domain* of f and denoted by Ess f. For instance,

Ess j = Ess j^3 = Ess *exp* = U ; Ess *log* = P; Ess j^{-1} = P + N; Ess *cos* = V.

Ess $j^2 = \{0\}$; in fact, 0 is the only essential number of any even power. Indeed, $j^2\,0 = 0$ and $j^2\,x' = (x')^2 \neq 0$ for any $x' \neq 0$. But if $x \neq 0$, then x is not essential for j^2 since $j^2(-x) = j^2\,x$.

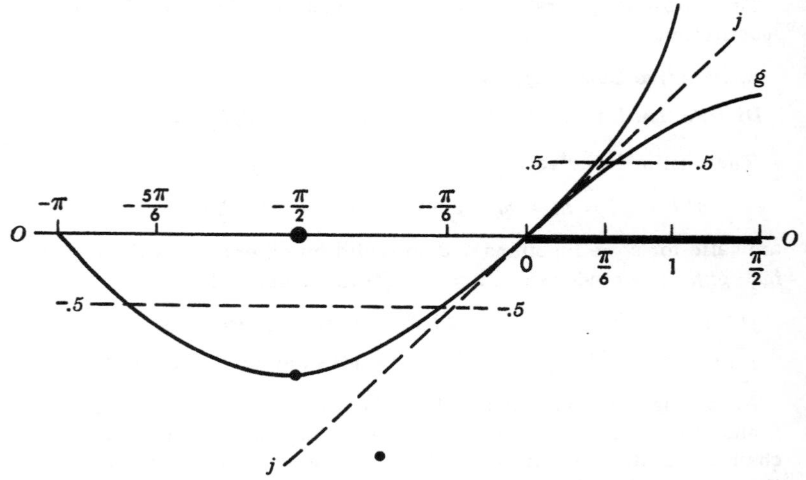

Fig. 28

Let g *ad hoc* denote the function *sin* (on $\{ -\pi \leq j \leq \pi/2 \}$). Its graph is shown in Fig. 28. The essential domain of g consists of the number $-\pi/2$ and all numbers that are > 0 and $\leq \pi/2$. If a horizontal line (such as *.5)* has exactly one point in common with g, then this point $(\pi/6)$ belongs to Ess g. So does the projection $-\pi/2$ of the point that the line *–1* has in common with g. If a horizontal line (such as *–.5)* has more than one point in common with g, then the projections of these points $(-\pi/6$ and $-5\pi/6)$ do not belong to Ess g.

If f is any function, by the *inverse* of f is meant in this book

(19) the class of all pairs (f x, x) for any x belonging to Ess f.

Thus the inverse is defined *for every function f*, and it is a *function;* that is to say, a class of pairs of numbers not containing two pairs with equal first, and unequal second, members. Indeed, the pair (f x, x) belongs to (19) only if x is essential for f, that is, if there is no number x$'$ $\neq$ x such that f x$'$ = f x. Hence the class (19) does not contain two pairs (f x, x) and (f x, x$'$) for x$'$ $\neq$ x.

The point at the altitude x above the point f x on O is the reflection in the line j of the point at the altitude f x above the point x. Hence *the graph of the inverse of f is the reflection in j of the function f* (on Ess f). In Fig. 28, g intersects some horizontal lines in two points. Consequently, the reflection in j of the entire curve g (not drawn in Fig. 28) intersects some vertical lines in two points and is not a simple curve nor the graph of a function. Only the reflection in j of the restriction g (on Ess g) is simple. This reflection (the graph of the inverse of g) consists of an isolated point and an arc drawn in Fig. 28.

The function j^2 (on Ess j^2) consists of the single pair (0,0) and cos (on Ess cos) is empty. Hence the inverse of j^2 consists of (0,0), and the inverse of cos is the vacuous function.

For reasons that will become clear presently, the inverse of f will be denoted by $j//f$, just as the reciprocal of f is denoted by $1/f$. Examples of mutually inverse functions include:

$j//j^3 = j^{1/3}$, and $j//j^{1/3} = j^3$; $j//exp_2 = log_2$ and $j//log_2 = exp_2$;
$j//j^2$(on U − N) $= j^{1/2}$ and $j//j^{1/2} = j^2$ (on U − N);
$j//j^2$ (on U − P) $= -j^{1/2}$ and $j//-j^{1/2} = j^2$ (on U − P).

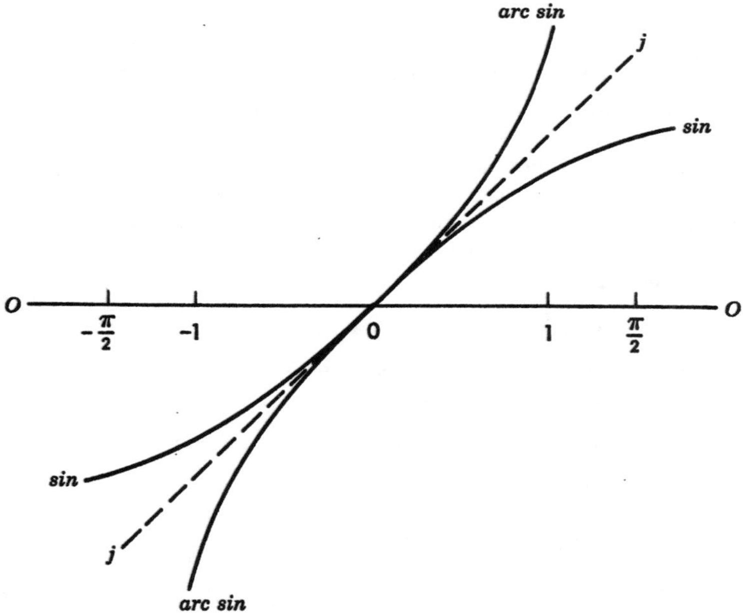

Fig. 29

Fig. 29 exhibits the mutual inverse of the function sin (on $\{-\pi/2 \leq j \leq \pi/2\}$), which is called the arcsine function and is denoted by $arcsin$. For any function, such as g in Fig. 28, that is an extension of sin (on $\{-\pi/2 \leq j \leq \pi/2\}$) and a restriction of sin, the inverse is (as is $j//g$) a restriction of $arcsin$. Thus $arcsin$ is what might be called an *inextensible* inverse of a restriction of *sin* (namely, of the restriction of *sin* to the interval between $-\pi/2$ and $\pi/2$). But there are inextensible inverses of other restrictions of *sin* that are different from *arcsin*. For instance, as the reader should prove and illustrate by drawings,

$$j//sin\ (\text{on}\ \{3\pi/2 \leq j \leq 5\pi/2\}) = arcsin + 2\pi$$

and

$$j//sin\ (\text{on}\ \{\pi/2 \leq j \leq 3\pi/2\}) = 2\pi - arcsin.$$

The reason why, in analogy to the symbol $1/f$ for the reciprocal of f, the symbol $j//f$ is here used for the inverse of f is that, in analogy to $(1/f) \cdot f = 1$ (on Dom $1/f$) and $f \cdot (1/f) = 1$ (on Dom $1/f$), one has

(20) $(j//f)\,f = j$ (on Ess f) and $f(j//f) = j$ (on Dom $j//f$),
Dom $(j//f)$ = Ran f (on Ess f) and Ran $(j//f)$ = Ess f.

Clearly, for any x in Ess f, the value f x belongs to Dom $(j//f)$: and the value of $j//f$ for f x is x. This completes the proof of the first formula (20). The reader should prove the second and verify both formulas geometrically. Since the graphs of f (on Ess f) and of $j//f$ are symmetric with regard to j, the construction explained in the example of Fig. 26 yields a part of the line j if either of the two functions is substituted into the other.

To find, for a given function f, the inverse of f, proceed as follows: first determine f(on Ess f), then, if this function is denoted by f^*, determine a function g such that $f^*g = j$. In Examples 1–4, $f^* = f$ so that only the second problem arises.

Example 1. To find $j//j^3$, determine g in such a way that $j^3 g = j$. Now $j^3 g = g^3$. If g^3 is to be equal to j, then g must be $j^{1/3}$. Indeed, one easily verifies that $j//j^3 = j^{1/3}$.

Example 2. To find $j//(2j + 3)$, solve $(2j + 3)\,g = j$, that is, $2g + 3 = j$. Clearly, $j//(2j + 3) = \frac{1}{2}(j - 3)$.

Example 3. To find $j//\frac{j+1}{j-1}$ determine g so that $\frac{j+1}{j-1}\,g = j$, that is, $\frac{g+1}{g-1} = j$. Clearly, $g = \frac{j+1}{j-1}$. Thus $\frac{j+1}{j-1}$ is its own inverse.

Example 4. To find $j//(j + exp_2)$ one has to solve $(j + exp_2)g = j$, that is, $g + exp_2 g = j$ for g. This is an exceedingly difficult problem.

Example 5. Consider the function $\frac{1}{2}\left(exp_2 + \frac{1}{exp_2}\right)$. Only one number is essential for this function, namely, 0. However, if f denotes *ad hoc* the restriction of the preceding function to U – N, then Ess f =

Dom f. To find $j//f$, determine a function g whose range consists of positive numbers and for which $fg = j$; that is to say, $\frac{1}{2}(\exp_2 g + 1/\exp_2 g) = j$. Set, *ad hoc*, $h = \exp_2 g$. Then the problem amounts to determining h so that $\frac{1}{2}(h + 1/h) = j$; that is to say, $h^2 - 2j \cdot h + 1 = 0$. Clearly, $h = j + \sqrt{j^2 - 1}$. Hence $g = \log_2\left(j + \sqrt{j^2 - 1}\right)$.

Example 6. To find $j//(j^3 - j)$, one has to determine Ess $(j^3 - j)$. One can prove (see Exercise 24, p. 166) that this class consists of the numbers $> 2/\sqrt{3}$ and the numbers $< -2/\sqrt{3}$. To determine values of the inverse function, one has to solve cubic equations.

Just as one writes $f - g$ for $f + (-g)$ and f/g for $f \cdot (1/g)$, one can write $f//g$ for $f(j//g)$. For instance,

$$\exp_2//\exp_2 = j;\ \log_2//\log_2 = j \text{ (on P)};\ j^m//j^{2n-1} = j^{m/(2n-1)};$$

$$\sec^2//\tan = (1 + \tan^2)//\tan = 1 + j^2 \text{ or } \sec^2\arctan = 1 + j^2;$$

$$\cos//\sin = \sqrt{1 - \sin^2}//\sin = \sqrt{1 - j^2} \text{ or } \cos\arcsin = \sqrt{1 - j^2}.$$

The reader should graphically substitute into *cos* the arcsine curve (i.e., the reflection in j of the sine curve between $-\pi/2$ and $\pi/2$; the domain of *arcsin* is the interval between -1 and 1). The result is the upper half of the unit circle.

EXERCISES

35. For each of the following functions, determine the value for x, the domain, and the range, and construct the graph:

(a) $j^2\sin$; (b) $\sin j^2$; (c) $\log \text{abs}$; (d) $\text{abs} \log$; (e) $\sin j^{-1}$.

36. Draw any simple curve f and construct these graphs of (a) $f(j + 2)$; (b) $f(2j)$; (c) $f(-2j)$; (d) $(2j + 1)f$. How are these graphs related to that of f?

37. Carry out the following substitutions:
(a) $(f - g)h$; (b) $(f - 3)h$; (c) $(f - f3)h$; (d) $(f - f3)3$; (e) $(f - f3)4$; (f) $(f_1 + f_2)gh$; (g) $(f_1 + f_2)(g_1 \cdot g_2)h$; (h) $\left(\frac{f-3}{j-3}\right)g$; (i) $\left(\frac{f - f3}{j-3}\right)g$; (j) $\left(\frac{f - fga}{j - ga}\right)g$.

38. Show that $j^2 f = f^2$ and $j^{-1} f = 1/f$.

39. On cross-section paper, using 1/8 inch as the unit, construct $\sin(3j) + \sin(\pi j)$ on $\{0 \leqq j \leqq 24\}$. This curve illustrates the acoustical

phenomenon of beats (waves which sectionwise reinforce and destroy each other).

40. If $f = \frac{a\,j + b}{c\,j + d}$ and $g = \frac{p\,j + q}{r\,j + s}$ (where a, b, c, d, p, q, r, s stand for constant functions), find: (a) fg; (b) gf; (c) $j//f$. Find conditions under which (d) $fg = gf$; (e) $j//f = f$.

41. Express without reference to numerical variables the equality $\sin(2x) = 2 \sin x \cdot \cos x$ and similar laws for $\sin(3x)$, $\cos(2x)$, $\tan(2x)$. Then express in terms of numerical variables the equality $\log_{10}(2j) = \log_{10} 2 + \log_{10}$ and similar laws for $\exp_2(5j)$ and $\cot(2j)$.

★42. Give examples of three functions f, g, h for which $f(g \cdot h) = fg \cdot fh$ *(a)* is, *(b)* is not, valid.

43. Show that fg is the class of all pairs of numbers (x, z) with the following property: there is a number y such that the pair (x, y) belongs to g and the pair (y, z) belongs to f.

44. Show that $ff = f$ is valid for the following functions: *(a)* j, 3, *abs*; *(b)* any extension of the function j (on $\{-1 \leqq j \leqq 2\}$) whose range is $\{-1 \leqq j \leqq 2\}$; *(c)* other functions.

45. Construct the curve $\arctan = j//\tan$ (on $\{-\pi/2 \leq j \leq \pi/2\}$) and $\arccos = j//\cos$ (on $\{0 \leq j \leq \pi\}$). Prove that $\arccos + \arcsin = \pi/2$, where the number on the right side stands for the constant function.

46. Show that, for any integer n,

$$\arcsin + 2n\pi \quad \text{and} \quad 2n\pi - \arcsin$$

are inverses to restrictions of *sin*.

47. *(a)* Draw the graph of $j//(j + \exp_2)$.

(b) Compute the inverse of $\frac{1}{2}(\exp_2 + 1/\exp_2)$ (on U – P). How is it related to the inverse in Example 5? Draw the graphs of both functions and both inverses.

(c) Compute and draw $j//\frac{1}{2}(\exp_2 - 1/\exp_2)$.

(d) For which integers p and q has $j^{p/q}$ a significant inverse?

★48. A function f is called *strictly increasing (strictly decreasing)* if $f\,a < f\,b$ (or $f\,b < f\,a$) for any two numbers in Dom f such that $a < b$. Show that for such a function Dom f = Ess f and that $j//f$ is strictly increasing (strictly decreasing).

★49. Show that while $j//(j//f)$ may be a restriction of f, one has $j//(j//(j//f)) = j//f$ for any function f.

9. POLYNOMIALS, RATIONAL FUNCTIONS, ELEMENTARY FUNCTIONS

3, $\sqrt[3]{2}\cdot j+5$, $j^2+3\cdot j+2$, and $j^{17}-1$ are examples of polynomial functions, briefly, polynomials. A function p is called a *polynomial* if and only if there is a non-negative integer n and n + 1 constant functions $a_0, a_1, \ldots, a_n$ such that $p = a_n \cdot j^n + \ldots + a_1 \cdot j + a_0$ or $\mathrm{a}_n j^n + \ldots + \mathrm{a}_1 j + \mathrm{a}_0$, according to the convention on p. 86. The numbers $\mathrm{a}_0, \ldots, \mathrm{a}_n$ (the values of the constant functions in the first definition) are called the *coefficients* of p. If $\mathrm{a}_n \neq 0$, then n is called the degree of p — briefly, deg p. The constant functions $\neq O$ are the polynomials of degree 0, whereas O is called a polynomial without a degree.

If p and q are polynomials, then so are $p+q$, $p\cdot q$ and $p\,q$.

$$\deg(p+q) \leq Max(\deg p, \deg q);\ \deg(p\cdot q) = \deg p + \deg q;$$

$$\deg(p\,q) = \deg p \cdot \deg q.$$

(Here, p and q are function variables whose scope is the class of all polynomials. They are what might be called *polynomial variables.)*

A polynomial may also be defined as a function that can be obtained from constant functions and j by a finite number of additions and multiplications.

A function is called *rational* if it can be obtained by the operations just mentioned and, besides, transitions to the reciprocal. Clearly, a function is rational if and only if it is the quotient of two polynomials. Examples are

$$(21)\ \frac{j^2+3j+2}{j+2} = (j+1)\ (\text{on } U - \{-2\}) \text{ and } \frac{j^2-1}{j-1} = (j+1)(\text{on } U - \{1\}).$$

In algebra, one is primarily interested in the system of the coefficients of a polynomial, called a *polynomial form.* But it would not be wise to denote the polynomial form of j^2+3j+2 simply by 1,3,2. Apart from other difficulties, the form of $j^{17}-1$ would then consist of 16 zeros between 1 and −1. It is more advisable to supply each coefficient with, as it were, *a place card carrying the the number of its position.* As such place cards we will use asterisks. Thus the form of j^2+3j+2 should be written $1*^2, 3*^1, 2*^0$. However, to conform (as far as it is possible) to the traditional notation of forms, we will omit (where this seems desirable)

(a) the symbol $*^0$;

(b) the number 1 in $*^1$, and write $*$ instead of $*^1$;

(c) all coefficients 0 together with their place cards;

(d) all coefficients 1 on places other than the lowest, whereas their place cards must be retained; e.g., $*^2$ instead of $1*^2$;

Finally one may use + shaped commas between the terms, and - shaped commas before negative coefficients; as in

$*^2+3*+2$ (or $1*^2+3*^1+2*^0$) and $*^2-1$ (or $1*^2+0*^1-1*^0$);

$*^{17}-1$ is the form of $j^{17}-1$.

Addition and multiplication of polynomial forms are defined in analogy to those of polynomial functions. Thus

$(* + 1) \cdot (* - 1) = *^2 - 1$, $(* + 1) \cdot (* + 2) = *^2 + 3 * + 2$, $(* + 1) + (*^2 - 1) = *^2 + *$.

In particular, the sum of, say, the forms $*^2$, $-3*$, and 2 is the form $*^2 - 3 * + 2$, wherefore the + and - shaped commas may be regarded as plus and minus signs. More generally,

$$(a_0 + a_1 * + \dots + a_m *^m) \cdot (b_0 + b_1 * + \dots + b_n *^n)$$
$$= a_0 b_0 + (a_0 b_1 + a_1 b_0) * + \dots + a_m b_n *^{m+n}.$$

Algebraists define also *rational forms* $\dfrac{a_0 + \cdots + a_m *^m}{b_0 + \dots + b_n *^n}$ and operate with them somewhat as with fractions. A rational form $\dfrac{a_m *^m + \dots + a_0}{1}$ is said to be polynomial, since it behaves like the polynomial form $a_m *^m + \dots + a_0$; just as the rational number $\frac{a}{1}$ is called integral, since it behaves like the integer a. Just as $\frac{4}{6}$ and $\frac{2}{3}$ are called equal because $4 \cdot 3 = 6 \cdot 2$, the rational forms

$$(22) \qquad \frac{*^2 + 3 * + 2}{* + 2} \quad \text{and} \quad \frac{*^2 - 1}{* - 1}$$

are called equal because

$$(*^2 + 3 * + 2) \cdot (* - 1) = (* + 2) \cdot (*^2 - 1).$$

Yet the rational functions (21) corresponding to the functions (22) are not equal.

The class $\{ sin = O \}$ contains infinitely many numbers (namely, all multiples of π), whereas, if p is a polynomial, $\neq O$, the class $\{ p = O \}$ cannot contain more than deg p numbers. It follows that *sin* is not a polynomial; and one readily proves that neither is *sin* a rational function. Also the other trigonometric functions, the so-called arc functions, (i.e., the inverses of restrictions of the trigonometric functions) and the exponential and logarithmic functions to various bases can be proved to be nonrational.

A function is called *elementary* if it can be obtained from constant functions, exp_2 and log_2, the trigonometric and the arc functions by a finite number of additions, multiplications, and substitutions. Since $j^c = exp_2 (c\ log_2)$, it follows that j^c is elementary for any c. All rational functions are elementary.

The function $j//(j + exp_2)$ is not elementary, but the proof of this fact is difficult. Other nonelementary functions are the elliptic integrals whose graphs were approximately constructed in Chapter II.

EXERCISE

50. Denote by *trg* the function assuming for any number x that is not an odd multiple of π the value *tan* (x/2). Show that there are six rational functions such that the substitution of *trg* into them results in *sin, cos, tan, cot, sec,* and *csc*. If *arctrg* denotes *j//trg*, how are *arcsin, arccos*, etc. related to *arctrg* ?

10. REMARKS CONCERNING THE TRADITIONAL FUNCTION NOTATION IN PURE MATHEMATICS

Remark 1. In the classical and the current literature, no distinction is maintained between functions and numbers (e.g., between a constant function and its value) – a distinction achieved in this book by the simple device of using italics for functions and function variables, and roman type for numbers and numerical variables. In the traditional literature, both the constant function *3* and its value, the number 3, are denoted by 3 in roman type; while italics are reserved for single letters when used as mathematical symbols; e.g., both the function variable c (whose scope is the class of all constant functions) and the numerical variable c (whose scope is the class of all numbers) are denoted by *c* in italic type. On the other hand, multi-letter-designations of specific functions, such as the logarithm or the arctangent, are not usually italicized. For the convenience of the reader, the situation is summarized in the following table.

Numbers and numerical variables	Functions and function variables	Function values
3, 0, $\sqrt{3}$, e	*3*, *log*, *arctan*	*log* 3, *arctan* x
a, c, x, y	*c*, *f*, *g*	*f* 3, *f* a, *g* x,
	are traditionally printed	
3, 0, $\sqrt{3}$, e	3, log, arctan	log 3, arctan x
a, *c*, *x*, *y*	*c*, *f*, *g*	$f(3)$, $f(a)$, $g(x)$

Throughout this book, references to the literature will follow the traditional usage.

Remark 2. The traditional literature lacks symbols for the identity and the nth power functions. These are referred to as "the function *x*" and "the function x^n" – that is, by the same symbols that designate the values of those functions for x. Thus most texts use (often in one and the same sentence) italic *x* (1) as a numerical variable; (2) as the designation of the identity function (in this book denoted by j); (3) as the designation of what herein is denoted by $j\,x$, the value that j assumes for x. One can read "The function x associates with any number x the number x", "The function x is the class of all pairs (x, x) for any number x" and the like. Correspondingly, x^n denotes both j^n and $j^n\,x$.

Remark 3. Traditionally, in analogy to the notation x^n for (among other things) j^n, also the functions *cos, arctan, log,* etc. (which do possess symbols) are referred to by their values as "the function cos x", etc.

In the course of one of his unrecorded travels, Gulliver met islanders who applied their arithmetic mainly, if not exclusively, to groups of shells (used as coins). Their arithmetical vocabulary consisted of

1) 4, read "four"; 5, read "five"; ... called numbers; (they wrote 4s for four shells, 5s for five shells ...)

2) symbols I, II, and III read "one shell", "two shells" and "three shells" – also referred to as numbers.

Mathematicians on the island, dissatisfied with the discrepancy between the two types of symbols, established uniformity by making references to "the number 4s", "the number 5s", etc. in analogy to "the number II". They would have avoided various difficulties (described at the foot of this page and on p. 195) if they had rather introduced new symbols, say 1, 2, and 3, read "one," "two," and "three" (for the numbers themselves), in contradistinction to I, II, and III for one shell, two shells, and three shells. They might have retained I, II, III as synonyms for 1s, 2s, and 3s.

Remark 4. According to Section 7, one can define the sum of the functions *cos* and j as the function assuming for any x the value $(cos + j)\,x = cos\,x + j\,x = cos\,x + x$. Traditionally, one of two usages is adopted. Either one defines the sum of "the function cos x" and "the function x" as "the function cos x + x" assuming for any number a the value $\cos a + a$, thus using a numerical variable other than x, namely, a. Or one uses the numerical variable x and, for the mere purpose of the definition, introduces a symbol, s, for the sum and defines the latter as "the function $s(x)$" such that $s(x) = \cos x + x$ for any x. Thereafter, one abandons the symbol s and refers to the sum function by its value, that is, as "the function cos x + x." For the sake of uniformity, one refrains even from writing (cos + log) x and writes only cos x + log x.

Gulliver's islanders had difficulties defining the sum of two numbers. Since they could not help writing

$$\text{II} + \text{I} = \text{III}, \quad \text{II} + 4s = 6s, \quad \text{II} + \text{III} = 5s,$$

they refrained from writing 4 + 5 = 9 or (4 + 5)s = 9s and, for the sake of uniformity, wrote only 4s + 5s = 9s.

Remark 5. In the classical literature, the identity function, as a result of its anonymity, has been altogether neglected. Its neutral role in substitution, described in the formulas

$$(18) \qquad jf = f \quad \text{and} \quad fj = f,$$

is rarely mentioned. How would this property be expressed in the classical notation? Obviously, not by $x(f) = f$ or $x(f(x)) = f(x)$. Rather, one would have to write:

(18′) If $j(x) = x$ for any x, then $j(f(x)) = f(x)$ and $f(j(x)) = f(x)$ for any x, or the like.

Not only is (18′) cumbersome in comparison with (18), but the antecedent of the implication (18′) is an *ad hoc* definition of a symbol for the identity function. Now, of course, this symbol would not have to be just *j*. One might express (18) by saying

(18″) If $h(x) = x$ for any x, then $h(f(x)) = f(x)$ and $f(h(x)) = f(x)$ for any x.

The point is that even in the classical theory, notwithstanding its countless references to "the function x," the formulation of the most important property of that very function necessitates the introduction, at least *ad hoc*, of a more adequate symbol than x. Under a serious indictment, the classical notation pleads guilty!

Remark 6. Some modern books contain symbols (that seem to go back to E.H. Moore) such as $f(.)$ and $\cos(.)$ or $f(*)$ and $\cos(*)$ for the functions f and cos, whereas the values that those functions assume for x are denoted by $f(x)$ and $\cos x$. In this way, the traditional ambiguity of $f(x)$ and $\cos x$ is indeed eliminated. But what about x^2 or $x^2 + x$? Only by using symbols such as j and j^2 can one distinguish $j^2(.)$ and $(j^2 + j)(.)$ from $j^2(x)$ and $(j^2 + j)(x)$. However, if one uses a symbol for the identity function, then he does not need suffixes such as (.), since f, *cos*, j^2, $j^2 + j$ denote functions, and $f\,x$, $\cos x$, $j^2 x = x^2$, $(j^2 + j)\,x = x^2 + x$ denote their values for x.

Remark 7. The symbol *j* for the identity function and the letter x in the traditional formulas are by no means simply interchangeable. Remark 5 demonstrates that in (18) one may not replace *j* by x. Nor can one always replace x by *j*. Such a replacement is completely out of the question if x is a numerical variable used conjunctively or imperatively, as in

"The class of all x such that $x^2 - 1 = 0$ contains two numbers",
"Find x such that $x^2 - 1 = 0$".

If, in the traditional formulas

(23) $\cos^2 x + \sin^2 x = 1$ and $x^2 \geq \sin^2 x$ for any x,

x were replaced by *j*, the nonsensical legend "for any *j*" would have to be omitted. In $\cos^2 j + \sin^2 j = 1$, the 1 would have to be italicized since the sum of the functions on the left side equals the constant function *1* and not the number 1. (We would also italicize the symbols of the two trigonometric functions.) After that is done, *j* may be omitted in the entire first formula and in the right side of the second. The formulas thus obtained,

$$sin^2 + cos^2 = 1 \quad \text{and} \quad j^2 \geq sin^2,$$

express relations between specific functions that are tantamount to properties of any number expressed in (23).

Also outside of mathematics, facts about classes can be described in various ways. For instance, one may say

(+) In the decisive meeting, X declared a certain law constitutional, for any member X of the U.S. Supreme Court,

or one may say

(++) The U.S. Supreme Court unanimously declared that law constitutional.

Remark 8. The functions

denoted in this book by:	*sin*,	*sin*(*j* + *3*),	*sin*(2*j*);
are traditionally called:	sin x,	sin(x + 3),	sin(2x) or sin 2x;

Some beginners are disturbed by the fact that *j* appears only twice in the first line, whereas x occurs in each example below. But similar asymmetries occur in the traditional expressions. Only one example in the second line contains an additive annex to x. Nobody habitually denotes the first function by sin(x + 0) or sin (1x). Of course, one might – just as he might write *sin j* instead of *sin*, for that matter – but it would be superfluous, wherefore one prefers not to add 0, not to multiply by 1, and not to substitute *j*.

After a little practice, the beginner will not miss *j* in *sin* any more than 0 or 1 in sin x.

Remark 9. Many texts indicate the equality of two functions (in this book expressed by $f = g$) by writing $f(x) \equiv g(x)$, read "$f(x)$ identically equal to $g(x)$." The distinction herein maintained between functions and their values for x makes such a symbol superfluous. For instance, *sin* ≠ *O* even though, for infinitely many x (namely, for all multiples of π), *sin* x = *O* x = 0.

Remark 10. Consider the statement

(24) $2 + 3x + x^2 + \sin x$ is not a polynomial.

The literature abounds in references to x in such statements as "the variable." Exactly how is this x traditionally used?

Not indicatively: replacement of x by 0 or 1 would result in the false statements that the functions 2 and $6 + \sin 1$ are not polynomials.

Not imperatively: no problem is formulated in (24); no number is to be found.

Nor conjunctively in the simple sense of the word: replacement of x by 0 and 1 yields the numbers 2 and $6 + \sin 1$ which are not elements of the function.

Nor does x in (24) seem to possess the most characteristic property of numerical variables (and in fact, of variables in general): the property of being replaceable by other letters without any change of the meaning. It is at least doubtful whether the statement (24) with t or a instead of x would be understood without any additional remark.

Notwithstanding the countless references to x in statements of the type (24) as "a variable", it appears that x therein is really used rather as a designation of the identity function (denoted by j in this book), which is a specific function just as is *cos*, and is not a variable of any kind.

However, the traditional use of x as a symbol for the identity function is not consistent. On the one hand, even if one interprets x in "the function $\log x$" as the identity function, he cannot very well do the same for the first x in "the function $\log x \sin x$."

On the other hand, $2 + 3x + x^2$ is sometimes used as an abbreviation for the class of all pairs of numbers $(x, 2 + 3x + x^2)$, where x is a numerical variable. But inasmuch as in the abbreviation x is used as a numerical variable, the letter has a peculiar connotation not adhering, e.g., to t or a. If in a traditional treatise (24) is expressed in terms of a, the statement

"$2 + 3a + a^2 + \sin a$ is not a polynomial"

is amplified by the remark "in a" or "as a function of a" or "as a function of the variable a." The above-mentioned special connotation of the letter x will not here be elaborated since in this book x is not being used in this obscure fashion.

Remark 11. The notions "a function of x," "a polynomial in a," "a function of the numerical variable t" and the like have not been defined in this chapter. In Chapter VII, where the step from numerical variables into the completely different world of variable quantities will be taken, sound concepts such as "a function of the pressure" and "a function of the variable quantity u" will be clearly defined. There, it will appear that the notion "a function of a numerical variable" is the result of a misunderstanding.

Remark 12. In polynomial and rational forms, algebraists traditionally use the letter x instead of the asterisk used on p. 97, that is, as a *place card for coefficients*. As a result, rational functions in calculus and the corresponding forms in algebra are currently denoted by the same symbols; e.g.,

traditionally, both $\frac{j^2 - 1}{j - 1}$ and $\frac{*^2 - 1}{* - 1}$ are denoted by $\frac{x^2 - 1}{x - 1}$.

This is unfortunate, since functions and forms do not follow the same laws. For instance, the forms (22) are equal, the functions (21) are not, and the discrepancies (even in the realm of polynomials) are much more pronounced where generalizations of the number concept are taken into consideration.

For instance, in the realm of integers reduced modulo 2, the functions j, j^2, j^3, ... are equal whereas the forms $*$, $*^2$, $*^3$, ... are regarded as pairwise unequal.

But even the one mentioned above is bound to confuse at least the thinking beginner, who is told

a) in *arithmetic*, that $\frac{x^2 - 1}{x - 1} = x + 1$ holds for any number $x \neq 1$;

b) in *algebra*, that $\frac{x^2 - 1}{x - 1} = x + 1$ (without any qualification); in fact, he is expected to answer dozens of questions by writing down formulas of this type;

c) in *calculus*, that $\frac{x^2 - 1}{x - 1}$ and $x + 1$ are different functions, inasmuch as "the variable x" may assume the value 1 in the latter but not in the former (and differences of this kind are crucial in defining derivatives of functions).

True, x in $x^2 + 3x + 2$ is called "the *variable*" in calculus (see Remark 10), and "the *indeterminate*" in algebra. But the definition of an indeterminate is rather obscure. The following is a quotation from one of the best current books. "Let 'x' be any symbol.... Nothing is assumed known about x.... One forms sums, products, and differences of x with [numbers] and with itself.... x^n is defined as $x \cdot x \ldots x$ to n factors". It is not clear how one can add numbers to something about which nothing is assumed known, nor what multiplication of x with itself may mean under these circumstances.

The analysis of the basic concepts contained in the present chapter seems to clarify the situation.

a) *Arithmetic:* $\frac{x^2 - 1}{x - 1} = x + 1$, for any number x, $\neq 1$, where x (in roman type) may be replaced by any other lower-case letter in roman type and, in the formula, by the designation of any number $\neq 1$. Here, x is a numerical variable.

b) *Algebra:* $\frac{*^2 - 1}{* - 1} = * + 1.$

This formula expresses the equality of two rational forms – systems of numbers whose order is indicated by the symbols $*$, $*^2$, ... One operates with these systems of numbers according to definite rules.

c) *Calculus:* $\frac{j^2 - 1}{j - 1} = j + 1$ (on $U - \{1\}$), where j is the identity function, defined as the class of all pairs (x, x) for any x, and 1 is the constant function of the value 1. Functions can be added, multiplied (see Section 7) and restricted to subclasses of their domains, as $j + 1$ to $j + 1$ (on $U - \{1\}$) (see p. 80). Formula (c) is free of variables of any kind. It relates specific functions, just as the formula $15/3 = 5$ relates specific numbers.

The arithmetical formula

$$\frac{x^2 - a^2}{x - a} = x + a \text{ for any x and any a such that } x \neq a$$

contains two numerical variables, x and a, which, without any change of the meaning, may be replaced by any two non-identical letters, for instance, by a and x.

The statements made in calculus and algebra

$\frac{j^2 - a^2}{j - a} = j + a$ (on $U - \{a\}$) for any a, and $\frac{*^2 - a^2}{* - a} = * + a$ for any a (where a is the constant function of the value a) contain one numerical variable, a.

In statements about any quadratic function, $a j^2 + b j + c$, or any quadratic form, $a *^2 + b * + c$, the coefficients a, b, c (often referred to as "the constants a, b, c" or "the parameters a, b, c") are the only numerical variables.

EXERCISES

51. In setting up the traditional symbols for functions (i.e., their values for x) observe the following three rules:

(1) Avoid sums, products, and quotients of functions such as $(\log \cdot \tan + f/\sin)\, x$, and replace them by the sums, products, and quotients of the values, as in $\log x \tan x + f(x)/\sin x$. In contrast, write powers of functions, such as $\cos^2 x$ in preference to powers of values, $(\cos x)^2$.

(2) In a product, write the factors side by side, as in log x tan x without separating them by dots unless either the second factor is constant (as in log x · 2 or 3 · 2) or the former factor ends in x and the latter factor is a power function or is replaceable by any constant function (as in log x · x, sin x · x^2, cos x · c). If possible, avoid this situation by permuting the factors.

(3) In expressing values or in substituting functions that are free of sums and products, use parentheses after single-letter symbols, as in $f(x)$, $g(\log x)$, $\Gamma(\sqrt{x})$, $J_0(x)$; and don't use them after multi-letter-symbols, as in cos x, log sin x, log log log x, sin $f(x)$. Insert between parentheses all sums and products that are substituted, as in $\log(x^2 + \sin x)$, arcsin (sin x cos x), $f(x + a)$, $g(3a)$.

Set up the traditional symbols for the following functions (not all of which are given in the simplest possible way):

$3\,tan$, $tan\,3$, $tan(3j)$, $tan(3 \cdot j)$, $tan(3 + j)$, $tan\,j$, $tan \cdot j$, $j\,tan$, $tan \cdot j^{-1}$, $log\,cos$, $log \cdot cos$, $(log + j) \cdot cos$, $(log + j)cos$; $(log \cdot sin)j^2$, $exp(j + f)$, $(j + f)exp$, $\frac{f}{g} \cdot sin$, $\frac{f}{g}\,sin$;

and express in the traditional way the law

$$j \cdot tan \cdot j = tan \cdot j^2 .$$

Express without reference to x the following functions

log arctan x, $e^{\sin x \cos x}$, $x^2 \cdot \sqrt{\cos x}$, $f(e^{g(x)})$, $f(g(h(x)))$.

52. Express in the traditional notation (which often necessitates the use of implications and the introductions of *ad hoc* symbols) the following laws:

$(f + g)h = fh + gh$, $jj = j$, $f(gh) = f(gh)$, $(j \cdot arcsin)sin = sin \cdot j$.

Express each of the following laws in a single formula:

if $p(x) = f(x)\,g(x)$, then $p(\log x) = f(\log x)\,g(\log x)$;

if $r(x) = f(x) \sin x$, then $r(\arcsin x) = x f(\arcsin x)$;

if $q(x) = \dfrac{x}{f(x)}$, then $q(\cos x) = \dfrac{\cos x}{f(\cos x)}$.

CHAPTER V

ON LIMITS

1. APPROXIMATE EQUALITIES

If $y \sim 2$ *and* $z \sim 6$, *then* $y + z \sim 8$; $y - z \sim -4$; $y \cdot z \sim 12$; $y/z \sim 1/3$; $z/y \sim 3$.

Here, $y \sim 2$ indicates that y is *close* to 2. But without an explanation as to *what* numbers are close to 2, to 6, to 8, etc., the above statement is vague and useless. In contrast, the following statement is precise.

If $\quad 2 - a < y < 2 + a$ *and* $6 - b < z < 6 + b$,
then $\quad 8 - (a + b) < y + z < 8 + (a + b)$
and $\quad$ (*if* $|a| < 2$ *and* $|b| < 6$),
$$12 - (6a + 2b - ab) < y \cdot z < 12 + (6a + 2b + ab).$$

The last two inequalities are obtained by adding and multiplying the first two. Obviously, $y + z$ differs from 8 by less than .01 if $a + b < .01$; e.g., if $a < .005$ and $b < .005$, or if $a < .002$ and $b < .008$. Here and in what follows, two numbers x_1 and x_2 are said to *differ by at most* d (> 0) if and only if

$$x_1 - d < x_2 < x_1 + d \text{ or, which is equivalent, if } x_2 - d < x_1 < x_2 + d;$$

in still other words, if $-d < x_1 - x_2 < d$ or, which is equivalent, if $-d < x_2 - x_1 < d$.

$y \cdot z$ differs from 12 by less than d if $6a + 2b + ab < d$; for instance, if $a < d/18$ and $b < d/6$, since in this case $6a < d/3$, $2b < d/3$, and $ab < d^2/108$, which is also $< d/3$ (unless $d > 36$, in which case one obtains $6a + 2b + ab < d$ even for $a = 1$ and $b = 1$).

In the precise statement, the number d may be chosen arbitrarily small, that is, as small as one pleases. Thus, $y + z$ is arbitrarily close to 8, and $y \cdot z$ is arbitrarily close to 12, provided that y is sufficiently close to 2 and z sufficiently close to 6. What proximity is sufficient has just been specified: for any d (> 0),

if $2 - d/2 < y < 2 + d/2$ and $6 - d/2 < z < 6 + d/2$, then $8 - d < y + z < 8 + d$;

if $2 - d/18 < y < 2 + d/18$ and $6 - d/6 < z < 6 + d/6$, then $12 - d < y \cdot z < 12 + d$.

These are, of course, not necessary proximities. In the first case, e.g.,

if $2 - d/3 < y < 2 + d/3$ and $6 - 2d/3 < z < 6 + 2d/3$,
then also, $\quad 8 - d < y + z < 8 + d$.

The facts just established may be expressed by writing

If $y \underset{\text{suf}}{\sim} 2$ *and* $z \underset{\text{suf}}{\sim} 6$, *then* $y + z \underset{\text{arb}}{\sim} 8$ *and* $y \cdot z \underset{\text{arb}}{\sim} 12$.

Conversely, from $y + z \underset{\text{suf}}{\sim} 8$ it cannot be concluded that either $y \underset{\text{arb}}{\sim} 2$ or $z \underset{\text{arb}}{\sim} 6$. For instance, if $y = 3$ and $z = 5$, then $y + z = 8$, yet y is far from 2 and z is far from 6.

It is easy to prove that

if $y \underset{\text{suf}}{\sim} 2$ and $z \underset{\text{suf}}{\sim} 6$, then $y - z \underset{\text{arb}}{\sim} -4$ and $y/z \underset{\text{arb}}{\sim} 1/3$

and to substantiate this statement by inequalities. Similarly, it can be proved that

if $y \underset{\text{suf}}{\sim} 0$ and $z \underset{\text{suf}}{\sim} 6$, then $y + z \underset{\text{arb}}{\sim} 6$, $y \cdot z \underset{\text{arb}}{\sim} 0$, $y/z \underset{\text{arb}}{\sim} 0$.

But all that can in this case be proved about z/y is that that quotient is either a large positive number or the negative of a large positive number (according to whether y is positive or negative).

If $y \underset{\text{suf}}{\sim} 0$ and $z \underset{\text{suf}}{\sim} 0$, then $y + z \underset{\text{arb}}{\sim} 0$ and $y \cdot z \underset{\text{arb}}{\sim} 0$.

But in this case no general statement at all can be made about either y/z or z/y, as is illustrated in the following examples:

If	$y = .01$	.0001	.001
and	$z = .005$	.005	–.00001
then	$y/z = 2$	.02	–100

EXERCISES

1. How close should y be to 1/2, and z to 6 in order that:

(a) $y + z$ and 6.5 differ by less than .01;

(b) $z - y$ and 5.5 differ by less than c (for any number $c > 0$);

(c) $y \cdot z$ and 3 differ by less than .02;

(d) z/y and 12 by less than .1;

(e) y/z and 1/12 by less than b (for any number b);

(f) all five preceding conditions are satisfied simultaneously?

In case (c) as well as in case (e) give two different examples of sufficient proximities.

2. Let a and b be any two numbers. What proximity of y and z is sufficient to guarantee that the difference between

(a) $y + z$ and $a + b$ be less than .01;

(b) $y - z$ and $a - b$ be less than c (> 0);

(c) $y \cdot z$ and $a \cdot b$ be at most c (> 0)?

3. If n is a positive integer and c is any positive number, how close to 2 must x be chosen in order that $x^2 - 1$ differ from 3 by less than *(a)* 10^{-n}, *(b)* c, *(c)* .001, *(d)* .03? Notice that in this and the following exercises the questions read "How close... *must* x be chosen...?" Distinguish the cases where $x > 2$ and where $x < 2$.

4. How close to 2 must x be chosen in order that $(x^4 - 1)/5$ differ from 3 by less than *(a)* 10^{-n}, *(b)* c, c > 0, *(c)* .001, *(d)* .03?

5. Show that, if n is sufficiently large, then 2^n exceeds: *(a)* 100, *(b)* 1,000,000, *(c)* 10^m, for any positive integer m. How large must n be chosen if 2^{-n} is to be less than: *(a)* .01, *(b)* .000001, *(c)* 10^{-m}, for given m, *(d)* any small positive number d?

2. THE LIMIT OF *f* AT *a*

If x differs from 2 by less than d, where $|d| < 2$, then $(2 - d)^2 - 1 < x^2 - 1 < (2 + d)^2 - 1$. Hence $x^2 - 1$ differs from 3 by less than $4d + d^2 = d(4 + d)$ and, if $d < 1$ (so that $4 + d < 5$) by less than 5d. Consequently, for any positive number c (< 5), $x^2 - 1$ differs from 3 by less than c provided x differs from 2 by less than c/5. Indeed,
if $2 - c/5 < x < 2 + c/5$,
then $(2 - c/5)^2 - 1 < x^2 - 1 < (2 + c/5)^2 - 1 = 3 + 4c/5 + c^2/25$, which is less than 3 + c, since c < 5 implies that $c^2/25 = (c/5) \cdot (c/5) < c/5$. Thus

(1*) $$\text{If } x \underset{\text{suf}}{\sim} 2, \text{ then } x^2 - 1 \underset{\text{arb}}{\sim} 3.$$

Traditionally, (1*) is expressed by writing $\lim_{x \to 2} (x^2 - 1) = 3$ (read "the limit, as x approaches 2, of $x^2 - 1$ is 3").

Clearly, by the same token, if $z \underset{\text{suf}}{\sim} 2$, then $z^2 - 1 \underset{\text{arb}}{\sim} 3$ and if $t \underset{\text{suf}}{\sim} 2$, then $t^2 - 1 \underset{\text{arb}}{\sim} 3$, and hence

$$\lim_{z \to 2} (z^2 - 1) = 3 \quad \text{and} \quad \lim_{t \to 2} (t^2 - 1) = 3.$$

All these synonymous formulas describe a property of the function $j^2 - 1$, which in this book will be expressed by saying that $j^2 - 1$ has the limit 3 at 2 and by writing

(1) $\lim_2 (j^2 - 1) = 3$ (read "limit at 2 of $j^2 - 1$ is equal to 3").

Similarly

(2) $\lim_2 \frac{1}{5} (j^4 - 1) = 3, \ \lim_3 (j^2 - 1) = 8, \ \lim_3 \frac{1}{5} (j^4 - 6) = 15.$

Even these first examples show that the concept of limit has nothing to do with the maxima and minima of a function as the word limit might suggest. If L is used instead of limit, then

$$\mathsf{L}_2 (j^2 - 1) = 3 \quad \text{and} \quad \mathsf{L}_3 \frac{1}{5} (j^4 - 6) = 15.$$

More generally, for any positive integer n and any number a

(3) $$\lim_{a} j^{n} = a^{n}.$$

Indeed, if $0 < d < a$ and x differs from a by less than d, then

$$(a-d)^n - a^n < x^n - a^n < (a+d)^n - a^n .$$

But $(a+d)^n - a^n$

$$= [(a+d)^{n-1} + a(a+d)^{n-2} + a^2(a+d)^{n-3} + \cdots + a^{n-1}] \cdot d,$$

which is less than $n(a+d)^{n-1}d$, since the factor between brackets is the sum of $(a+d)^{n-1}$ and $n-1$ terms $< (a+d)^{n-1}$. From $d < a$ it follows that $n(a+d)^{n-1}d < n(2a)^{n-1}d$; and if $d < 1$, then also $n(a+d)^{n-1}d < n(a+1)^{n-1}d$. Hence for any number c, if $d < a$ and $d < c/n(2a)^{n-1}$, or if $d < 1$ and $d < c/n(a+1)^{n-1}$, then $a-d < x < a+d$ implies $x^n - a^n < c$ and, similarly, $-c < x^n - a^n$. Thus, if $x \underset{suf}{\sim} a$, then $x^n \underset{arb}{\sim} a^n$, as is claimed in (2).

Very concise expressions of the preceding results can be formulated in terms of the concept of the restriction of a function.

For instance, (1) is rendered

$$(j^2 - 1)\,(\text{suf near } 2) \underset{arb}{\sim} 3.$$

More precisely, for any number c (< 5), the restriction

$$(j^2 - 1)\,(\text{on } \{2 - \tfrac{1}{5}c < j < 2 + \tfrac{1}{5}c\})$$

lies between the constant functions $3 - c$ and $3 + c$ or, which is equivalent, between the analogous restrictions of these functions. Hence one might write

(1′) $$3 - c < j^2 - 1 < 3 + c \text{ on } \{2 - \tfrac{1}{5}c < j < 2 + \tfrac{1}{5}c\},$$

where it is understood that for all three functions in the inequalities restrictions to the same class are to be considered.

Similarly,

(3′) $$a^n - c < j^n < a^n + c \quad \text{on } \left\{ -\frac{1}{n(2a)^{n-1}}c < j - a < \frac{1}{n(2a)^{n-1}}c \right\}.$$

In fact, (1) and (3) may be considered as digests of (1′) and (3′). If h_1 and h_2 denote, *ad hoc*, the functions assuming for c the values

$h_1 c = \frac{1}{5}c$ and $h_2 c = \dfrac{1}{n(2a)^{n-1}}c$, then

(1″) $$3 - c < j^2 - 1 < 3 + c \text{ on } \{2 - h_1c < j < 2 + h_1c\}, \text{ and}$$

(3″) $$a^n - c < j^n < a^n + c \text{ on } \{-h_2c < j - a < h_2c\}.$$

More generally (for any function *f* and any number a), if for all x that are $\underset{suf}{\sim}$ a (but $\neq$ a), the values *f* x are $\underset{arb}{\sim}$ b, then one says that *f* has the limit b at a or that the limit of *f* at a equals b, and one writes

(4) $$\lim_a f = b \quad \text{or} \quad \underset{a}{\mathsf{L}} f = b.$$

(4), which means

$$f \underset{\text{arb}}{\sim} b \text{ on } \{ a \neq j \underset{\text{suf}}{\sim} a \},$$

may be considered as a digest of the statement that for some function h whose domain is P and whose range is a subclass of P (that is, some function defined for all positive numbers and assuming positive values):

$$b - c < f < b + c \text{ on } \{ a - h\,c < j < a + h\,c \}.$$

In (1″) and (3″), the functions h_1 and h_2 are $\frac{1}{5}\,j$ and $\frac{1}{n(2a)^{n-1}}\,j$.

If $f = j^{-1}$ and a = 2, it is easily seen that $\lim_2 j^{-1} = \frac{1}{2}$. However, if $f = j^{-1}$ and a = 0, then there is no number to which the values of the function would be arbitrarily close for all numbers that are sufficiently close to 0. This fact is expressed by saying: *The function* j^{-1} *has no numerical limit at* 0, or $\lim_0 j^{-1}$ *does not exist*. Near 0 (where the hyperbola j^{-1} has a gap), one branch of j^{-1} rises above, the other sinks below, any preassigned level. Similarly, the function $\frac{1}{j-2}$ has no limit at 2. Nevertheless it is possible to write down the symbols $\lim_0 j^{-1}$ and $\lim_2 \frac{1}{j-2}$. They suggest that there are limits of j^{-1} at 0 and of $\frac{1}{j-2}$ at 2, which is not the case. Those symbols (like 2/0) do not designate anything.

Beginners frequently ask: If $\lim_a f = b$, is b *precisely* or *approximately* the limit of f at a (or the limit of f x as x approaches a)? The answer is: *precisely*. In the case of $j^2 - 1$, if x is sufficiently close to 2, then the value for x, that is $x^2 - 1$, is arbitrarily close to the precise number 3 but not arbitrarily close to any other number. For instance, $x^2 - 1$ is not $\underset{\text{arb}}{\sim}$ 3.01 if x $\underset{\text{suf}}{\sim}$ 2. For, if it were, then $x^2 - 1$ would differ from 3.01, in particular, by less than .001 if x $\underset{\text{suf}}{\sim}$ 2. But this is not the case. If x differs from 2 by less than .0001, then $x^2 - 1$ differs by less than .001 from 3 (Cf. Exercise 3). Consequently $x^2 - 1$ differs from 3.01 by more than .009 and not by less than .001.

What is approximate is the equality of the values of the function $j^2 - 1$ close to 2, on the one hand, and the precise number 3, on the other.

Some beginners think: *Obviously* $x^2 - 1$ is arbitrarily close to 3 if x is sufficiently close to 2, because $x^2 - 1$ is *precisely* equal to 3 if x is *precisely* equal to 2. But whether a function has what is called a limit at a, and if so, what that limit is are questions that concern the behavior of the function *near* a and have nothing to do with its behavior strictly *at* a (for instance, with whether or not its domain includes a, and if so, what the function value at a is).

Whether or not there are certain fish near a definite spot in the ocean has nothing to do with whether or not such fish are *at* that spot, which may be a rock.

Example 1. Consider the square of the signum function assuming the value 0 for 0, the value 1 on P and the value −1 on N. Clearly sgn^2 assumes the value 1 on P + N and the value 0 at 0. Everywhere near 0 the values are 1. Consequently,

$$\lim_0 sgn^2 = 1 \quad \text{and yet} \quad sgn^2\ 0 = 0.$$

Example 1 illustrates the significance of the words "but $\neq 0$" in the definition of the limit of *f* at a. If these words were omitted, *f* would have the limit b at a only if also the value of *f at* a were arbitrarily close to b (and, in fact, equal to b). The resulting definition would be logically possible but different in content from the one given above (which is universally accepted). For, according to the former, sgn^2 would not have a limit at 0, whereas, according to the latter, it has the limit 1.

Example 2. Consider $2^{-1/x^2}$ for x near 0. If x is close to 0, then

(a) x^2 is a positive number that is very close to 0;

(b) $1/x^2$ is a very large positive number;

(c) $2^{1/x^2}$ is a tremendously large positive number;

(d) $2^{-1/x^2}$ (the reciprocal of $2^{1/x^2}$) is very, very close to 0.

In fact, it is not hard to show that, if x is sufficiently close to 0, then $2^{-1/x^2}$ is as close to 0 as one pleases; in symbols, $\lim_{x \to 0} 2^{-1/x^2} = 0$. But one cannot say "this is obvious, because if x is *precisely* equal to 0, then $2^{-1/x^2}$ is *precisely* equal to 0." For, if, in $2^{-1/x^2}$, x is replaced by 0, the result is the meaningless symbol $2^{-1/0^2}$ and not a number.

$2^{-1/x^2}$ is the value for x of the function obtained by substituting $-j^{-2}$ into exp_2, that is, the function $exp_2(-j^{-2})$. The number 0 does not belong to the domain of this function (since no value of the function is defined for 0). Yet, as was shown above, the function has a limit at 0, namely 0; in symbols, $\lim_0 [\, exp_2\, (-j^{-2})] = 0$.

Fig. 30

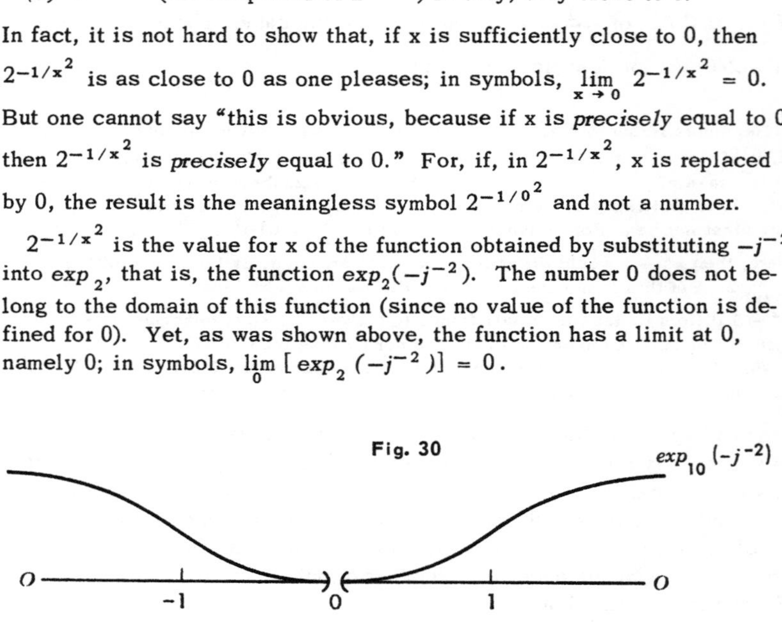

In Fig. 30, there is the graph of $exp_2(-j^{-2})$. This curve consists (somewhat like the hyperbola j^{-1}) of two branches – one branch to the left, and one to the right, of 0 – and does not intersect the vertical line through 0 at all. (The number 0 does not belong to the domain of the function.) But whereas the two branches of j^{-1} do not approach any point above 0, those of the curve $exp_2(-j^{-2})$ do. They approach the point with altitude 0, since, according to *(d)*, all points of the graph near 0 have altitude very, very close to 0. These statements illustrate geometrically that, whereas j^{-1} has *no* limit at 0, the function $exp_2(-j^{-2})$ has a limit at 0, namely the limit 0.

Beginners sometimes confuse "having the limit 0" and "having *no* limit". These concepts are in a relation comparable to that between a man having a purse with no money in it and a man having no purse.

Example 3. Consider the function obtained by pairing with any x the largest integer $\leqq$ x. It has no limit at 2 (or at any integer). If x is any number less than 2, then no matter how close x is to 2, the function has the value 1 for x, whereas for 2 and any number between 2 and 3, the function has the value 2.

A function f is said to be *continuous at* a (Bolzano, 1817) if

(1) a belongs to the domain of f;

(2) f has a limit at a;

(3) the limit of f at a is equal to f a, the value of f at a.

In symbols, f is continuous at a if

$$\lim_{a} f = f\,a \quad \text{or} \quad \lim_{x \to a} f\,x = f\,a \quad \text{or} \quad \lim_{z \to a} f\,z = f\,a \ldots .$$

A function that is continuous at every number of its domain is called *continuous*. Examples of continuous functions are the exponential functions exp_b (for any positive base b), the sine and cosine functions and, for any n, the functions j^n. Indeed,

$$\lim_{a} j^n = j^n\,a = a^n \quad \begin{cases} \text{for any a, if } n > 0, \\ \text{for any a} \neq 0, \text{ if } n < 0. \end{cases}$$

For instance, j^{-1} is continuous (that is to say, continuous at any number in Dom j^{-1}) even though j^{-1} is not continuous at 0 (which does not belong to Dom j^{-1} and where, besides, j^{-1} has no limit).

The function in Example 2 is not continuous at 0 since 0 does not belong to its domain. But it has an extension that is continuous at 0, namely the function pairing $2^{-1/x^2}$ with any x $\neq$ 0 *and* 0 *with* 0; in other words, the class consisting of

$$(0, 0) \text{ and all pairs } (x, 2^{-1/x^2}) \text{ for any } x \neq 0.$$

The one point in the graph of that extension that does not belong to the curve in Fig. 30 fills the gap between the two branches of the latter curve. Clearly, no other extension of the function in Example 2 is continuous at 0. If, e.g., the pair (0, 1) is adjoined to it, the resulting extension is not continuous at 0 since its limit at 0 and its value at 0 are not equal.

The graph of any function is a simple class of points. The graph of a continuous function whose domain is connected (i.e., an interval, a ray, or U) is often called a simple *curve*.

EXERCISES

6. Which of the following denote numbers and which are meaningless symbols? *(a)* $\lim_{-1} \frac{3}{1 - j^{-1}}$; *(b)* $\lim_{0} \frac{3}{1 - j^{-1}}$; *(c)* $\lim_{1} \frac{3}{1 - j^{-1}}$;

(d) $\lim_{1} \exp_{10} j^{-1}$ or $\lim_{x \to 1} 10^{1/x}$; *(e)* $\lim_{0} \exp_{10} j^{-1}$.

7. Where does each of the following functions fail to have a limit? *(a)* $\log_{10}$; *(b)* *tan*; *(c)* $\frac{1}{1 - j^{-1}}$; *(★d)* $\frac{1}{1 + \exp_2 j^{-1}}$; *(e)* the class of all pairs $(x, x - [x])$, where $[x]$ is the largest integer $< x$.

★8. The function h mentioned in the explanation of $\lim_{a} f = b$ was illustrated by h_1 and h_2 in the cases (1) and (3). Determine it in the examples (2). In all these cases $h = k\, j$ for some number k. What is h for $\lim_{0} j^{1/2} = 0$ and for $\lim_{0} j^{2/3} = 0$?

9. If $f\,x$ is arbitrarily close to b for any number x that is sufficiently close to a *and greater than* a, then b is called the *right-side limit* of f at a; in symbols, $\lim_{a+0} f = b$. The *left-side limit*, $\lim_{a-0} f$, is defined in an analogous way. Verify the following statements:

(a) $\lim_{0+0} sgn = 1$ and $\lim_{0-0} sgn = -1$.

(b) If $f\,x$ is the largest integer $< x$, then for every integer n

$$\lim_{n+0} f = n \quad \text{and} \quad \lim_{n-0} f = n - 1 .$$

(c) If f is the function of Exercise 7 *(d)*, then $\lim_{0-0} f = 1$ and $\lim_{0+0} f = 0$.

(d) If $\lim_{a} f = b$, then $\lim_{a+0} f = \lim_{a-0} f = b$, and conversely.

10. Show the continuity of the following functions:

(a) of *cos* and *sec* at 0; (Hint: Use the inequality that *sin* x < x.)

(b) of log_{10} at 1, and at any number a > 0.

11. For what reasons are j^{-1} and the functions in Examples 1, 2, and 3 not continuous at 0?

3. THE LIMITS OF SOME COMPOUND FUNCTIONS

Having studied the definition of the number $\lim_a f$, the beginner expects to be given a rule by virtue of which, whenever a function f and a number a are given, $\lim_a f$ can actually be computed Such a sweeping rule, however, does not exist. Only restricted rules can be formulated.

Suppose that

$$\lim_a g = b \text{ and } \lim_a h = c \quad \text{(the same a in both cases)}.$$

This means:

$$\text{if } x \underset{suf}{\sim} a, \text{ then } g\,x \underset{arb}{\sim} b \text{ and } h\,x \underset{arb}{\sim} c.$$

It follows that

$$g\,x + h\,x \underset{arb}{\sim} b + c \quad \text{and} \quad g\,x \cdot h\,x \underset{arb}{\sim} b \cdot c.$$

Now $g\,x + h\,x$ is the value for x of the sum $g + h$; and $g\,x \cdot h\,x$ that of the product $g \cdot h$. Consequently, if $x \underset{suf}{\sim} a$, then $(g + h)x \underset{arb}{\sim} b + c$ and $(g \cdot h)x \underset{arb}{\sim} b \cdot c$; that is to say, the limit of $g + h$ at a is b + c, that of $g \cdot h$ is b · c. Thus one arrives at the following

Law About Limits of Sums and Products (First Form).

If $\lim_a g = b$ *and* $\lim_a h = c$, *then* $\lim_a (g + h) = b + c$ (with an analogue for multiplication). Since $b = \lim_a g$ and $c = \lim_a h$, the above law may be formulated in the following

Second Form: $\lim_a (g + h) = \lim_a g + \lim_a h$ and

$\lim_a (g \cdot h) = \lim_a g \cdot \lim_a h$, whenever $\lim_a g$ and $\lim_a h$ exist.

Here, no reference is made to the names, b and c, of the limits of g and h at a. On the other hand, a name for $g + h$ might be introduced, say f, whereby the law would assume the

Third Form. if $f = g + h$, then $\lim_a f = \lim_a g + \lim_a h$, provided that $\lim_a g$ and $\lim_a h$ exist.

This statement is a restricted rule by which $\lim_a f$ can actually be computed, namely, in the cases where f is the sum of two functions whose limits at a are known. In these cases, the rule stipulates that one should simply add the known limits.

One might, of course, introduce b, c, f, and a fourth name, and give the law the fourth and fifth forms:

If $\lim_a g = b$ and $\lim_a h = c$, and $f = g + h$, then $\lim_a f = b + c$;

If $\lim_a g = b$, $\lim_a h = c$, $\lim_a f = d$, and $f = g + h$, then $d = b + c$.

Some persons grasp a law such as the one just discussed more easily if for certain objects names are introduced that could be dispensed with, whereas others prefer formulations without superfluous names and symbols. One point, however, ought to be clear to *every* student regardless of his personal preference in the matter: *the equivalence of the five preceding formulations* of one and the same law. In fact, the recognition of this equivalence is an important step on the way to an understanding of many mathematical laws.

For instance, from $\lim_2 j^2 = 4$ and $\lim_2 j^3 = 8$ it follows that $\lim_2 (j^2 + j^3) = 12$ and $\lim_2 (j^2 \cdot j^3) = 32$. From $\lim_0 exp_{10}(-j^{-2}) = 0$ and $\lim_0 sgn^2 = 1$ it follows that $\lim_0 [exp_{10}(-j^{-2}) + sgn^2] = 1$. Of course, these results can also be obtained, without reference to the Law about the Limits of Sums, by direct application of the limit concept to the sum function.

It should be noted that $g + h$ may have a limit at a, although neither g nor h has. For instance, neither sgn nor $1 - sgn$ has a limit at 0. Yet their sum (which is the constant function 1) has the limit 1 at 0.

If both g and h are continuous at a, then $\lim_a g = g\,a$ and $\lim_a h = h\,a$. Hence, by the above rules, $\lim_a (g + h) = g\,a + h\,a$, which is $(g + h)\,a$. Since, at a, the limit of $g + h$ and the value of $g + h$ are equal, the function $g + h$ is continuous at a. Similarly, $g \cdot h$ is continuous at a. In other words, *if both g and h are continuous at* a, *then so are their sum and their product.*

Next, suppose that $\lim_a h = b$ and $\lim_b g = c$. This means[1] that, if $x \underset{suf}{\sim} a$, then $h\,x \underset{arb}{\sim} b$ and $gh\,x \underset{arb}{\sim} c$. Consequently,

(5) if $\lim_a h = b$ and $\lim_b g = c$, then $\lim_a (gh) = c$.

(5) may also be expressed in the form

if $\lim_a h = b$ and g has a limit at a, then $\lim_a (gh) = \lim_b g$.

---

[1] Unless $h\,x = b$ for some numbers x arbitrarily close to a, and $g\,b \neq c$. For instance, if $g = 1 - sgn^2$ and $h = O$, then $\lim_0 h = 0$, $\lim_0 g = 0$; yet, $gh = 1$ (the constant function of value 1), and thus $\lim_0 gh = 1$. This case must be ruled out to make the following conclusions valid.

Without introducing the name, b, of the limit of h at a, one can give (5) the following concise formulation:

Law About Limits of Results of Substitution. The limit of gh at a *is equal to the limit of g at the limit of h at* a; in a formula,

$$(5') \qquad \lim_{a} (gh) = \lim_{\lim_a h} g \cdot$$

For instance, knowing that

$$\lim_{1} \frac{j^2 - 1}{j - 1} = 2 \quad \text{and} \quad \lim_{2} \frac{j^2 - 4}{j - 2} = 4,$$

one can, by virtue of the above rule, compute the limit at 1 of the function obtained by substituting $\frac{j^2 - 1}{j - 1}$ into $\frac{j^2 - 4}{j - 2}$, that is, of

$$\left[\left(\frac{j^2 - 1}{j - 1}\right)^2 - 4\right] \Big/ \left[\frac{j^2 - 1}{j - 1} - 2\right].$$

According to the rule, that limit equals 4. This result can be confirmed by bringing the last quotient into the form

$\frac{j^4 - 6j^2 + 8j - 3}{(j - 1)^3}$, which is equal to $(j + 3)$ (on U $- \{ 1 \}$) and therefore has the limit 4 at 1.

★If h has a limit at a and g is continuous at the number $\lim_a h$, then the preceding law implies

$$(5'') \qquad \lim_{a} (gh) = g(\lim_{a} h).$$

For instance, log_{10} is continuous. Hence, if h has a limit at a that is positive (and thus belongs to Dom log_{10}), then (5′) implies

$$(6) \qquad \lim_{a} log_{10}\, h = log_{10} \lim_{a} h, \text{ if } \lim_{a} h > 0.$$

Moreover, it is easily seen that, if h is continuous at a and g is continuous at h a, then gh is continuous at a.

EXERCISES

12. What can be said about the limits at 0 of $g + h$, $g - h$, $g \cdot h$, gh and hg if: *(a)* $g = j^{-1}$ and $h = j^2 - j^{-1}$; *(b)* $g = j^{-2}$ and $h = j^{-2} + j^{-1}$; *(★c)* $g = j^3$ and $h = sgn$?

13. Give an example of two functions g and h whose product has a limit at a, although neither g nor h has a limit at a. Show that, if $\lim_a (g \cdot h) = c$ and $\lim_a g = b \neq 0$, then $\lim_a h = c/b$. What can be concluded from $\lim_a (g + h) = c$ and $\lim_a g = b$?

14. Show *(a)* that $\lim_0 (sgn\, j^2) = 1$ although $\lim_a sgn$ does not exist; *(b)* that $\lim_0 (sgn^2 j^3) = 1$.

15. Prove

(a) $\lim\limits_{0} f = \lim\limits_{a} [f(j-a)]$; (b) $\lim\limits_{a} f = \lim\limits_{a+b} [f(j-b)]$;

(c) $\lim\limits_{a} [f(c\,j)] = \lim\limits_{ca} f$, for any number c; simplify $\lim\limits_{a} [f(cj+c')]$;

(d) $\lim\limits_{a} [g(a + \sin \cdot (j-a))] = g\,a$; similarly, if h is any function all of whose values are between 0 and 1, then

$$\lim_{a} [g(a + h \cdot (j-a))] = g\,a.$$

4. THE LIMITS OF QUOTIENTS OF FUNCTIONS

As one can readily verify, from $\lim\limits_{a} g = b$ and $b \neq 0$ it follows that $\lim\limits_{a} \frac{1}{g} = \frac{1}{b}$. Since h/g is the product of h and $1/g$, the rule for products implies that

if $\lim\limits_{a} g = b$ and $b \neq 0$ and $\lim\limits_{a} h = c$, then $\lim\limits_{a} \frac{h}{g} = \frac{c}{b}$.

Furthermore, if $\lim\limits_{a} g = 0$, then $1/g$ has no limit at a. (Examples: $g = j$ or j^2 or $\sin$ for $a = 0$; or $g = j - 2$ for $a = 2$) Neither has h/g a limit at a if $\lim\limits_{a} g = 0$ and $\lim\limits_{a} h \neq 0$. (Example: $g = \sin$ and $h = \cos$ for $a = 0$).

What is – this is the only remaining case – the limit at a of h/g if both g and h have the limit 0 at a? As suggested by the remarks on page 108, a general statement as to the value of $\lim\limits_{a} (g/h)$ is impossible. Indeed, for $a = 0$:

1. if $h = 6j$ and $g = 2j$, then $\lim\limits_{0} (h/g) = 3$;
2. if $h = 2j$ and $g = 6j$, then $\lim\limits_{0} (h/g) = \frac{1}{3}$;
3. if $h = 3j^2$ and $g = 6j$, then $\lim\limits_{0} (h/g) = 0$;
4. if $h = 6j$ and $g = 3j^2$, then h/g has no limit at 0.

To find $\lim\limits_{a} (h/g)$ if g and h are given specific functions such that $\lim\limits_{a} g = \lim\limits_{a} h = 0$ is a problem which, if at all soluble, requires a special investigation of the particular g and h, and usually some mathematical invention. For instance, to find

$$\lim_{.75} \frac{j^2 - .75^2}{j - .75} \qquad (7)$$

one may conceive the idea of factoring $j^2 - .75^2$ into $j + .75$ and $j - .75$. Then it is clear that the quotient, being equal to $(j + .75)$

(on U − {.75}), sufficiently close to .75, assumes values arbitrarily close to 1.5. Hence the limit (7) is 1.5.

Now the idea of factoring $j^2 - .75^2$ into $j + .75$ and $j - .75$ is utterly simple; and, in presence of the denominator $j - .75$, it might be called most "natural." But all that should not obscure the more basic fact that the afore-mentioned idea is *a mathematical invention.* There is nothing logically cogent about factoring $j^2 - .75$ at all, or into $j + .75$ and $j - .75$, in particular. Logically, the factoring into $(j^2 - .75^2)^{1/3}$ and $(j^2 - .75^2)^{2/3}$ is equally correct. In solving other problems, the latter factoring is even the appropriate idea. But it would be of no avail in an attempt to find (7); in fact, it would complicate that problem.

It is of the utmost importance that the beginner realize this situation. Otherwise he might be misled into the belief that he lacks an understanding of the limit concept or that that concept is altogether beyond his grasp — just because he cannot determine some specific limits. The task of solving a specific limit problem presupposes

(1) the understanding of the limit concept, which is a matter of learning and can be expected from any student of science and engineering;

(2) a creative act, be it ever so modest, which cannot be expected from everyone.

While some persons possess a natural skill in this direction, others need guidance (by example) to acquire, on a modest scale, the art of conceiving the "right idea". Emphasis on the creative nature of any solution of a limit problem may keep these students from becoming discouraged.

More generally, by using the formula (valid for every positive integer n)

$$j^n - a^n = (j - a) \cdot (j^{n-1} + a\,j^{n-2} + \cdots + a^{n-2}\,j + a^{n-1}),$$

one obtains

(8) $\lim\limits_{a} \dfrac{j^n - a^n}{j - a} = n \cdot a^{n-1}$ for any positive integer n and any a,

since, except at a, the quotient equals the second factor in the preceding formula, and this factor is the sum of n terms each having the limit a^{n-1} at a.

In Fig. 31, there are the graphs of the functions $\dfrac{j^3 - 1}{j - 1}$ and $\dfrac{j^2 - (1/4)}{j - (1/2)}$. They look like the parabola $j^2 + j + 1$ and the line $j + \dfrac{1}{2}$ except for gaps at 1 and $\dfrac{1}{2}$ respectively. Since the two branches of either curve approach each other near the gaps, at the altitudes 3 and 1, these numbers are the limits of the quotients.

Now let r be any rational number > 0, say p/q, where p and q are positive integers. Then $\dfrac{j^r - a^r}{j - a} = \dfrac{(j^{1/q})^p - (a^{1/q})^p}{(j^{1/q})^q - (a^{1/q})^q}$. Here, both the

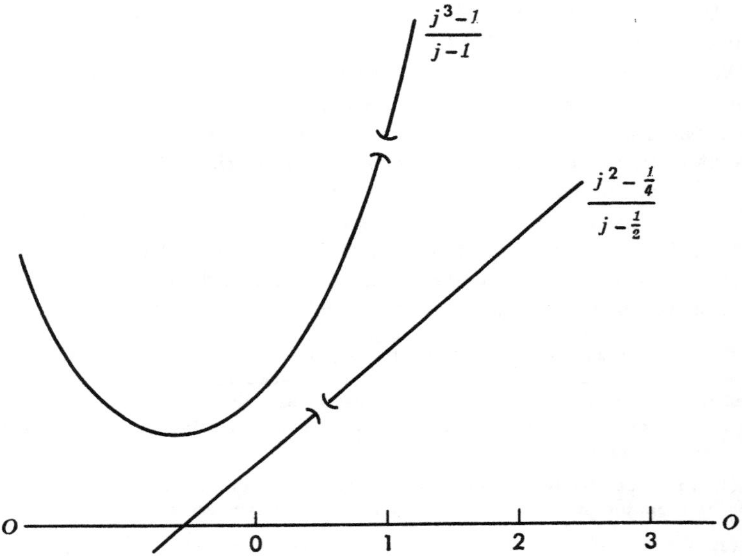

Fig. 31

numerator and the denominator can be divided by the common factor $j^{1/q} - a^{1/q}$. The remaining factor in the numerator

$$(j^{1/q})^{p-1} + a^{1/q}(j^{1/q})^{p-2} + (a^{1/q})^2\,(j^{1/q})^{p-3} + \cdots + (a^{1/q})^{p-1},$$

has the limit $p(a^{1/q})^{p-1}$ at a. Similarly, the remaining factor in the denominator has the limit $q(a^{1/q})^{q-1}$ at a. Consequently, if $a \neq 0$, the limit at a of the quotient is

$\frac{p}{q}\,a^{(p-q)/q} = ra^{r-1}$. Thus, if $a \neq 0$, formula (8) remains valid if n is replaced by the positive rational number r.

If $s = -r$, then

$$\frac{j^s - a^s}{j - a} = \frac{-1}{a^r j^r} \cdot \frac{j^r - a^r}{j - a}.$$

Hence the limit at a of the quotient equals $\frac{-1}{a^{2r}} \cdot r \cdot a^{r-1} = -ra^{-r-1}$ $= sa^{s-1}$. Summarizing, one can formulate the following important law

(8′) $\lim_a \frac{j^t - a^t}{j - a} = ta^{t-1}$ for any rational number t and any $a \neq 0$.

In algebra, the student is told to "simplify" expressions such as $1/(\sqrt{x} + \sqrt{a})$ by rationalizing the denominator (that is to say, by freeing it of roots). This is done by multiplying both denominator and numerator by $\sqrt{x} - \sqrt{a}$. In $\frac{\sqrt{x} - \sqrt{a}}{x - a}$, the denominator *is* free of roots so that, from the algebraic viewpoint, the expression leaves nothing to be desired. But to find its limit as x approaches a, that is, $\lim_{a} \frac{j^{1/2} - a^{1/2}}{j - a}$, by the procedure that led to formula (8′), one has to conceive the idea of rationalizing the *numerator* although by doing so he introduces roots in the denominator and thus algebraically "complicates" the fraction. Frequently, to find the limit of a function one must do the opposite of what in algebra is called simplifying.

A totally different idea is used to find $\lim_{0} \frac{\sin}{j}$. In Fig. 32, an arc from A to B of a circle has been drawn. Let its radius OA = OB be the unit of length, and call $\widehat{AB}$ the length (in this unit) of the arc. The ratio of $\widehat{AB}$ to 2π (the perimeter of the entire circle) is equal to that of the area of the sector to π (the area of the circle). Thus the area of the sector in square units equals $\frac{1}{2}\widehat{AB}$. The sector contains the triangle OCB (whose area is $\frac{1}{2} \sin \widehat{AB} \cdot \cos \widehat{AB}$) and is contained in the triangle OAD (whose area is $\frac{1}{2} \tan \widehat{AB}$, provided $0 < \widehat{AB} < \frac{1}{2}\pi$). It follows that

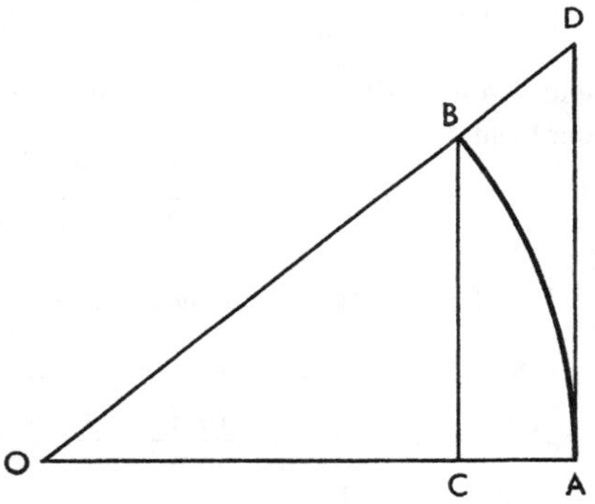

Fig. 32

$$\sin \widehat{AB} \cdot \cos \widehat{AB} < \widehat{AB} < \tan \widehat{AB} = \frac{\sin \widehat{AB}}{\cos \widehat{AB}},$$

or $$\sin \cdot \cos < j < \frac{\sin}{\cos} \text{ on } \{0 < j < \pi/2\}.$$

Division by $\sin \widehat{AB}$ (or $\sin$) yields

$$\cos AB < \frac{\widehat{AB}}{\sin \widehat{AB}} < \frac{1}{\cos \widehat{AB}} \text{ or } \cos < \frac{j}{\sin} < \frac{1}{\cos} \text{ on } \{0 < j < \pi/2\}.$$

Now if $x \underset{suf}{\sim} 0$, then $\cos x \underset{arb}{\sim} 1$ and $\frac{1}{\cos x} \underset{arb}{\sim} 1$. Lying between the numbers $\cos x$ and $1/\cos x$ (which are arbitrarily close to 1), the quotient $x/\sin x$ itself is as close to 1 as one pleases if $x \underset{suf}{\sim} 0$. Consequently, the same is true for the quotient $\frac{\sin x}{x}$ (see Fig. 26). Hence the important formula

(9) $$\lim_0 \frac{j}{\sin} = 1 \text{ and } \lim_0 \frac{\sin}{j} = 1.$$

EXERCISES

16. Find: (a) $\lim_2 \frac{j^3 - 8}{j^2 - 4}$; (b) $\lim_2 \frac{j^{-1} - 1/2}{j - 2}$; (c) $\lim_a \frac{j^{-2} - a^{-2}}{j - a}$; (d) $\lim_a \frac{j^{-1/2} - a^{-1/2}}{j - a}$; (e) $\lim_1 \frac{j^{1/3} - 1}{j - 1}$; (f) $\lim_a \frac{j^m - a^m}{j^n - a^n}$ where m and n are positive integers. Prove that the last limit is equal to the quotient

$$\lim_a \left[\frac{j^m - a^m}{j - a}\right] \Big/ \lim_a \left[\frac{j^n - a^n}{j - a}\right].$$

17. For a rational number t, find $\lim_a \frac{j - a}{j^t - a^t}$, (a) using (8′); (b) by a direct proof, similar to that of (8′).

18. Find: (a) $\lim_a \frac{\sin(j - a)}{j - a}$; (b) $\lim_0 \frac{\sin(c\,j)}{j}$; (c) $\lim_0 \frac{\cos - 1}{j^2}$ (Hint: Express cos in terms of $\sin(\frac{1}{2}\,j)$); (d) $\lim_0 \frac{\tan}{j}$ (Hint: Replace tan by $\frac{\sin}{\cos}$).

19. If $\lim_a \frac{g - ga}{j - a} = b \neq 0$, and $\lim_a \frac{h - ha}{j - a} = c$,

show that $\lim_a \frac{h - ha}{g - ga} = \frac{c}{b}$.

20. It is unlikely that a beginner would find $\lim_{x \to 0} \frac{10^x - 1}{x}$, that is, $\lim_0 \frac{\exp_{10} - 1}{j}$, or $\lim_1 \frac{\log_{10}}{j - 1}$. But he should try; for he should experience the difficulty of these problems, before they are solved in the next section.

5. THE EXPONENTIAL AND LOGARITHMIC FUNCTIONS

In Fig. 14 there are several exponential curves. They all have the altitude 1 at 0, as has the line $j + 1$ (which, of course, is not one of them). According to their relation to that line, the exponential curves to various bases > 1 fall into two classes:

a *lower* class consisting of the curves that intersect $j + 1$ also in a point with a positive projection, as does exp_2;

an *upper* class consisting of the curves that intersect $j + 1$ also in a point with a negative projection, as does exp_4.

For any given point $x \neq 0$ on O, it is easy to determine the base b for which the curve exp_b intersects $j + 1$ just above x. For the number b with that property, the altitude at x of the curve exp_b (that is, b^x) and of the line $j + 1$ (that is, $x + 1$) must be equal; that is to say, $b^x = x + 1$. Consequently, $b = (x + 1)^{1/x}$. For instance, if $x = .5$, then $b = 1.5^2 = 2.25$ (hence $exp_{2.25}$ belongs to the lower class); if $x = -.25$, then $b = .75^{-4} = 3.16...$ (hence $exp_{3.16...}$ belongs to the upper class).

There is a number between 2 and 4 (more precisely, between 2.25 and 3.16...) which separates the bases in the lower and in the upper class. It is the number repeatedly referred to in Chapters II-IV that has been denoted by e. The curve exp_e (dotted in Fig. 14) has neither a point with positive nor a point with negative projection in common with $j + 1$. The line $j + 1$ is not a chord of exp_e; it is its tangent at 0.

To find the value of e approximately, consider the bases of exponential curves intersecting $j + 1$ at points with projections very close to 0.

Above the positive points .1, .01, .001, ... , the line intersects the exponential curves to the bases 1.1^{10}, 1.01^{100}, 1.001^{1000}, ...

Above the negative points −.1, −.01, −.001, ... , the line intersects the curves to the bases $.9^{-10}$, $.99^{-100}$, $.999^{-1000}$, ...

From 7-place logarithm tables, one can determine the logarithms of these numbers to 3 decimals:

$$\begin{aligned}
log_{10}\, 1.1^{10} &= 10\, log_{10}\, 1.1 &&= .432;\\
log_{10}\, 1.01^{100} &= 100\, log_{10}\, 1.01 &&= .434;\\
log_{10}\, 1.001^{1000} &= 1000\, log_{10}\, 1.001 &&= .434;\\
log_{10}\, .9^{-10} &= -10\, log_{10}\, .9 &&= .458;\\
log_{10}\, .99^{-100} &= -100\, log_{10}\, .99 &&= .436;\\
log_{10}\, .999^{-1000} &= -1000\, log_{10}\, .999 &&= .435.
\end{aligned}$$

Thus log_{10} e lies between .434 and .435. More accurate approximations yield $log_{10}\, e = 0.4343...$ and $e = 2.71828...$.

The function exp_e and its inverse, the function log_e, are of paramount importance in calculus. They are referred to as the *natural* exponential and logarithmic functions and will be denoted simply by

$$exp \quad \text{and} \quad log.$$

Thus $$exp\, x = e^x, \text{ for every } x$$

and $log\, x$, for every $x > 0$, is the exponent y for which $e^y = x$.

Clearly, $$e^{log\, x} = x \quad \text{or} \quad exp\, log\, x = x \text{ for every } x > 0,$$

$$log\, e^x = x \quad \text{or} \quad log\, exp\, x = x \text{ for every } x.$$

By its definition, e is the number from which the numbers $(x + 1)^{1/x}$ differ arbitrarily little if x is sufficiently close to 0. If the letter *h* (in an *ad hoc* use) denotes the function assuming, for any $x \neq 0$, the value $(x + 1)^{1/x}$, then

$$e = \lim_0 h = \lim_{x \to 0} (x + 1)^{1/x}.$$

*By formula (6) (based on the continuity of the function log_{10}), from $e = \lim_0 h$ it follows that $log_{10}\, e = \lim_0 log_{10}\, h$. Now $log_{10}(x + 1)^{1/x} = \frac{1}{x} log_{10}(x + 1)$ or $log_{10}\, h = \dfrac{log_{10}(j + 1)}{j}$. Thus

$$log_{10}\, e = \lim_0 \frac{log_{10}(j + 1)}{j}.$$

Here, clearly, the base 10 might be replaced by any other positive number. If it is replaced by e, then, on the left side, $log_e\, e = 1$, and it follows that

$$\lim_0 \frac{log\,(j + 1)}{j} = 1 \quad \text{or} \quad \lim_{x \to 0} \frac{log\,(x + 1)}{x} = 1.$$

If in the latter equality x is replaced by $e^t - 1$ (or if in the former the substitution $exp - 1$ is applied), then one obtains

$$(10) \qquad \lim_0 \frac{j}{exp - 1} = 1 \quad \text{or} \quad \lim_{t \to 0} \frac{t}{e^t - 1} = 1$$

and hence also

$$\lim_0 \frac{exp - 1}{j} = 1 \quad \text{or} \quad \lim_{t \to 0} \frac{e^t - 1}{t} = 1.$$

EXERCISES

21. Show that $exp - 1$ and $log(j + 1)$ are inverse functions, and plot their graphs.

22. Show that

(a) $exp = exp_{10}(.4343\ldots j)$ and $exp_{10} = exp\,(2.30\ldots j)$;

(b) $log = 2.30\ldots log_{10}$ and $log_{10} = .4343\ldots log$;

(c) $exp = exp_b\left(\frac{1}{log\, b} j\right)$ and $exp_b = exp\,(log\, b \cdot j)$.

(d) Set up two-place tables of *log* n for the integers n = 1, 2, ... , 10 (see Exercise 4 of Chapter II).

23. Plot the graphs of the so-called hyperbolic sine and cosine, that is, the functions defined as follows:

$$sinh = \frac{1}{2}\left(exp - \frac{1}{exp}\right) = \frac{1}{2}\,(exp - j^{-1}\,exp) \text{ and}$$

$$cosh = \frac{1}{2}\left(exp + \frac{1}{exp}\right) = \frac{1}{2}\,(exp + j^{-1}\,exp),$$

associating with any number x the values $\frac{1}{2}(e^x - e^{-x})$ and $\frac{1}{2}(e^x + e^{-x})$, respectively, and denoted by *sinh* x and *cosh* x. Prove the following relations,

$$cosh^2 - sinh^2 = 1 \text{ and } sinh(2j) = 2\,sinh \cdot cosh,$$

and try to set up further analogues of well-known trigonometric formulas. Prove that *sinh* has an inverse, *j//sinh,* denoted by *arcsinh* and that

$$arcsinh = log(j + \sqrt{j^2 + 1}).$$

Then show that *cosh* has no significant inverse but that its restriction to P, as well as its restriction to N, has an inverse and that *j//cosh* (on P) $= log(j + \sqrt{j^2 - 1})$ and *j//cosh* (on N) $= log(j - \sqrt{j^2 - 1})$. By geometric considerations, prove that the last two functions must be each other's negative. Is this the case for the above two logarithms?

6. "INFINITE" LIMITS AND LIMITS AT "INFINITY"

None of the functions j^{-2}, $-j^{-2}$, j^{-1}, and $sin\, j^{-1}$ has a numerical limit at 0. Yet, near 0, their graphs behave quite differently. On both sides of 0, the curve j^{-2} rises beyond 10^4, 10^6,... above all points closer to 0 than .01, .001,... In fact, the altitude of j^{-2} is arbitrarily large above all points that are sufficiently close to 0 – a fact usually described by writing

$$\lim_{x \to 0} 1/x^2 = \infty \text{ or } \lim_{z \to 0} 1/z^2 = \infty, \ldots$$

(read "the limit of $1/x^2$ as x approaches 0 is infinite,"...).

The property of the function j^{-2} underlying these synonymous statements may be expressed in the formula

$$\lim_0 j^{-2} = \infty \text{ (read "the limit at 0 of } j^{-2} \text{ is infinite").}$$

In this formula, it is not intended to equate two numbers. Neither $\lim\limits_0 j^{-2}$ nor ∞ is a number. The intention is to symbolize a certain type of behavior of the (finite) values of j^{-2} near 0; namely, the fact that $1/x^2$ is above any level if x is sufficiently close to 0 -- a behavior not by any means shared by all functions without a numerical limit at 0. For example, $-1/x^2$ sinks below any level if x is sufficiently close to 0, which will be symbolized by $\lim_{0} -j^{-2} = -\infty$.

For the hyperbola j^{-1}, one branch rises, the other one sinks near 0. Thus j^{-1} has neither the limit ∞ nor the limit $-\infty$ at 0. All that might be said is that j^{-1} has the right-side limit ∞, and the left-side limit $-\infty$ at 0; in formulas,

$$\lim_{0+0} j^{-1} = \infty \qquad \text{and} \qquad \lim_{0-0} j^{-1} = -\infty.$$

Still another type of behavior is exemplified by the function $sin\ j^{-1}$ assuming the value $sin(1/x)$ for any $x \neq 0$. Obviously, Ran $sin\ j^{-1} = \{-1 \leq j \leq 1\}$. The value of $sin\ j^{-1}$ is 1 for $2/\pi$, $2/5\pi$, $2/9\pi,\ldots$; and -1 for $2/3\pi$, $2/7\pi$, $2/11\pi,\ldots$. Thus to the right of 0, as close to 0 as one pleases, $sin\ j^{-1}$ assumes the values 1 and -1 (and, in fact, any value of its range) which rules out a numerical right-side limit as well as the right-side limit ∞ or $-\infty$. The same is true on the left side.

The symbol ∞ (which does not designate a number) is also used to describe certain types of behavior of functions for large numbers of their domains. For instance, for sufficiently large x, the number $1/x$ is arbitrarily close to 0; in symbols

$$\lim_{x\to\infty} 1/x = 0, \quad \lim_{t\to\infty} 1/t = 0,\ldots \text{ or } \lim_{\infty} j^{-1} = 0;$$

$$\lim_{\infty} j^{-2} = 0, \quad \lim_{\infty} (j^{-1/2} + 1) = 1, \quad \lim_{\infty} [(j^{-1} + 1) \cdot (j^{-2} + 3)] = 3,$$

$$\lim_{\infty} exp_{1/2} = 0, \quad \lim_{-\infty} j^{-1} = 0, \quad \lim_{-\infty} exp_2 = 0, \quad \lim_{-\infty} j^{-2} = 0, \quad \lim_{\infty} j^2 = \infty.$$

The last equality expresses the fact that, if x is a sufficiently large (finite) number, x^2 is larger than any (finite) number.

In contrast, the sine function has neither a numerical limit nor the limit ∞ at infinity; $\lim_{\infty} sin$ is meaningless, as is $\lim_{0+0} sin\ j^{-1}$.

EXERCISES

24. Find the following limits at infinity:

(a) $\lim\limits_{\infty} \left[j \cdot \left(\frac{1}{3} j^{-1} - \frac{1}{2} j^{-2} + \frac{1}{6} j^{-3}\right)\right]$; (b) $\lim\limits_{\infty} \mathit{sin}\, j^{-1}$;

(c) $\lim\limits_{-\infty} \mathit{sin}\, j^{-1}$; (d) $\lim\limits_{\infty} \mathit{exp}\, j^{-1}$; (e) $\lim\limits_{-\infty} \mathit{exp}\, j^{-1}$; (f) $\lim\limits_{\infty} j^{-1} \cdot \mathit{sin}$.

25. Show that if c is any number $\neq 0$, then none of the functions

$$\mathit{sin}\,(cj),\quad \mathit{sin}\, c\left(j + \frac{1}{2}\right),\quad \mathit{sin}\left[c\left(j - \frac{1}{2}\right)\right]$$

has a limit at infinity.

26. Show that

(a) $\lim\limits_{0} f = b$ implies $\lim\limits_{\infty} (f\, j^{-1}) = b$, and $\lim\limits_{-\infty} (f\, j^{-1}) = b$;

(b) $\lim\limits_{\infty} (f\, j^{-1}) = b$ implies $\lim\limits_{0+0} f = b$, and

$\lim\limits_{-\infty} (f\, j^{-1}) = b$ implies $\lim\limits_{0-0} f = b$;

(c) $\lim\limits_{\infty} f = b$ implies $\lim\limits_{0+0} (f\, j^{-1}) = b$;

(d) $\lim\limits_{\infty} [j \cdot \mathit{log}\,(1 + j^{-1})] = 1$. (Hint. Apply the preceding results to one of the formulas at the end of Section 5.)

(e) Similarly, prove $\lim\limits_{\infty} [j^{-1}\,(\exp j^{-1} + 1)] = 1/2$ and

$$\lim_{\infty} \frac{b\, j^{-1}}{\mathit{exp}\,(b\, j^{-1}) - 1} = 1.$$

7. THE LIMIT OF SEQUENCES

Just as U in this book denotes the class of all numbers, Y will hereinafter designate *the class of all positive integers.* A function with the domain Y is called a complete sequence – briefly, a *sequence.*

The term *incomplete sequence* may be used for functions whose domains are subclasses of Y, consisting, e.g., of the first 5 integers, or of the integers > 5 or of the even integers.

If f is a function for which Dom f contains all positive numbers, then f (on Y) is a sequence. For instance,

the function	i.e., the class of pairs	is a sequence usually described by
j (on Y)	(1,1), (2,2),..., (n,n)...	1, 2, ..., n,...
$2j - 1$ (on Y)	(1,1), (2,3),..., (n,2n − 1),...	1, 3, ..., 2n − 1,...
3 (on Y)	(1,3), (2,3),..., (n,3),...	3, 3, ..., 3,...
j^2 (on Y)	$(1,1^2), (2,2^2),\ldots, (n,n^2),\ldots$	$1^2, 2^2, \ldots, n^2, \ldots$
j^{-1} (on Y)	$(1,1), \left(2,\frac{1}{2}\right),\ldots, \left(n, \frac{1}{n}\right),\ldots$	$1, \frac{1}{2}, \ldots, \frac{1}{n}, \ldots$
$\frac{1}{2}\, exp_2$ (on Y)	$(1,1), (2,2),\ldots, (n,2^n),\ldots$	$1, 2, \ldots, 2^n, \ldots$
sin (on Y)	(1, sin 1), (2, sin 2),..., (n, sin n)...	sin 1, sin 2,..., sin n,...
$sin(\pi j)$ (on Y)	(1,0), (2,0),..., (n,0),...	0, 0, ..., 0,...
$sin(\frac{1}{2}\pi j)$ (on Y)	$(1,1), (2,-1)\ldots; (n, (-1)^{n-1}),\ldots$	$1, -1,\ldots, (-1)^{n-1},\ldots$
f (on Y)	(1, f 1), (2, f 2),..., (n, f n),...	f 1, f 2,..., f n,...

Clearly, if $\lim_{\infty} f = b$, then $\lim_{\infty}$ [f (on Y)] = b. But $\lim_{\infty}$ [f (on Y)] = b implies merely that if $\lim_{\infty} f = b'$ then $b' = b$; it does not imply that f has a limit at ∞. For instance,

$$\lim_{\infty} [sin\ (\pi\ j)\ (\text{on Y})] = 0 \text{ even though } \lim_{\infty} [sin\ (\pi\ j)] \text{ does not exist.}$$

The concept of the limit of a sequence goes back to Wallis (1656).

The sequence (1, f 1), (2, f 1 + f 2),..., (n, f 1 + f 2 + ... + f n), ... will hereinafter be denoted by $\sum_1 f$. For instance,

$$\sum_1 j:\ (1,1), (2, 1+2),\ldots, (n, 1+2+\ldots+n),\ldots$$

$$\sum_1 (2\,j - 1):\ (1,1), (2, 1+3),\ldots, (n, 1+3+\ldots+2n-1),\ldots$$

Clearly, $\sum_1 j = [\frac{1}{2}\, j \cdot (j + 1)]$ (on Y), $\sum_1 (2\,j - 1) = j^2$ (on Y).

If $\sum_1 f$ (on Y) has a limit at ∞, then this limit is denoted by $\sum_1^{\infty} f$ and called *the sum of the series* f 1 + f 2 + ... + f n + ... Conversely, the sum of the numbers f 1, f 2,..., f n,... ad infinitum (in this order) is defined as $\lim_{\infty} \sum_1 f$ (on Y) if this limit exists, and is left undefined if there is no such limit. For instance, $\lim_{\infty} \sum_1 exp_{\frac{1}{2}}$ (on Y) = 1; hence

$$\sum_1^{\infty} exp_{\frac{1}{2}}\ (\text{on Y}) = \frac{1}{2} + \left(\frac{1}{2}\right)^2 + \ldots + \left(\frac{1}{2}\right)^{n-1} + \ldots = 1.$$

On the other hand, neither $\sum_1 j^{-1}$ (on Y) nor $\sum_1 sin\left(\frac{1}{2}\pi j\right)$ (on Y) has a limit at ∞. Therefore

$$\sum_1^\infty sin\left(\frac{1}{2}\pi j\right) = 1 + (-1) + 1 + (-1) + \dots + (-1)^{n-1} + \dots$$

and

$$\sum_1^\infty j^{-1} = 1 + \frac{1}{2} + \frac{1}{3} + \dots + \frac{1}{n} + \dots$$

remain undefined symbols.

EXERCISES

27. Using formula (E_n) on p. 46 and Exercises 10 and 12 in Chapter III, find functions on Y which are equal to

$\sum_1 j^2$, $\sum_1 (2j - 1)^2$, $\sum_1 j^3$.

28. For any function f, define a function Δf by
$\Delta f = f - f(j-1)$;
thus the values of Δf on Y are

$$\Delta f\,1 = f\,1 - f\,0,\ \Delta f\,2 = f\,2 - f\,1,\ \dots,\ \Delta f\,n = f\,n - f\,(n-1).$$

For instance, $\Delta j^2 = j^2 - j^2(j-1) = j^2 - (j-1)^2 = 2j - 1$.
Find $\Delta\,1$, $\Delta\,j$, $\Delta\,j^3$, $\Delta\left(\frac{1}{2}\,j\cdot(j-1)\right)$, and $\Delta\,log$.
Prove the following formulas:

$$\Delta \sum_1 f = f, \qquad \sum_1 \Delta f = f - f\,0,$$

which are analogous to the Reciprocity Laws of Calculus,

$$\mathbf{D}\int_1 f = f \quad \text{and} \quad \int_1 \mathbf{D} f = f - f\,1.$$

CHAPTER VI

THE BASIC CONCEPTS OF CALCULUS

In Chapters I – III, the two basic problems of calculus were solved accurately for step lines and polygons, and approximately for simple curves. The results about functions and limits obtained in Chapters IV and V yield accurate solutions for certain curves and functions.

1. THE DERIVATIVE

The slope of the chord of j^2 between .75 and b (between a and b) is $\frac{b^2 - .75^2}{b - .75}$ for any b $\neq$ .75 and $\left(\frac{b^2 - a^2}{b - a}\right.$ for any b $\neq$ a$\left.\right)$.

If b $\underset{suf}{\sim}$ .75 (and b $\underset{suf}{\sim}$ a), then this slope is $\underset{arb}{\sim}$ 1.5 (and $\underset{arb}{\sim}$ 2 a), or

$$\lim_{b \to .75} \frac{b^2 - .75^2}{b - .75} = 1.5 \quad \left(\text{and } \lim_{b \to a} \frac{b^2 - a^2}{b - a} = 2a\right).$$

Thus the slope of the tangent to j^2 at .75 (and at a) is

$$\mathbf{D}\, j^2\, .75 = 1.5 \quad (\text{and } \mathbf{D}\, j^2\, a = 2a).$$

The numerical variable b may be replaced by any other letter, as in

$$\lim_{x \to .75} \frac{x^2 - .75^2}{x - .75} = 1.5, \quad \lim_{t \to .75} \frac{t^2 - .75^2}{t - .75} = 1.5, \ldots .$$

All these synonymous formulas express the property $\lim\limits_{.75} \frac{j^2 - .75^2}{j - .75} = 1.5$ of the function $\frac{j^2 - .75^2}{j - .75}$, which might be called the *difference quotient of* j^2 *at* .75 (See Fig. 17, p. 34.)

More generally, the slope of the chord of the curve f between a and b is the difference quotient $\frac{f\,b - f\,a}{b - a}$ of f between a and b, that is, the value for b of the function $\frac{f - fa}{j - a}$, which will hereinafter be called *the difference quotient of f at* a. In view of j a = a and $f\,j = f$, it may be written in the more symmetrical forms

$$\frac{f - fa}{j - ja} \quad \text{and} \quad \frac{fj - fa}{j - a}.$$

If this difference quotient has a limit at a, then this limit is the slope, $\mathbf{D}\, f$ a, of the tangent of f. Thus

$$\mathbf{D}\, f\, a = \lim_{b \to a} \frac{f\,b - f\,a}{b - a} = \lim_{x \to a} \frac{f\,x - f\,a}{x - a} = \ldots \text{ or}$$

(1) $\mathbf{D}\, f\, \mathrm{a} = \lim\limits_{\mathrm{a}} \dfrac{f - f\mathrm{a}}{j - \mathrm{a}} = \lim\limits_{\mathrm{a}} \dfrac{f - f\mathrm{a}}{j - j\mathrm{a}} = \lim\limits_{\mathrm{a}} \dfrac{fj - f\mathrm{a}}{j - \mathrm{a}}.$

For instance, according to pp. 122 and 120,

(2) $\mathbf{D}\, sin\, 0 = \lim\limits_{0} \dfrac{sin - 0}{j - 0} = \lim\limits_{0} \dfrac{sin}{j} = 1,$

(3) $\mathbf{D}\, j^{\mathrm{r}}\mathrm{a} = \mathrm{r}\, j^{\mathrm{r}-1}\, \mathrm{a}$ for any rational exponent r, and any a in Dom j^{r}.

On the other hand, the difference quotient of the function *abs* at 0 is $\dfrac{abs}{j}$ and assumes the value 1 on P, and the value −1 on N. The symbol $\lim\limits_{0} \dfrac{abs}{j}$, and consequently the symbol **D** *abs* 0, is meaningless. Therefore, the function *abs* has no derivative at 0 and, indeed, the curve *abs* has a corner at 0.

f having *no* derivative at a and *no* tangent at a must not be confused with *f* having the derivative 0 *(zero)* at a, that is, having a *horizontal* tangent at a. Whereas *abs* has a corner at 0, the curves j^2 and j^3 have horizontal tangents at 0. The line *4* (as well as any horizontal line) has the slope 0 and a horizontal tangent (namely, itself) everywhere.

Formula (1) makes the slope of the curve *f* at a, one of the two basic concepts of calculus introduced in Chapter I, computable. But one may take a different view of the subject. He may feel that, in the first chapters, the tangent to *f* at a was not clearly introduced; that, in contrast to the chord of *f* between a and b, the tangent to *f* at a was only intuitively discussed (see, e.g., the construction on p. 35) and not rigorously defined. Anyone taking this view may consider formula (1) as a definition of **D** *f* a.

The two points of view can be contrasted as follows:

The less critical student says:	The more critical student says:
(1) I define $\mathbf{D}\, j^2\, .75$ as the slope of the tangent to j^2 at .75.	(1) I don't quite know what is meant by a tangent to a curve.
(2) I draw that tangent *t*; it is the line that the chords of j^2 between .75 and b approach as b approaches .75.	(2) Neither, consequently, can I draw the tangent to j^2 at .75; nor do I know precisely what is the slope of that tangent.
(3) Like any nonvertical line, *t* has a slope, say s. By the construction of *t*, its slope s must be the limit of the slopes of the chords approaching *t*.	(3) I do know the chords of j^2. The slope of that chord between .75 and b is $\dfrac{\mathrm{b}^2 - .75^2}{\mathrm{b} - .75} = .75 + \mathrm{b}$ for any $\mathrm{b} \neq .75$.
(4) That limit is 1.5. Hence s = 1.5 and $\mathbf{D}\, j^2\, .75 = 1.5$.	(4) I can prove that these slopes approach the limit 1.5 as b approaches .75.

(continued)

(5) I will utilize (3) and (4) in defining the slope of the tangent to j^2 at .75 (of which so far I have had no clear idea). I will say that D j^2 .75 is the aforementioned limit, 1.5.

(6) Finally, I will define: the tangent t to j^2 at .75 is the line having the slope D j^2 .75 (that is, 1.5) and, at .75, the same altitude as j^2.

In a discussion, the less critical student asks the other one: "Why that backward reasoning? I can prove that the slope of t is not < 1.5 nor > 1.5. Knowing that Abraham Lincoln was not assassinated before 1865 nor after 1865, can I not conclude that he was assassinated in 1865?"

Answer: "Knowing that Lincoln was assassinated, you certainly can. But knowing that Franklin D. Roosevelt was not assassinated before 1945 nor after 1945 you cannot conclude that he was assassinated in 1945, can you?"

We suggest the following three steps for computing D f a, say, D sin a:

Step I. Write down the difference quotient of f at a, or between a and x ($\neq$ a), that is,

$$\frac{f - fa}{j - a} = \frac{sin - sin\, a}{j - a} \quad \text{or} \quad \frac{f\,x - f\,a}{x - a} = \frac{sin\, x - sin\, a}{x - a}.$$

Step II (the most important step). Bring the difference quotient into a form that reveals the approximate value of the quotient for x $\sim$ a, which in the old form was not apparent. This is the most difficult step because there is no general rule by which such a transformation can always be achieved (see pp. 118, 119, and 121) and each problem requires a special study. For some functions (e.g., for j^r where r is rational (see p. 119) the difference quotient can be transformed into a function without the denominator $j - a$; for other functions, into a product in which one factor is a quotient with the denominator $j - a$, but nevertheless has a limit at a that can easily be found. For instance, for the function sin, the following transformations – the first based on trigonometry, the second on a regrouping of factors – prove to be revealing:

$$\frac{sin\, x - sin\, a}{x - a} = \frac{2\, sin\, \frac{x-a}{2} \cdot cos\, \frac{x+a}{2}}{x - a} = cos\, \frac{x+a}{2} \cdot \frac{sin\, \frac{x-a}{2}}{\frac{x-a}{2}}$$

or

$$\frac{sin - sin\, a}{j - a} = cos\, \frac{1}{2}(j + a) \cdot \frac{sin\, \frac{1}{2}(j - a)}{\frac{1}{2}(j - a)}.$$

The second factor (with $j - a$ in the denominator) can be obtained by substituting $\frac{1}{2}(j - a)$ into $\frac{sin}{j}$, which at 0 has the limit 1.

Step III. Obtain the limit of the difference quotient in the new form in which this problem is feasible. In the example, the factors of the difference quotient have the limits

$$\lim_a \left[\cos \frac{1}{2}(j + a)\right] = \cos a \quad \text{and} \quad \lim_a \frac{sin \frac{1}{2}(j - a)}{\frac{1}{2}(j - a)} = \lim_0 \frac{sin}{j} = 1.$$

Hence, $\lim_a \frac{sin - sin\ a}{j - a}$ is the product of the two limits, that is, of cos a and 1. Thus the following generalization of (2):

$$\text{(4)} \qquad \mathbf{D}\, sin\ a = \lim_a \frac{sin - sin\ a}{j - a} = \cos a.$$

Similarly, to find **D** *exp* a:

I. Write down the difference quotient of *exp* between a and x, that is,

$$\frac{e^x - e^a}{x - a}.$$

II. In this case, transform the quotient by factoring it as follows:

$$\frac{e^x - e^a}{x - a} = e^a \cdot \frac{e^{x - a} - 1}{x - a},$$

which is suggested by the result $\lim_0 \frac{exp - 1}{j} = 1$ (p. 124).

III. Find the limit, as x approaches a, of the transformed difference quotient. The limit is the product of e^a and the limit of $\frac{e^{x - a} - 1}{x - a}$, that is,

$$\text{(5)} \qquad \mathbf{D}\, exp\ a = e^a \cdot 1 = e^a.$$

In particular, **D** *exp* 0 = 1. Indeed, on p. 123 the curve $exp = exp_e$ was defined as that exponential curve which has the line $j + 1$ as its tangent at 0. **D** exp_b 0 and **D** exp_b a will be found later. (See p. 237.)

In Chapters I and II, the derivative **D** *f* of a curve *f* was introduced as the curve whose altitude is equal to the slope **D** *f* x at any x where *f* has a derivative. From formulas (3), (4), (5) it follows that

(3′) $\mathbf{D}\, j^r = r\, j^{r-1}$, for any rational r,

(4′) $\mathbf{D}\, sin = \cos$,

(5′) $\mathbf{D}\, exp = exp$.

On the basis of their experience with elementary functions, mathematicians originally believed that every continuous function f had a derivative $\mathbf{D}\, f\, x$ for most numbers x belonging to Dom f – in geometrical language, that the points where a continuous curve has no tangent are exceptional. There exist even incorrect proofs of this statement (Ampère, 1806). Bolzano constructed a continuous curve without a tangent at any point but never published this great result. In the 1870's, Weierstrass defined continuous functions without derivatives anywhere by adding infinitely many cosine curves each having a greater frequency, but a smaller amplitude, than the preceding curve. One of his examples is

$$\frac{1}{2}\cos(13\,\pi\, j) + \frac{1}{2^2}\cos(13^2\,\pi\, j) + \frac{1}{2^3}\cos(13^3\,\pi\, j) + \ldots + \frac{1}{2^n}\cos(13^n\,\pi\, j) + \ldots .$$

In the interval between 0 and 1, on the 13 oscillations of amplitude $\frac{1}{2}$, Weierstrass superimposes first 13^2 oscillations of amplitude $\frac{1}{4}$, then 13^3 oscillations of amplitude $\frac{1}{8}$, and so on. The result of each superposition is a curve that has a tangent at every point but is, as it were, more tremulent than the preceding curve. The limit of the curves so obtained can be shown to be the graph of a function that is continuous everywhere without having a derivative anywhere.

EXERCISES

1. Find $\mathbf{D}\cos 0$, $\mathbf{D}\tan 0$, $\mathbf{D}\log 1$, $\mathbf{D}\, j^{\frac{1}{2}}\, a$, $\mathbf{D}\, j^{-1}\, x$.

2. Find $\mathbf{D}\cos$, $\mathbf{D}\tan$, $\mathbf{D}\, j^{\frac{1}{2}}$ $\mathbf{D}\, j^{-1}$.

3. Show that $\lim_{h\to 0} \dfrac{f(a+h) - f\,a}{h}$, that is, $\lim_{0} \dfrac{f(j+a) - fa}{j}$ (a limit that is frequently used as the definition of $\mathbf{D}\, f\, a$) indeed equals $\lim_{a} \dfrac{f - fa}{j - a}$. Plot some curve f, and then the graph of $\dfrac{f(j+2) - f\,2}{j}$.

4. $\mathbf{D}\,\mathbf{D}\, f$ (the derivative of the function $\mathbf{D}\, f$) is called *the second derivative of f* and denoted by $\mathbf{D}^2 f$. (The square of $\mathbf{D}\, f$ would be denoted by $(\mathbf{D}\, f)^2$.) Similarly, $\mathbf{D}^3 f = \mathbf{D}\,\mathbf{D}^2 f$, and so on. Find:

(a) $\mathbf{D}^2 j$, $\mathbf{D}^k j^n$ for any k, in particular $k = n$ and $k = n + 1$;
(b) $\mathbf{D}^k \sin$, $\mathbf{D}^k \exp$; (c) show that $\mathbf{D}^3 f$ equals the second derivative of $\mathbf{D}\, f$.

5. If $\lim_a h = 0$ and $\lim_a g = 0$, if h and g have derivatives at a, and if $\mathbf{D}\, g\, a \neq 0$, then $\lim_a \dfrac{h}{g} = \dfrac{\mathbf{D}\, h\, a}{\mathbf{D}\, g\, a}$. Prove this statement, which is a simple case of what is known as *l'Hospital's Rule*. (See Exercise 19, p. 122).

★6. Show that if f has a derivative at a, then f is continuous at a.

2. ANTIDERIVATIVES

The number $\sqrt{9}$ might be called *a particular square root* of 9; so might $-\sqrt{9}$. The class of all numbers whose square equals 9 will be denoted by $\{\pm\sqrt{9}\}$ and called the *square root class* of 9. Clearly, $\{\pm\sqrt{9}\} = \{\sqrt{9}, -\sqrt{9}\}$.

Common properties of the particular square roots are conveniently expressed in terms of the self-explanatory numerical variable $\pm\sqrt{9}$ with the scope $\{\pm\sqrt{9}\}$, which might be referred to as *indefinite square root.*

The use of these concepts and symbols is illustrated by the following statements about square roots:

	particular	indefinite	square root class
true	$\sqrt{9} > 2$, $-\sqrt{9}+1 < \sqrt{9}$, $\lvert -\sqrt{9}\rvert = 3$, $\sqrt{9}+(-\sqrt{9}) = 0$, $\sqrt{9}\cdot(-\sqrt{9}) = -9$.	$(\pm\sqrt{9})^4 = 81$, $2 < \lvert\pm\sqrt{9}\rvert < 4$, $1 \pm \sqrt{9}$ is a root of the equation $x^2 - 2x - 8 = 0$.	$\{\pm\sqrt{9}\}$ contains two elements; if n belongs to $\{\pm\sqrt{9}\}$, then so does $-n$.
false	$-\sqrt{9} > 2$, $(-\sqrt{9})\cdot(-\sqrt{9}) = -9$	$\lvert\pm\sqrt{9}\rvert > 4$	$\{\pm\sqrt{9}\}$ contains three elements.

The distinction between (a) particular roots, (b) indefinite roots, and (c) root classes (a distinction that might well be utilized in expositions of arithmetic and algebra) has a perfect analogue in the theory of antiderivatives.

a. Particular Antiderivatives. The function that assumes the value 0 for 0 and whose derivative is *cos* will be denoted by $\mathbf{D}^{-1}_{(0,0)}$ *cos*. It equals the function *sin* and will be referred to as a *particular* antiderivative of *cos*, more specifically, as *the* (0, 0)-*antiderivative of cos*. More generally, for any number b, the function assuming the value b for 0 and having the derivative *cos* is a particular antiderivative of *cos*. It will be called the (0, b)-antiderivative of *cos*. It is defined by

$$\mathbf{D}\,\mathbf{D}^{-1}_{(0,b)}\ cos = cos \quad\text{and}\quad \mathbf{D}^{-1}_{(0,b)}\ cos\ 0 = b.\ \text{Clearly, } \mathbf{D}^{-1}_{(0,b)}\ cos = sin + b.$$

Similarly, the function $\mathbf{D}^{-1}_{(a,b)}$ *cos* assumes the value b for a. It equals $sin + b - sin\ a$. Consequently,

$$\mathbf{D}^{-1}_{(a',b')}\ cos = \mathbf{D}^{-1}_{(a,b)}\ cos \text{ if and only if } b' - sin\ a' = b - sin\ a.$$

The symbol $\mathbf{D}^{-1}_{(a,b)}$ is used (just as is **D**) in such a way that only the immediately following function is within its reach. Thus the value that $\mathbf{D}^{-1}_{(a,b)}$ *cos* assumes for a may, without ambiguity, be denoted by

$\mathbf{D}^{-1}_{(a,b)}$ *cos* a.

b. Antiderivative Classes. The class of all particular antiderivatives of *cos* (i.e., of all functions having the derivative *cos*) will be denoted by

$\{D^{-1}\ cos\}$. It will be referred to as the *antiderivative class* of *cos*. More generally, $\{D^{-1} g\}$ is the class of all functions having the derivative g. Thus

(6) $$f \text{ is in } \{D^{-1}\ g\} \text{ if and only if } D\ f = g.$$

To test whether the first of two given functions, f and g, is an antiderivative of the second, all one has to do is to determine $D\ f$ and to compare $D\ f$ with g. In this way, one can verify, for instance, that j^9 is an antiderivative of $9\ j^8$; and one can refute that *exp* is an antiderivative of j^6.

c. Indefinite Antiderivatives. Properties that are common to all antiderivatives of *cos* can be conveniently expressed in terms of a function variable with the scope $\{D^{-1}\ cos\}$. A self-explanatory symbol (which does not require any legend) is $D^{-1}\ cos$, as in $D^2D^{-1}\ cos = -\ sin$; that is to say, in this formula, $D^{-1}\ cos$ may be replaced by any element of the class $\{D^{-1}\ cos\}$, that is, by any function having the derivative *cos*; e.g., by *sin* and by *sin* + 5. More generally,

(7) $$DD^{-1}g = g \text{ for any function } g.$$

In view of the definition of $D^{-1}\ g$, (7) is an immediate consequence of (6). The derivative of any antiderivative of g (that is, of any function having the derivative g) equals g.

The use of the concepts and symbols introduced above is illustrated by the following statements about antiderivatives.

	particular	indefinite	antiderivative class
true	$D^{-1}_{(0,2)}\ cos\ \pi/2 = 3$, $1 \leq D^{-1}_{(0,2)}\ cos \leq 3$, $D^{-1}_{(0,3)}\ cos = 3 + D^{-1}_{(0,0)}\ cos$.	$D^2D^{-1}\ cos = -\ sin$, $\left[D^{-1}\ cos\right]_c^{c'} = \int_c^{c'} cos.$	$\{D^{-1}\ cos\}$ contains infinitely many functions. Jointly, the graphs of the functions in $\{D^{-1}\ cos\}$ cover the plane.
false	$D^{-1}_{(0,2)}\ cos = 2$, $D^{-1}_{(0,2)}\ cos\ \pi = 1$.	$1 \leq D^{-1} cos \leq 3$, $D^{-1}\ cos = sin$.	if f belongs to $\{D^{-1}\ cos\}$, then so does $-f$.

If f is a horizontal line, then $D\ f = O$. Conversely, if f is a straight line such that $D\ f = O$, then f is horizontal. In what follows, use will be made of the much more general statement that *no simple curve other than a horizontal line has the derivative* O. In Chapter X, this statement will be proved. In the remainder of the present chapter, it plays the role of a

Fundamental Hypothesis. If f is a function such that **D** $f = O$, *then f is constant.*

An immediate consequence of this hypothesis is the following

FUNDAMENTAL THEOREM CONCERNING ANTIDERIVATIVES. *If* f_1 *and* f_2 *belong to* $\{\mathbf{D}^{-1} g\}$, *then* $f_1 - f_2$ *is a constant function.*

Indeed, if f_1 and f_2 belong to $\{\mathbf{D}^{-1} g\}$, then $\mathbf{D}\, f_1 = g$ and $\mathbf{D}\, f_2 = g$. Hence $\mathbf{D}\, f_1 - \mathbf{D}\, f_2 = O$. A simple reasoning (see Exercise 4(b), p. 227) shows that $\mathbf{D}\,(f_1 - f_2) = \mathbf{D}\, f_1 - \mathbf{D}\, f_2$. Hence $\mathbf{D}\,(f_1 - f_2) = O$, wherefore, according to the Fundamental Hypothesis, $f_1 - f_2$ is constant.

If k_1 and k_2 belong to $\mathbf{D}^{-1} h$, the value of the constant function $k_1 - k_2$ can be found by evaluating $k_1 - k_2$ for any a belonging to the (common) domain of k_1 and k_2. The value is $k_1\,\mathrm{a} - k_2\,\mathrm{a}$. Hence $k_1 - k_2 = k_1\,\mathrm{a} - k_2\,\mathrm{a}$, and one has the following

COROLLARY OF THE FUNDAMENTAL THEOREM. If k_1 and k_2 belong to $\mathbf{D}^{-1} h$ then $k_1 - k_1\,\mathrm{a} = k_2 - k_2\,\mathrm{a}$ and $k_1\,\mathrm{b} - k_1\,\mathrm{a} = k_2\,\mathrm{b} - k_2\,\mathrm{a}$.

In connection with antiderivative classes, some general remarks about classes of functions will be useful. Two such classes are called *equal* if and only if any function belonging to either class belongs also to the other. Because of frequent references to the class of all constant functions, it is convenient to have a symbol for this class. Script $\mathcal{C}$ will be used as such. For instance,

(8) $$\{\mathbf{D}^{-1} O\} = \mathcal{C}.$$

In the calculus of antiderivatives one frequently has to add one and the same function to every function of a class, for instance, *sin* to every constant function (that is, to every function of the class $\mathcal{C}$). It will be convenient to denote the class of all resulting functions by $sin + \mathcal{C}$. In this way one can write, for instance,

(9) $$\{\mathbf{D}^{-1} cos\} = sin + \mathcal{C};$$ and more generally,

(10) $$\{\mathbf{D}^{-1} g\} = \mathbf{D}^{-1} g + \mathcal{C}.$$

A very convenient form of expressing (9) is

(9') $$\mathbf{D}^{-1} cos + \mathcal{C} = sin + \mathcal{C}$$

In (9') two classes of functions are equated: the class of all functions obtained by adding any particular antiderivative of *cos* and any constant function; and the class of all functions that are the sum of *sin* and a constant function.

Obviously, one must not, in (9′), "cancel" $\mathcal{C}$ nor even contract the two $\mathcal{C}$'s into one. The cancellation would result in the false statement that any antiderivative of *cos* is *sin;* the contraction, in the statement that a function and a class of functions are equal.

It is further clear that

(11) $$\mathcal{C} = a + \mathcal{C} \text{ for any constant function } a.$$

The concepts introduced in this section permit consistent and unambiguous formulations of the laws concerning antiderivatives – formulations in which every symbol has a clearly defined meaning. In fact, there is a variety of correct formulations, as the following examples illustrate.

A. *Statements claiming that a certain function belongs to a certain class of functions.*

1. Statements referring only to *specific functions,* as
sin is in $\{\mathbf{D}^{-1}\,cos\}$; $sin + 3$ is in $\{\mathbf{D}^{-1}\,cos\}$.

2. Statements referring to *function variables,* as
for any c, $sin + c$ is in $\{\mathbf{D}^{-1}\,cos\}$; $\mathbf{D}^{-1}\,cos$ is in $\{\mathbf{D}^{-1}\,cos\}$.

B. *Formulas expressing an equality.*

1. The equality of *two functions,* as
$\mathbf{D}^{-1}_{(0,3)}\,cos = sin + 3$, $\mathbf{D}^{-1}_{(0,b)}\,cos = sin + b$ for any b.

2. The equality of *two classes of functions,* as
$\{\mathbf{D}^{-1}\,cos\} = sin + \mathcal{C}$ and $\{\mathbf{D}^{-1}\,cos\} = \mathbf{D}^{-1}\,cos + \mathcal{C}$ and $\mathbf{D}^{-1}\,cos + \mathcal{C} = sin + \mathcal{C}$.

Of course not permissible are formulas in which functions and classes of functions containing more than one function are equated. Examples of such incorrect formulas are
$\{\mathbf{D}^{-1}cos\} = sin + c$ for any constant function c, and $\mathbf{D}^{-1}\,cos = sin + \mathcal{C}$.
The first claims that the *class* of all functions having the derivative *cos* equals any *function* obtained by adding *sin* and a constant function; the second, that any *function* having the derivative *cos* equals the *class* of all functions obtained by adding *sin* to any constant function.

Only formulas such as $\mathbf{D}\{\mathbf{D}^{-1}g\} = g$ and $\mathbf{D}\,\mathcal{C} = O$ would (in most cases) be harmless since the classes on the left side contain only one element, namely, the functions g and O.

Finally, it is incorrect to write

$$\mathbf{D}^{-1}\,cos = sin + c, \text{ for any constant function c,}$$

since it is not true that *any* antiderivative of *cos* is equal to any function $sin + c$; for instance, $\mathbf{D}^{-1}_{(0,1)}\,cos \neq sin + 5$.

If A and B are classes of numbers and k is a number, let $k \cdot A$ denote the class of all numbers $k \cdot a$ for any a belonging to A; and C, the class consisting of the two numbers 1 and −1. Then, in analogy to (8), $\{\pm\sqrt{1}\} = C$. Moreover, the following statements about square roots (in which c is a numerical variable with the scope C) are analogous to the corresponding statements about antiderivatives.

A_1. 3 is in $\{\pm\sqrt{9}\}$, $3 \cdot (-1)$ is in $\{\pm\sqrt{9}\}$.

A_2. $3 \cdot c$ is in $\{\pm\sqrt{9}\}$, $\pm\sqrt{9}$ is in $\{\pm\sqrt{9}\}$.

B_1. $\sqrt{9} = 3$.

B_2. $\{\pm\sqrt{9}\} = 3 \cdot C$, $\{\pm\sqrt{9}\} = \pm\sqrt{9} \cdot C$, $\pm\sqrt{9} \cdot C = 3 \cdot C$.

The analogues of the incorrect statements are

$\{\pm\sqrt{9}\} = 3 \cdot c$, $\pm\sqrt{9} = 3 \cdot C$, and $\pm\sqrt{9} = 3 \cdot c$.

Actually, algebraists use the symbol ± 1 sometimes for the class consisting of 1 and −1, sometimes for a numerical variable with that class as scope. Similarly, in the traditional calculus, "indefinite antiderivative" of g sometimes means the class $\{D^{-1}g\}$, sometimes the function variable $D^{-1}g$ with that class as scope.

EXERCISES

7. For $j + 1$, determine the (0,0)-, the (0,3)-, the (2,3)-, and the (a,b)-antiderivatives.

8. Determine, where it can be done, a′ and b′ in such a way that: $D^{-1}_{(a,b)}\, g = D^{-1}_{(a',0)}\, g = D^{-1}_{(0,b')}\, g$ for $g = \cos$, $j + 1$, j^2, 3, $\exp$, $\log$.

3. THE INTEGRAL

Let f be a simple curve whose projection on the line O is from 0 to 1. Let $r_n^{(f)}$ be the regular step line adapted to f and having n steps (p. 40). The area under it is

$$\int_0^1 r_n^{(f)} = \frac{1}{n} \cdot \left[f\left(\frac{1}{2n}\right) + f\left(\frac{3}{2n}\right) + \dots + f\left(\frac{2n-1}{2n}\right)\right]. \tag{12}$$

This average of the values of f at n equi-spaced points between 0 and 1 lends itself in many cases (e.g., in Examples 1,2,3 below) to an almost algebraic determination as well as to a computation by certain machines.

If n is sufficiently large, the area under the step line is as close as one pleases to $\int_0^1 f$, the area under the curve f from 0 to 1; that is to say,

$$\int_0^1 f = \lim_{n\to\infty} \int_0^1 r_n^{(f)} = \lim_{n\to\infty} \frac{1}{n} \cdot \left[f\left(\frac{1}{2n}\right) + f\left(\frac{3}{2n}\right) + \dots + f\left(\frac{2n-1}{2n}\right)\right]. \tag{13}$$

Formulas (12) and (13) make the area under a curve or the integral computable. One who takes the view that, in the first chapters of this book, that basic concept of calculus was only intuitively discussed may take (12) or (13) as a *definition* of $\int_0^1 f$.

Example 1. $\int_0^1 \exp = e - 1$, because

1. $$\int_0^1 r_n^{(\exp)} = \frac{1}{n} \cdot [e^{1/2n} + e^{3/2n} + \dots + e^{(2n-1)/2n}]$$
$$= \frac{1}{n} \cdot e^{1/2n} \cdot [1 + e^{1/n} + (e^{1/n})^2 + \dots + (e^{1/n})^{n-1}]$$
$$= \frac{1}{n} \cdot e^{1/2n} \cdot \frac{(e^{1/n})^n - 1}{e^{1/n} - 1} = (e-1) \cdot e^{1/2n} \cdot \frac{1/n}{e^{1/n} - 1};$$

and, according to (6),

2. $$\lim_{n \to \infty} \left[(e-1) \cdot e^{1/2n} \cdot \frac{1/n}{e^{1/n} - 1}\right] =$$
$$\lim_\infty \left[(e-1) \cdot \exp\left(\frac{1}{2}\, j^{-1}\right) \cdot \frac{j^{-1}}{\exp j^{-1} - 1}\right]$$
$$= (e-1) \cdot \lim_\infty \left(\exp \frac{1}{2}\, j^{-1}\right) \cdot \lim_\infty \frac{j^{-1}}{\exp j^{-1} - 1} = (e-1) \cdot 1 \cdot 1 = e - 1,$$

where the equality sign before the last is based on (10), p. 124.

Example 2. $\int_0^b j^2 = \frac{b^3}{3}$, because, according to Exercise 10, p. 48,

1. $$\int_0^b r_n^{(j^2)} = \frac{b}{n} \cdot \left[\left(\frac{b}{2n}\right)^2 + \left(3\frac{b}{2n}\right)^2 + \dots + \left((2n-1) \cdot \frac{b}{2n}\right)^2\right]$$
$$= \frac{b^3}{4n^3}\, [1^2 + 3^2 + \dots + (2n-1)^2]$$

2. $$\lim_{n \to \infty} \frac{b^3}{4n^3} \cdot \left(\frac{4}{3}n^3 - \frac{1}{3}n\right) = \lim_\infty \frac{b^3}{4} \cdot \left(\frac{4}{3} - \frac{1}{3}\, j^{-2}\right) = \frac{b^3}{3}.$$

(13) can be generalized to

$$(14)\quad \int_a^b f =$$

$$\lim_{n\to\infty} \frac{1}{n} \cdot \left[f\left(a+\frac{d}{2}\right)+f\left(a+3\frac{d}{2}\right)+\dots+f\left(a+(2n-1)\frac{d}{2}\right)\right]\cdot(b-a)$$

where $d = (b-a)/n.$

In defining $\int_a^b f$ it is not necessary to take into account *all* regular step lines adapted to f. For a continuous function f, it would be sufficient to consider, e.g.,

$$r_1^{(f)},\ r_3^{(f)},\ r_9^{(f)},\ \dots,\ r_{3^n}^{(f)},\ \dots .$$

Each value of f computed in determining the area under one of these step lines is utilized in determining the area under the subsequent step line (see Exercise 6, p. 43) – another advantage in computation by machines.

On the other hand, also regular step lines that are not adapted to f and even non-regular step lines with sufficiently narrow steps crossing f may be taken into consideration. (A step, i.e., a horizontal segment, is said to *cross* f if it has the same altitude as has f above some point of the projection of the step.) Even the areas under such step lines differ from $\int_0^1 f$, as defined in (13) or (14), as little as one pleases (cf. Exercise 16). Customarily, this last property, rather than (14), is used as the definition of $\int_a^b f$; that is to say, $\int_a^b f$ is defined as the number from which the areas of all step lines with sufficiently thin steps crossing f differ as little as one pleases, if there is such a number. Clearly, a number satisfying this strong condition must be equal to that on the right side of (14). But the latter can be determined even for some non-continuous functions f for which there is no number satisfying the strong condition.

Example 3. Let f, *ad hoc*, denote the extension of j^{-1} assuming the value 0 for 0. Then, according to (14), $\int_{-1}^1 f = 0$ (a formula studied by Cauchy), since $j^{-1}(-x) = -j^{-1}x$ and hence $\int_{-1}^1 r_n^{(f)} = 0$ for any n. (In each such sum, the positive and the negative terms cancel each other.) Yet, among the regular step lines whose steps are as thin as one pleases and cross f *off center*, one can easily exhibit some under which the areas are, say, > 1000, others under which the areas are $< -10{,}000$, etc. Thus there is no number from which the areas under *all* regular steplines with sufficiently thin steps would differ by as little as one pleases.

The customary definition of $\int_a^b f$ does not lend itself as directly to computations by algebra or by machines as does (14). On the other hand, it yields a simple proof of the important

Addition Law concerning Domains. $\int_a^b f + \int_b^c f = \int_a^c f.$

This law is difficult to prove from the definition (14). The following

Addition Law concerning Integrands. $\int_a^b (f + g) = \int_a^b f + \int_a^b g$

is an almost immediate consequence of either definition. For instance,

$$\int_a^b [(j - m)^2 \cdot f] = \int_a^b (j^2 \cdot f) - 2m \int_a^b (j \cdot f) + m^2 \int_a^b f,$$

for any f and any number m.

Thus, in case that $\int_a^b f = 1$ and $\int_a^b (j \cdot f) = m$ (a case that is of importance in statistics) $\int_a^b [(j - m)^2 \cdot f] = \int_a^b (j^2 \cdot f) - m^2.$

The names of the two addition laws are based on the customary designation of the interval between a and b as the *domain of integration* and the function f as the *integrand* in the integral $\int_a^b f.$

Both definitions yield also the following important law:

If $g_1 \leq g_2$, then $\int_a^b g_1 \leq \int_a^b g_2.$

For instance, if $f \geq O$, then $\int_a^b f \geq \int_a^b O = 0$, and the following

First Fundamental Inequality. $Min_a^b\, g \cdot \int_a^b f \leq \int_a^b (g \cdot f) \leq Max_a^b\, g \cdot \int_a^b f,$

where $Min_a^b\, g$ and $Max_a^b\, g$ are the smallest and the largest number of Ran g (on $\{a \le j \le b\}$). Applying the First Inequality to $f = 1$, one obtains the

Second Fundamental Inequality. For any g and any a and b (> a),

$$(b - a) \cdot Min_a^b\, g \le \int_a^b g \le (b - a) \cdot Max_a^b\, g\,.$$

In Chapters I and II, the a-integral, $\int_a f$, of a simple curve f was defined as the curve whose altitude at any point b in Dom f equals $\int_a^b f$. From Example 2 it follows that the 0-area line of the parabola j^2 is the curve $\frac{1}{3}\, j^3$. Using the Addition Law concerning the Domains one readily concludes that, more generally,

$$\int_a j^2 = \frac{1}{3}\,(j^3 - a^3).$$

Similarly, generalizing Example 1, one finds

$$\int_0 exp = exp - 1.$$

$\int_a^b g - \int_a^b f$ is the excess of the area from a to b under g over that under f. This difference might be called the area *above f and below g* from a to b or, briefly, the area *from f to g* and from a to b. In view of the Addition and Multiplication Laws, this area can be found by determining $\int_a^b (g - f)$. For instance, the area from j^3 to j^2 and from a to b equals $\int_a^b (j^2 - j^3) = \left[\frac{1}{12}\,(4 - 3j) \cdot j^3\right]_a^b$. This area is 1/12 from 0 to 1; it is − 1/12 from 1 to 4/3, and 0 from 0 to 4/3.

Clearly, the area from g to f is the negative of that from f to g. The area under g from a to b is the area from O to g and from a to b.

The area of the trapezoid from $-j$ to 3 and from 0 to 2 is $\int_0^2 (3 + j) = 8$. The contribution of the rectangle from O to 3 is 6;

that of the triangle from $-j$ to O is 2. It should be noted that, since $-j$ is below O, the area under $-j$ (that is, the area from O to $-j$) is -2. However, what contributes to the area of the trapezoid is not the area from O to $-j$ but rather the area from $-j$ to O. More generally,

$$\int_a^b (g - f) = \int_a^b (g - O) + \int_a^b (O - f) \text{ and}$$

$$\int_a^b (g - f) = \int_a^b (g - h) + \int_a^b (h - f),$$

for any three curves, f, g, and h.

In Chapter IX, the area $\int_{-r}^{r} \sqrt{r^2 - j^2}$ under the upper semicircle will be found to be $\frac{1}{2}\pi r^2$ (in agreement with the elementary formula). Similarly, $\int_{-r}^{r} -\sqrt{r^2 - j^2} = -\frac{1}{2}\pi r^2$. The area from $-\sqrt{r^2 - j^2}$ to $\sqrt{r^2 - j^2}$ is, of course, $\frac{1}{2}\pi r^2 - \left[-\frac{1}{2}\pi r^2\right] = \pi r^2$. Suppose one moves along the former curve from $-r$ to r and then along the latter from r back to $-r$. During this entire round trip, the circular domain with the positive area is to the left of the traveler, and the trip is counterclockwise. The area from $\sqrt{r^2 - j^2}$ to $-\sqrt{r^2 - j^2}$ from $-r$ to r is $-\pi r^2$. Traveling along the former curve from $-r$ to r and then along the latter from r back to $-r$, one finds the domain with the negative area on the right side, and the trip is clockwise.

In the course of a motion along a curve f from a to b, then (vertically) to g, then along g from b to a, and back to f, one may encircle several domains. Those that lie to the left (the right) of the traveler make a positive (a negative) contribution to the area from f to g. For instance, suppose one moves along j^3 from 0 to 4/3, then down to j^2, and then along j^2 from 4/3 to 0. The loop between j^3 and j^2 from 0 to 1 (which is on one's left side) makes the contribution 1/12 to the area from j^3 to j^2 and from 0 to 4/3; the loop between 1 and 4/3 (which is on one's right side), the contribution $-1/12$. The latter loop is encircled clockwise, the former counterclockwise.

Besides the, as it were, *oriented* areas from f to g and from g to f, one may be interested in the area *between f and g*, defined as the sum of the absolute values of the areas of all loops between f and g. For instance, the area between O and *sin* from 0 to 2π equals $|2|+|-2| = 4$. To find the area between f and g, one must determine all loops between the two curves, which presupposes the determination of the points where the curves intersect. This is unnecessary if one wishes to find an oriented area.

EXERCISES

9. Find $\int_a^b (k_0 + k_1 j)$ in terms of a, b, k_0, and k_1, for any four numbers, and compare the result with that of plane geometry (see Exercise 11a, p. 48). If f is any function whose domain includes the interval $\{c \leq j \leq c + d\}$, there is only one straight line having the same altitudes above c and c + d as has f, namely, the chord of f between c and c + d. Determine the area from c to c + d of the trapezoid under that chord. Then divide the interval from a to b into n equal parts of lengths $d = (b - a)/n$. Replace the portion of f above each part by the chord and determine the area under the polygon of n sides thus inscribed into f.

10. For any three numbers k_0, k_1, k_2, the curve $k_0 + k_1 j + k_2 j^2$ will be referred to as a *vertical parabola*, more specifically, as the vertical parabola with the coefficients k_0, k_1, k_2 ("vertical," because its axis is vertical, whereas, e.g., the axis of the semiparabola $j^{1/2}$ is horizontal). Find the area from a to b under that vertical parabola in terms of a, b, k_0, k_1, k_2, for any five numbers. If f is a function whose domain includes the interval $\{c - d \leq j \leq c + d\}$, then there is only one vertical parabola having the same altitude above c − d, c, and c + d as has f. Determine its coefficients in terms of c, d, and the three values of f, and compute the area from c − d to c + d under that parabola.

Then consider a function f whose domain includes the interval $\{a \leq j \leq b\}$. Divide the interval from a to b into 2n segments of length $d = (b - a)/2n$. Replace the portion of f above the 1st pair, the 2nd pair, ... , the nth pair of these segments by the vertical parabola having the same altitude above the division points as has f, and determine the area under the "parabolic polygon," $s_n^{(f)}$, thus inscribed in f. The result is

$$\int_a^b s_n^{(f)} = \frac{d}{3} \cdot \{[f(a) + 4f(a + d) + f(a + 2d)]$$
$$+ [f(a + 2d) + 4f(a + 3d) + f(a + 4d)] + \ldots$$
$$+ [f(a + (2n - 2)d) + 4f(a + (2n - 1)d) + f(b)]\}$$

or, if the terms are rearranged to simplify the computation,

$$\int_a^b s_n^{(f)} = \frac{d}{3}\{ f(a) + f(b)$$
$$+ 2[f(a+2d) + f(a+4d) + \dots + f(a+(2n-2)d)]$$
$$+ 4[f(a+3d) + f(a+5d) + \dots + f(a+(2n-1)d)]\},$$

where $d = (b-a)/2n$.

The important formula $\int_a^b f = \lim_{n\to\infty} \int_a^b s_n^{(f)}$ or $\int_a^b f \sim \int_a^b s_n^{(f)}$

for large n is known as *Simpson's Rule.* The computation of

$\int_a^b s_n^{(f)}$ presupposes the determination of $2n+1$ values of f at equi-

spaced places between a and b, just as does the computation of

$\int_a^b r_{2n}^{(f)}$. But, in general, $\int_a^b s_n^{(f)}$ is closer to $\int_a^b f$ than is $\int_a^b r_{2n}^{(f)}$.
Illustrate this by computing

(a) $\int_0^1 s_2^{(j^3)}$ and $\int_0^1 r_4^{(j^3)}$; *(b)* $\int_0^1 s_2^{(j^4)}$ and $\int_0^1 r_4^{(j^4)}$

and by comparing the results with $\int_0^1 j^3 = 1/4$ and $\int_0^1 j^4 = 1/5$,

respectively. Of course, there are exceptions; e.g., the polygon p with two sides whose vertices have the altitudes $p\,0 = 0$, $p\,\frac{1}{2} = \frac{1}{4}$ and $p\,1 = 1$.

Compare $\int_0^1 p$ with $\int_0^1 r_2^{(p)}$ and $\int_0^1 s_1^{(p)}$.

11. Compute $\int_0^b j^3$. Hint: To obtain the needed expression for the sum of the cubes of the first n odd numbers, notice that $1^3 + 3^3 + 5^3 + \dots + (2n-1)^3 = [1^3 + 2^3 + 3^3 + \dots + (2n)^3] - 8\cdot[1^3 + 2^3 + \dots + n^3]$, and use the expression for $1^3 + 2^3 + \dots + n^3$ obtained in Exercise 12, p. 49.

Corroborate the last result by computing, for any positive number $t < 1$, the area under Fermat's step line (with infinitely many steps) associated with j^3 ($= j^{3/1}$) whose first step on the right side is from tb to b (see Exercise 15, p. 49). Then compute the limit of the areas under these step lines as t approaches 1. In the same way show that

$\int_0^b j^{p/q} = \frac{q}{p+q} \cdot b^{(p+q)/q}$ if $p/q > -1$ and

$\int_a^\infty j^{p/q} = \frac{-q}{p+q}\ a^{(p+q)/q}$ if $p/q < -1$, where $a > 0$. On the basis of these results, determine for any rational number r the a-integrals $\int_a j^r$.

Compute $\int_0^b \cos$. Hint: Using the trigonometric expression for sin 2a and (repeatedly) that for *sin* b − *sin* a, prove that

$$\cos a + \cos 3a + \dots + \cos (2n-1)\, a = \frac{\sin 2na}{2 \sin a} \quad \text{or}$$

$$\underset{1}{\Sigma} \cos\ [\, a(2j - 1)\,] = \frac{\sin (2a\, j)}{2 \sin a} \quad \text{(see p. 128).}$$

In particular, show that $\int_0^{\pi/2} \cos = 1$, as suggested by graphical methods (p. 38). Then find $\int_a \cos$. (Here *sin* 2a stands for *sin* (2a), etc.)

12. Find the following areas: *(a)* from $-j^{1/2}$ to $j^{1/2}$ and from 1 to 4; *(b)* from j^3 to j and from −2 to 3; *(c)* from $\sqrt{\arctan + \log}$ to $j^2 + \sqrt{\arctan + \log}$ and from 1 to 4; *(d)* between $2j^3$ and j; *(e)* between O and $j^3 - 5j$.

13. Find the a-integrals $\int_a (c\, j)$, $\int_a (c\, j^2)$, $\int_a \sin$, $\int_a \exp$.

14. From the symmetry of the curves *log* and *exp* with regard to the line j, show that

$$\int_0^b \exp = e^b - 1 \text{ implies } \int_1^{e^b} \log = (b-1)e^b + 1 \text{ and hence}$$

$$\int_1^c \log = (\log c - 1) \cdot c + 1 \text{ and } \int_1 \log = (\log - 1) \cdot j + 1.$$

In a similar way, use the results of Exercises 12 and 13 to compute

$\int_0^c j^{1/3}$ and $\int_0^c \arccos$. Then find $\int_0^c j^{1/2}$ (see Exercise 5, p. 43). What follows from Exercise 16, p. 51?

★15. Assume that f is a continuous function assuming between $-a$ and a only non-negative values. Show that the following equality and two inequalities are valid by virtue of fundamental laws concerning integrals:

$$\int_{-a}^{a} (j^2 \cdot f) = \int_{-a}^{-d} (j^2 \cdot f) + \int_{-d}^{d} (j^2 \cdot f) + \int_{d}^{a} (j^2 \cdot f)$$

$$\geq \int_{-a}^{-d} (j^2 \cdot f) + \int_{d}^{a} (j^2 \cdot f) \geq d^2 \left[\int_{-a}^{-d} f + \int_{d}^{a} f \right] \text{ and}$$

that therefore, for any such function f and $0 < d < a$,

$$\int_{-a}^{-d} f + \int_{d}^{a} f \leq \frac{1}{d^2} \int_{-a}^{a} (j^2 \cdot f).$$

This is a special case of an inequality due to Bienaymé-Tchebisheff. For any function f satisfying the assumptions and having the property that $\int_{-10^6}^{10^6} (j^2 \cdot f)$ does not exceed, say, 2 one thus finds that

$$\int_{-10^6}^{-10} f + \int_{10}^{10^6} f \leq \frac{2}{100} = .02 \text{ and hence } \int_{-10^6}^{10^6} f - \int_{-10}^{10} f \leq .02;$$

that is to say, for such a function practically the entire area from -10^6 to 10^6 is concentrated in the portion from -10 to 10. (An example of such a function is $1/(exp\ abs)$ assuming for any x the value $1/e^{|x|} = e^{-|x|}$.)

★16. Let f be a function with the following property: For every positive integer k, there is a number $d_k > 0$ such that, if two numbers x' and x'' in Dom f differ by less than d_k, the values $f\,x'$ and $f\,x''$ differ by less than $1/k$. By dividing the interval from a to b into equal parts of lengths less than d_k, show that, regardless of where in each of these parts a point is selected, the average of the values of f at these points, multiplied by $b - a$, differs from $\int_a^b f$ by less than $(b - a)/k$. Then show that the same is true even if the interval was divided into unequal parts of lengths less than d_k.

4. THE FUNDAMENTAL RECIPROCITY LAWS OF CALCULUS

From results obtained in the two preceding sections and the fact that **D** c = *0*, for any constant function c, one concludes

$$\int_a j^2 = \frac{1}{3} j^3 - \frac{1}{3} a^3 \qquad \text{and} \qquad \mathbf{D}\left(\frac{1}{3} j^3 - \frac{1}{3} a^3\right) = j^2 ;$$

$$\int_a exp = exp - exp\ a \qquad \text{and} \qquad \mathbf{D}\ (exp - exp\ a) = exp ;$$

$$\int_a cos = sin - sin\ a \qquad \text{and} \qquad \mathbf{D}\ (sin - sin\ a) = cos .$$

Thus, if *f* is j^2 or *exp* or *cos*, and if, for any a, one determines first $\int_a f$ and then the derivative of the a-integral, the result if *f*. This bears out what the graphical and numerical methods in Chapters II and III suggested: that the reciprocity of area and slope established in Theorem I (p. 14) for horizontal lines, holds also for some curves. That it holds for all simple curves is claimed in the following

FIRST RECIPROCITY LAW. *For any continuous function f and any* a *and* b *belonging to Dom f.*

$$\mathbf{D} \int_a f = f, \text{ that is, } \mathbf{D} \int_a f\, b = f\, b.$$

Here, $\mathbf{D} \int_a f\, b$ denotes the value at b of $\mathbf{D} \int_a f$, the derivative of the function $\int_a f$.

To prove the law, if a and b are given, set *(ad hoc)*

$$q\ x = \frac{\int_a^x f - \int_a^b f}{x - b} .$$

Thus, for any x between a and b, the number *q* x is the difference quotient between b and x of the function $\int_a f$. The Addition Law concerning Domains implies

(15) $$q\ x = \frac{\int_b^x f}{x - b} .$$

According to the Fundamental Inequality,

$$(x - b) \cdot Min_b^x\ f \leq \int_b^x f \leq (x - b) \cdot Max_b^x\ f \text{ (for } x > b).$$

Consequently, (15) implies

$$Min_b^x\ f \leq q\ x \leq Max_b^x\ f.$$

If f is continuous at b, then for any number $c > 0$, there is a number d such that the restriction of f to the interval between b and b + d differs from the constant function $f\ b$ by less than c. In particular, for any number x between b and b + d, also $Max_b^x\ f$ and $Min_b^x\ f$ (and consequently $q\ x$) differ from f by less than c. In other words,

$$\text{if } x \underset{suf}{\sim} b, \text{ then } q\ x \underset{arb}{\sim} f\ b.$$

Since the derivative at b of $\int_a f$ equals the limit of $q\ x$ as x approaches b, it follows that

$$\mathbf{D} \int_a f\ b = \lim_{x \to b} q\ x = f\ b \text{ for any b and any a,}$$

as contended. Thus $\mathbf{D} \int_a f = f$ for any a, which completes the proof of the First Reciprocity Law.

The last formula can also be expressed by saying that $\int_a f$ belongs to $\{ \mathbf{D}^{-1} f \}$ or is an antiderivative of f. From $\int_a^a f = 0$, it follows that $\int_a f$ is the (a, 0)-antiderivative of f.

(16) $$\int_a f = \mathbf{D}_{(a,\ 0)}^{-1} f \quad \text{for any continuous } f.$$

The results mentioned at the beginning of this section can be restated as follows:

$$\mathbf{D}\left(\frac{1}{3}\, j^3\right) = j^2 \quad \text{and} \quad \int_a j^2 = \frac{1}{3}\, j^3 - \frac{1}{3}\, a^3 \quad \text{and} \quad \int_a^b j^2 = \left[\frac{1}{3}\, j^3\right]_a^b ;$$

$$\mathbf{D}\ exp = exp \quad \text{and} \quad \int_a exp = exp - exp\ a \quad \text{and} \quad \int_a^b exp = [\ exp\ ]_a^b ;$$

$\mathbf{D}\ sin = cos$ and $\int_a cos = sin - sin\ a$ and $\int_a^b cos = [\ sin\ ]_a^b$.

Thus if f is $\frac{1}{3}\ j^3$ or exp or sin, and one determines first $\mathbf{D}\ f$ and then the a-integral of this derivative, the result is $f - f$ a. Hence, any area curve of $\mathbf{D}\ f$ is parallel to (or identical with) the curve f. The integral from a to b of the derivative is $[f]_a^b$, that is, the rise f b $- f$ a of f from a to b.

This bears out another suggestion derived from the graphical and numerical methods: that the reciprocity of slope and area established in Theorem II (p. 14) for straight lines holds also for some curves. That it holds for all curves with continuous slope curves is claimed in the following

SECOND RECIPROCITY LAW. *If* $\mathbf{D}\ g$ *is continuous, then*

$$\int_a \mathbf{D}\ g = g - g\ \mathrm{a} \quad \text{and} \quad \int_a^b \mathbf{D}\ g = g\ \mathrm{b} - g\ \mathrm{a}.$$

Applying the First Reciprocity Law to the continuous function $\mathbf{D}\ g$, one finds that $\int_a \mathbf{D}\ g$ has the derivative $\mathbf{D}\ g$. Thus $\int_a \mathbf{D}\ g$ is an antiderivative of $\mathbf{D}\ g$. So is also g. Hence, by the Fundamental Theorem on Antiderivatives, $\int_a \mathbf{D}\ g - g$ is a constant function. The value of this difference at a is $\int_a^a \mathbf{D}\ g - g\ \mathrm{a} = -g$ a. Hence $\int_a \mathbf{D}\ g - g = -g$ a, or $\int_a \mathbf{D}\ g = g - g$ a.

Evaluating the functions on both sides for b one obtains $\int_a^b \mathbf{D}\ g = g\ \mathrm{b} - g$ a, which completes the proof of the Second Reciprocity Law.

As corollaries of the two Reciprocity Laws, one obtains the following generalizations of Theorems III and IV on p. 14.

If $f = \int_a g$ *for some number* a, *then* $g = \mathbf{D}\ f$.

If $g = \mathbf{D}\ f$, *then* $f = \int_a g$ *for some number* a.

In other words, if f is an integral of g, then f is an antiderivative of g, and vice versa.

The area under f from a to b, that is, $\int_a^b f$, is the value that the function $\int_a f$ assumes for b. According to (16), this value is equal to that of $D^{-1}_{(a,0)} f$ at b. Since any antiderivative of f differs from $D^{-1}_{(a,0)} f$ by a constant function, the rise from a to b of any antiderivative of f is equal to the value of $D^{-1}_{(a,0)} f$ at b. Thus the fundamental formula

$$(17) \qquad \int_a^b f = [D^{-1} f]_a^b ;$$

in words, the area under f from a to b is equal to the rise from a to b of any antiderivative of f. This statement expresses the

Calculus Solution of the Area Problem. *To determine the area or the integral* $\int_a^b f$, *find any antiderivative of* f *and compute its rise from* a *to* b.

EXERCISE

17. Define $\int^b f$ as the function assuming the value $\int_a^b f$ for any a and prove that $D \int^b f = - f$ (see Exercise 21, p. 18).

5. REMARKS CONCERNING THE CLASSICAL NOTATIONS IN CALCULUS

a. Derivatives. D for derivative seems to have been first used by the French mathematician Arbogast around 1800. Today, this symbol is quite common in treatises on operators and on differential equations. Problems therein are expressed in the form: Find $f(x)$ such that

$$D\, f(x) + 3f(x) = \sin x, \quad D^2 f(x) + f(x) = e^x, \ \ldots$$

Around 1900, Heaviside, for similar purposes, denoted the derivative by the letter p. In the traditional texts on calculus and in most scientific books applying calculus, however, three symbols prevail that go back to Leibniz, Cauchy, and Lagrange. The third is similar to (if not identical in meaning with) the symbol originally used by Newton.

(1) Following Leibniz, one denotes the derivative of f by $\frac{df(x)}{dx}$, as in

(18) $$\frac{d \sin x}{dx} = \cos x,$$

and one refers to the derivative as *differential quotient*. Just as $f(x)$ denotes both the function f itself and its value for x, the differential quotient of f symbolizes both the function **D** f and its value for x, that is, **D** f x. Clearly, in such a differential quotient, x cannot be consistently interpreted as a numerical variable. One cannot possibly write $\frac{df(2)}{d2}$. Even $\frac{df(a)}{dx}$ and $\frac{d \sin 2}{dx}$ look somewhat awkward. But the mere fact that the latter symbols (in contrast to the former) are not impossible seems to suggest that in a differential quotient the letter x is not in its two occurrences used in the same way.

To avoid the awkward symbols mentioned, one has invented the following notations for the values of differential quotients

$$\left(\frac{df(x)}{dx}\right)_{x=a} \quad \text{and} \quad \left(\frac{d \sin x}{dx}\right)_{x=2}.$$

Thus, in (18), the letter x is treated differently to the left and the right of the equality sign. For it is not necessary to write $(\cos x)_{x=2}$ for $\cos 2$.

The term "differential quotient" and Leibniz' symbol are reminiscent of the difference quotients as the limit of which the derivative of f is defined. However, they suggest a quotient of limits rather than a limit of quotients. Leibniz indeed regarded $\frac{df(x)}{dx}$ as a quotient of infinitely small numbers – a point of view not shared by Lagrange and Cauchy and definitely abandoned around 1870 under the influence of Weierstrass.

Newton had clearly stated in Book I of his Principia (1687) that those so-called quotients "are not really ratios of ultimate quantities but rather limits which the ratios of indefinitely decreasing quantities ever more closely approach [in the original: *limites ad quos quantitatum... decrescentium rationes semper appropinquant*]".

(2) At about 1840, Cauchy introduced the symbols $\mathbf{D}_x f(x)$ and $\mathbf{D}_x \sin x$. The prefix $\mathbf{D}_x$ really amounts to a contraction of Leibniz' $\frac{d}{dx}$, and in the 20th century differential quotients are indeed frequently written in the form $\frac{d}{dx} f(x)$, $\frac{d}{dx} \sin x$, etc. Obviously, also in Cauchy's symbol, the two letters x have to be treated differently. The value of the derivative of f for 3 is denoted by $(\mathbf{D}_x f(x))_{x=3}$ or by $\mathbf{D}_x f(3)$. According to Cauchy, $\mathbf{D}_x \sin x = \cos x$ and $\mathbf{D}_t \sin t = \cos t$. Occasionally one reads $\mathbf{D}_x \sin \pi = -1$ and $\mathbf{D}_t \sin \pi = -1$. No attempt will here be

made to clarify the somewhat obscure meaning of the letters x and t in these formulas since in this book the contents of those formulas are consistently rendered by **D** *sin* π = *cos* π = -1.

(3) Lagrange in 1787 introduced the term derivative (function dérivée) and the symbol $f'(x)$ (read "*f* prime of *x*") for both the function **D** *f* and its value **D** *f* x. The value of this function for 3 is denoted by $f'(3)$. Rarely is Lagrange's symbol applied to specific functions, as in $\sin' x$ and $(x^n)'$. The value of the former function for 3 might be denoted by $\sin' 3$; that of the latter would have to be symbolized by $(x^n)'_{x=3}$, since $(3^n)'$ would be incomprehensible. Clearly, $f'(g(3))$ is the number denoted in this book by **D** *fg* 3, while **D** *(fg)* 3 has to be written $[f(g(x))]'_{x=3}$.

Lagrange's "primes" have superseded (with one exception to be discussed in Chapter VII) Newton's original dots as in $\dot{f}$. The precise meaning of Newton's "pricked letters" will be discussed on p. 218.

b. Integrals. Integrals, denoted in this book by

$$\int_1^3 \textit{log}, \quad \int_a^b \textit{sin}, \quad \int_0^1 f, \quad \int_a^b g, \ \ldots$$

are traditionally referred to as *definite integrals* and are almost universally denoted by

$$\int_1^3 \log x\, dx, \quad \int_a^b \sin x\, dx, \quad \int_0^1 f(x)\, dx, \quad \int_a^b g(x) dx, \ \ldots,$$

symbols first used by Euler (1770) and in the above form by Fourier (1828). They are often attributed to Leibniz, who indeed introduced the sign $\int$ and symbols such as $\int \sin x\, dx$. The latter, just as is $\dfrac{d \sin x}{dx}$, is reminiscent of the expressions as the limit of which it is defined: $\int$ is a distorted S (for summation), and sin x dx recalls the terms $\sin x_1^* \cdot (x_{i+1} - x_i)$ of the sums. Of course, $\int_a^b \textit{sin}$ might as well be defined as the limit of sums of terms $\sin t_i^* \cdot (t_{i+1} - t_i)$. And indeed the traditional symbols

$$\int_a^b \sin x\, dx \text{ and } \int_a^b \sin t\, dt$$

are not only equal in value as are 2 + 3 and 5; they are identical in meaning as are 2 + 3 and II plus III. They (and other synonyms using other letters) designate what in this book is succinctly denoted by $\int_a^b \sin$.

What herein is called the a-integral of f and denoted by $\int_a f$ is traditionally written $\int_a^x f(t)\,dt$ or even $\int_a^x f(x)\,dx$, where the dummy appendix is particularly awkward. Hence $\frac{d}{dx}\int_a^x f(t)\,dt = f(x)$ renders $\mathbf{D}\int_a f = f$, the First Reciprocity Law.

c. Antiderivatives. The traditional notation for antiderivatives is among the curiosities in the history of mathematics. Suppose that in the formula

$$(17)\qquad \int_a^b f = [\mathbf{D}^{-1} f]_a^b\,,$$

which embodies the main result of calculus, one writes the integral in the traditional form and leaves a blank space for a symbol denoting any antiderivative of f, as in

$$\int_a^b f(x)\,dx = [\qquad]_a^b\,.$$

Suppose further, one tries to fill the blank space in such a way as to conceal the depth and power of formula (17) as much as possible. Obviously, no symbol for an antiderivative of f would be better suited for this purpose than $\int f(x)\,dx$. For, with this symbol filling the blank space, (17) assumes the form

$$(17')\qquad \int_a^b f(x)\,dx = \left[\int f(x)\,dx\right]_a^b,$$

and one of the greatest discoveries of all time looks like a truism.

$\int f(x)\,dx$ is indeed the symbol (essentially due to Leibniz) that is traditionally used for the antiderivative.

Some texts refer to antiderivatives as *indefinite integrals*, others as *primitive functions of f*. (The latter books use the term "indefinite integral" for the a-integrals of f.) In books on differential equations and operators, the symbols $\mathbf{D}^{-1}$ and $\mathbf{p}^{-1}$ for antiderivatives (corresponding to $\mathbf{D}$ and $\mathbf{p}$ for derivatives) are common. No standard symbol exists for the function denoted in this book by $\mathbf{D}^{-1}_{(a,\ b)} f$, the antiderivative of f that assumes the value b for a.

Nothing illustrates Leibniz' influence on the development of calculus as strikingly as the prevalence for over a quarter of a millenium of the symbol

$\int f(x)\,dx$ in the treatment of antiderivatives.

That traditional treatment is not quite correct. Of course, the general idea behind such standard formulas as

$$(19) \qquad \int \cos x\,dx = \sin x + c$$

is correct – the idea that any function having the derivative cos x is sin x plus some constant function, c. But (19) is not a connection of well-defined symbols as are, for instance, the traditional formulas

$\cos^2 x = 1 - \sin^2 x$ or $-f^2(x) \leq c^2$ for any function f and any constant function, c .

The former connects three specific functions, the latter, function variables. The inequality is valid if $f(x)$ is replaced by any specific function (such as cos x or e^x) and c by any constant function (such as that of value 5 or that of value 7). But (19) connects neither specific functions nor function variables. If $\int \cos x\,dx$ stands for *any function having the derivative* cos x, then c does not stand for any constant function. For instance, if $\int \cos x\,dx$ is replaced by sin x + 5, then c must be replaced by 5 and cannot be replaced by 7.

To transform (19) into a correct connection of three well-defined symbols, one may interpret $\int \cos x\,dx$ as *the class of all functions having the derivative* cos x, and c as *the class of all constant functions.* But then (19) does not express equality of functions; it expresses equality of two classes of functions in the sense that any function belonging to the class on the left side belongs to the class on the right side, and vice versa. Obviously, this interpretation must be supplemented by remarks concerning operations on classes of functions; in particular, by an explanation of what is meant by adding a particular function, such as sin x to a class of functions – explanations not usually given in texts on calculus.

It is believed that the present book offers the reader the first treatment of antiderivatives in which not only the ideas behind the formulas are correct but the formulas themselves are connections, according to articulate rules, of symbols each of which has a clearly defined meaning.

d. The Epochal Camouflage of Ideas in Leibniz' Notation.
The designation of antiderivatives as indefinite integrals and of derivatives as differential quotients has had two important consequences.

On the one hand, that notation camouflages the fundamental result of calculus and (as will be seen in Chapter VIII) some difficult operations on derivatives as truisms. In past centuries, undoubtedly, this very camouflage secured the basic ideas of calculus acceptance by many who would have shunned a method obviously surpassing their understanding. Without the protection of a plausible appearance, those great ideas might not even have survived – just as some bright butterflies would perish if, with folded wings, they did not assume the appearance of inconspicuous leaves.

On the other hand, the traditional notation has made it difficult to understand calculus. The symbol $\int_a^b f(x)\,dx$, while objectionable on account of its dummy part, is at least reminiscent of the sums of products as whose limit the integral is defined. The symbol $\int f(x)\,dx$, however, not only fails to remind one of the inverse of derivation – the operation he has to perform – but strongly suggests sums of products (the same sums of which the integral is reminiscent) with which the antiderivative concept has absolutely nothing to do.

The traditional symbols, introduced essentially by Leibniz, make it hard to distinguish definitions from theorems, technical difficulties from profound problems, and minor results from tremendous discoveries.

All in all, Leibniz' notation accounts for what, from a sociological point of view, are the two most striking facts in the history of calculus: that for centuries the use of that great theory has been enormously wide, and that even today its use is often merely mechanical.

EXERCISES

18. Translate the following formulas in the notations of Leibniz, Lagrange, and Cauchy:

(a) $\mathbf{D}\, sin = cos$; $\mathbf{D}\, sin\ a = cos\ a$; $\mathbf{D}\, cos\ 1 = -sin\ 1$; $\mathbf{D}\, exp\ x = exp\ x$; $\mathbf{D}\, j^5 = 5\, j^4$; $\mathbf{D}\, j^5\, 2 = 5\, j^4\, 2$; $\mathbf{D}\, 3 = O$; $\mathbf{D}\, j^{-1} = -j^{-2}$;

(b) $\int_0^1 cos = [\,sin\,]_0^1$; $\int_2^3 j^2 = \left[\frac{1}{3}\, j^3\right]_2^3$; $\int_0 exp = exp - 1$;

(c) the Second Reciprocity Law, and the Theorems on p. 14;

(d) $\{D^{-1} \cos\} = \sin + \mathcal{C}$; $\{D^{-1} j^3\} = \frac{1}{4} j^4 + \mathcal{C}$; $D^{-1}\exp + \mathcal{C} = \exp + \mathcal{C}$.

19. Translate into the notation used in this book the following statements:

(a) $\frac{d(x^4)}{dx} = 4x^3$; $\left(\frac{d \cos x}{dx}\right)_{x=1} = -\sin 1$; $\left(\frac{d e^x}{dx}\right)_{x=1} = e$; $\frac{dx}{dx} = 1$;

(b) $D_x x^2 = 2x$; $D_t \sin t = \cos t$; $\cos'\pi = -\sin \pi$;

(c) If $\frac{d f(x)}{dx} = g(x)$, then $f(x) = \int g(x)\, dx + c$.

(d) $\int_0^1 t\, dt = \frac{1}{2}$; $\int_0^1 x\, dx = \frac{1}{2}$; $\int_1^2 \sin^2 x\, dx = \int_1^2 (1 - \cos^2 t)\, dt$.

6. THE APPROXIMATION OF PRODUCT SUMS BY INTEGRALS AND OF DIFFERENCE QUOTIENTS BY DERIVATIVES

Chapters II and III contain accurate definitions of product sums or areas under step lines, and of difference quotients or slopes of chords. Integrals and derivatives were determined approximately by setting

$$\int_a^b f \sim \left[f\left(a + \frac{1}{2}d\right) + f\left(a + \frac{3}{2}d\right) + \ldots + f\left(a + \frac{2n-1}{2}d\right)\right] \cdot d \text{ for large n,}$$

where $d = (b - a)/n$;

$$Df\,a \sim \frac{f\,b - f\,a}{b - a} \text{ for } b \sim a.$$

Proximity is mutual. Hence, if one can determine $\int_a^b f$ and $D f a$ precisely – and by the methods developed in the present chapter (elaborated in Chapters VIII and IX) one frequently can – then he can find approximately product sums and difference quotients by reading the aforementioned formulas from right to left.

But why determine product sums and difference quotients approximately when they can be computed accurately? The answer is that the accurate computation – the addition of the areas under the various steps (each being the product of altitude and base) – even though it requires nothing but the most elementary arithmetic, usually is tedious and time consuming, whereas the approximate determination using calculus often is fast.

For instance, if one notices that a power curve, an exponential curve, a sine curve, or the like can be approximately passed through a step line (in the same sense in which in Chapter II a step line was approximately passed through a curve), then one approximates the area under the step line by the area under the curve. In some cases one can compute the latter area without even using paper and pencil.

Similarly, one replaces $\frac{f\,b - f\,a}{b - a}$ by $\mathbf{D}\,f\,a$; that is to say, one sets

(20) $$f\,b \sim f\,a + \mathbf{D}\,f\,a \cdot (b - a) \quad \text{for } b \sim a.$$

Doing this for every $b \sim a$ one finds

(21) $$f\,(\text{near a}) \sim fa + \mathbf{D}\,f\,a \cdot (j - a)\,.$$

The graph of the function on the right side is the tangent to f at a. Thus the procedure amounts to replacing, near a, the curve f by its tangent at a.

The approximate equalities (20) and (21) are unsatisfactory in that they lack any indication as to

(1) how close is the approximation of the curve by the tangent;

(2) how close b should be to a.

Obviously, no sweeping answer to these questions is possible (see Exercise 21).

(20) may be expressed by saying that, for $b \sim a$, and $c > 0$, the value $f\,b$ lies between

(22) $$f\,a + (\mathbf{D}\,f\,a - c) \cdot (b - a) \text{ and } f\,a + (\mathbf{D}\,f\,a + c) \cdot (b - a).$$

Also (22) is vague. But (22) yields an (as it were, qualitative) inference of great importance, valid for any curve f having a non-horizontal tangent at a, that is, such that $\mathbf{D}\,f\,a \neq 0$. If in this case in (22) one chooses $c = \frac{1}{2}\,\mathbf{D}\,f\,a$, he finds that $f\,b$ lies between

(23) $$f\,a + \frac{1}{2}\,\mathbf{D}\,f\,a \cdot (b - a) \text{ and } f\,a + \frac{3}{2}\,\mathbf{D}\,f\,a \cdot (b - a).$$

Which of these two numbers is greater depends upon the sign of $b - a$, that is, upon which of a and b is greater. But certainly one can formulate the following theorems for $b \sim a$.

THEOREM I.$^+$ *If* $a < b$, *then* $f\,a < f\,b$ *and if* $b < a$, *then* $f\,b < f\,a$, *provided* $\mathbf{D}\,f\,a > 0$.

THEOREM I.$^-$ *If* $a < b$, *then* $f\,b < f\,a$ *and if* $b < a$, *then* $f\,a < f\,b$, *provided* $\mathbf{D}\,f\,a < 0$.

Thus, in any case, if $\mathbf{D} f a \neq 0$ and if Dom f contains numbers near a on either side of a, then f assumes values $> f$ a on *one* side of a, and values $< f$ a, on the *other*. This fact may also be expressed in the following

THEOREM II. *If* a *belongs to* Dom $\mathbf{D}$ f, *and* Dom f *includes numbers near* a + 0 *and near* a – 0, *then* f *assumes near* a *values* $>$ f a *as well as values* $<$ f a *except possibly if* $\mathbf{D}$ f a = 0.

Newton applied these ideas to equations. For a given function f, a root of the equation f x = 0, i.e., a solution of the problem

(24) find x such that f x = 0

is a point where the curve f has the altitude 0, that is, a point common to the line O and the curve f. To approximate a root, Newton replaced f by the tangent (21) to f at a point, a, near the root and determined the point common to O and that tangent. Thus Newton replaced the problem (24) by the problem

(25) find x such that $fa + \mathbf{D} f a \cdot (x - a) = 0$.

The root of this linear equation, unless $\mathbf{D} f a = 0$, is

(26) $a - \dfrac{f a}{\mathbf{D} f a}$, which may be denoted by a_1.

Under conditions mentioned below, a_1 is closer to the root of (24) than is a. To approximate that root even more closely, Newton determined the point common to O and the tangent to f at a_1, that is,

$$a_1 - \frac{f a_1}{\mathbf{D} f a_1}, \text{ which may be denoted by } a_2;$$

and so on.

In many important cases, e.g., if $f x = x^n - c$ for two positive numbers n and c (not necessarily integers), iteration of this procedure rapidly approaches the root of (24) to which a is close – in the example, $\sqrt[n]{c}$. Thus Newton's procedure is an excellent method for the approximate determination of n-th roots. If $f = j^n - c$, then $\mathbf{D} f = n\, j^{n-1}$. Hence, if $a \neq 0$, then

$$a_1 = a - \frac{a^n - c}{n\, a^{n-1}}, \text{ that is, } \frac{(n-1)a^n + c}{n\, a^{n-1}} \text{ or } \left(1 - \frac{1}{n}\right) a + \frac{c}{n\, a^{n-1}}.$$

For instance, if n = 2, one computes

$$a_1 = \frac{1}{2}\left(a + \frac{c}{a}\right),\ a_2 = \frac{1}{2}\left(a_1 + \frac{c}{a_1}\right),\ \ldots$$

These numbers approach $\sqrt{c}$. If c = 2 and one chooses a = 1, he finds

$$a_1 = \frac{1}{2}(1 + \frac{2}{1}) = \frac{3}{2} = 1.5,$$

$$a_2 = \frac{1}{2}(1.5 + \frac{2}{1.5}) = \frac{17}{12} = 1.41\dot{6},$$

$$a_3 = \frac{1}{2}(1.41\dot{6} + \frac{2}{1.41\dot{6}}) = \frac{577}{408} = 1.41423..$$

Only a few general hints can here be given. If f is continuous in an interval and a and b are two numbers therein at which the values of f have opposite signs, then (24) has at least one root between a and b. Thus such a number, a, is within b − a from a root of (24). The chord of the curve f between a and b is the graph of

$$fa + \frac{f\,a - f\,b}{a - b}(j - a);$$

and the point common to O and this line is

$$a - \frac{f\,a}{\frac{f\,a - f\,b}{a - b}},$$ which may be denoted by b_1.

Clearly, b_1 lies between a and b. If, between a and b, $\mathbf{D} f$ assumes no values of oppositive sign, then (24) has only one root between a and b. If in that interval all values of $\mathbf{D}^2 f = \mathbf{D}\,\mathbf{D} f$ have the same sign as the number f a, then also a_1 lies between a and b, and the values of f *at* a_1 and b_1 have opposite signs. Thus the root of (24) lies between a_1 and b_1. (Fourier, 1830).

EXERCISES

20. Suppose one is testing the lifetime in seconds of 100 low-pressure arcs for a certain current. To describe the findings, one divides the time in intervals say of .2 seconds. One might find that 18 arcs have a lifetime between 0 and .2 seconds; 15, a lifetime between .2 and .4 seconds, etc. If each interval is characterized by its midpoint (the interval between 0 and .2 by .1, ... , that between 3.2 and 3.4 by 3.3), the results can be tabulated as follows:

Interval	.1	.3	.5	.7	.9	1.1	1.3	1.5	1.7	1.9	2.1	2.3	2.5	2.7	2.9	3.1	3.3
number of arcs	18	15	12	10	8	7	6	5	4	3	3	2	2	2	1	1	1

Draw the *histogram* (see Exercise 7, p. 24) and show that this step line may be approximated by the curve 100 *exp*$(-j)$ assuming for any

$x > 0$ the value $100\,e^{-x}$. By what integral would one replace the number of arcs with a lifetime between .2 and 2.4? The frequencies vary slightly from one set of 100 arcs to another, but the 0-area line of the curve fairly approximates the cumulative frequency polygon for any set of 100 arcs. Interpret the meaning of $\int_{3.2}^{4} exp\,(-j)$.

For the approximately $3 \cdot 10^{21}$ atoms in 1 gram of radium, the frequencies of the various lifetimes in years follow an exponential pattern similar to that of the electric arcs tabulated above, and reflected in the exponential decay of the total weight of the piece of radium. What is left of the 1 gram after t years is $e^{-4.33 \cdot 10^{-4} t}$ grams. When is $\frac{1}{2}$ gram left? (This time is called the *half life* of radium.) More generally, determine for what number the function $c\ exp(-a\,j)$ assumes the value c/2 and verify the result by computing the "half life" of the 100 electric arcs.

21. Let t be the line of slope 1 and the altitude 2 at 0. For any n, give an example of a function f_n such that $f_n\,0 = 2$ and $\mathbf{D} f_n\,0 = 1$ and that for 1/n the values of f_n and t differ by 1. (Clearly, t is the common tangent of all curves f_n). What is the significance of this exercise with regard to the vagueness of (21)?

22. If $\mathbf{D} f\,a \neq 0$, then according to (23), near a, f is between $f\,a + \frac{1}{2}\,\mathbf{D} f\,a \cdot (j - a)$ and $f\,a + \frac{3}{2}\,\mathbf{D} f\,a \cdot (j - a)$. What is the geometric meaning of this statement?

23. By Newton's method (a) approximate $\sqrt{3}$, $\sqrt{5}$, and $\sqrt[3]{10}$, starting with a = 2; (b) solve $\frac{1}{6}x^3 - x + .1 = 0$, starting with a = 1; (c) $e^x + x - 2 = 0$, starting with a = 0.

24. To get an insight into the pitfalls of Newton's method, consider the equation $\frac{1}{6}x^3 - x = 0$, whose roots are 0, $\sqrt{6}$, and $-\sqrt{6}$. Try to find 0 by Newton's method; that is to say, determine if possible a_1 and a_2 starting with the following values of a: (a) 1; (b) $\sqrt{6/5}$; (c) 1.2; (d) $\sqrt{2}$. In each of these four cases draw the graph of $\frac{1}{6}j^3 - j$ and the tangents at a, a_1 and a_2.

7. RELATIVE MAXIMA AND MINIMA

The importance of Theorem II lies in its implications concerning a problem that has attracted great attention to calculus and has stimulated the development of the calculus (especially of derivatives) more than any other question: the problem of *extrema* of a function – a collective term for maxima and minima.

f is said to possess a *maximum* (a *minimum*) at a if

$$f\,b \leq f\,a \quad (f\,b \geq f\,a) \text{ for every b in Dom } f.$$

$j^2 + 1$ possesses a minimum at 0. One refers to 0 as *the place of the minimum*; to 1, the value of the function at that place, as the *minimum* of $j^2 + 1$. This number is the smallest in Ran $(j^2 + 1)$. Since Ran $(j^2 + 1)$ does not contain a largest number, $j^2 + 1$ does not possess a maximum. For the function $-(j^2 - 1)^2$, whose graph looks somewhat like the curve in Fig. 20 (c) (p. 36), the maximum is 0, since the value of the function is 0 at 1 and -1 (the places of the maximum) and < 0 everywhere else.

Extrema as defined above are often called *absolute*, whereas f is said to possess a *relative maximum* (a *relative minimum*) at a if

$$f\,b \leq f\,a \quad (f\,b \geq f\,a) \quad \text{for every b in Dom } f \textit{ and near } \text{a}.$$

Clearly, if f possesses an absolute maximum at a, then it possesses also a relative maximum at a, but not necessarily vice versa. The function j^2 (on $\{-1 \leq j \leq 2\}$) has relative maxima at -1 and at 2; only the latter is absolute.

Clearly, the function j^{-2} possesses no maximum. Neither does the function assuming for any x the value $x - [x]$, where $[x]$ denotes the largest integer $\leq x$. Let, *ad hoc*, f denote this function. The value of f for any integer is 0. Obviously, Ran f consists of all numbers that are ≥ 0 and < 1. Thus, no number < 1 can be the maximum of f; nor can 1 or a number > 1 be the maximum, since nowhere does f assume a value ≥ 1.

g is said to possess a *proper* relative maximum at a if

$$g\,b < g\,a \text{ for every b belonging to Dom } g \text{ and near a and} \neq \text{a}.$$

$j + abs$ possesses no proper relative extremum, even though for every negative number the function has both a relative maximum and a relative minimum. At 0 it possesses a relative minimum but not a proper one.

The number c is said to be an interior element of Dom f – briefly, to be *interior* to Dom f – if Dom f includes an interval $\{c - d \leq j \leq c + d\}$ for some $d > 0$. The number a will be called an *initial* element of Dom f if Dom f includes an interval beginning at a, but no interval ending at a. Similarly, one defines a *terminal* element of Dom f. For instance, -1

is the initial, 1 the terminal element of Dom $\sqrt{1 - j^2}$; and −2 and 1 are initial, −1 and 2 terminal in Dom $\sqrt{-(j^2 - 1) \cdot (j^2 - 4)}$.

The theories of extrema of f at interior and at non-interior elements of Dom f will be treated separately.

a. Extrema at Interior Elements of Dom f. Theorem II (p. 160) can be restated as follows:

THEOREM III. *If* c *is interior to* Dom **D** *f, and if f possesses a relative extremum (that is, either a relative minimum or a relative maximum) at* c, *then* **D** f c = 0.

The equality **D** f c = 0 thus is a *necessary* condition for a relative extremum of f at an interior element of Dom f where f has a derivative – a condition that is satisfied if at such a place f possesses a relative extremum. Whether, conversely, for a number satisfying the condition, f actually possesses a relative extremum must in each case be investigated. For instance, **D** j^3 0 = 0, and yet j^3 has neither a relative maximum nor a relative minimum at 0. On the other hand, **D** $(-j^4)$ 0 = 0, and $-j^4$ has a maximum at 0; and **D** $(j^2 + 1)$ 0 = 0 and $j^2 + 1$ has a minimum at 0.

The inequality **D** f c $\neq$ 0, according to Theorem III, implies that f has no relative extremum at c. The inequality is so efficient in ruling out numbers as possible places for relative extrema that, in general, only few interior elements remain as possibilities – numbers of the following two types:

(1) *the numbers interior to* Dom f *belonging to the class* {**D** $f = 0$}

(2) *the numbers interior to* Dom f *that do not belong to* Dom **D** f; (clearly, if f has a derivative everywhere, there are no numbers of the second type). Besides these interior elements one must, of course, examine

(3) *the initial and terminal elements of* Dom f.

For a function a j^2 + b j + c, there is no number of types (2) and (3) and, if a $\neq$ 0, exactly one of type (1), since { 2a j + b = 0 } consists of the single number −b/2a. It is easy to verify by algebra that at −b/2a the function actually has a maximum if a < 0, and a minimum if a > 0.

The only numbers for which *sin* may (and actually does) possess relative extrema are the numbers belonging to { **D** *sin* = 0 } = { *cos* = 0 }, that is, the odd multiples of $\pi/2$. The function *exp*, for which **D** *exp* = *exp* > 0, has no extremum.

The functions *abs* and $j^{2/3}$ have minima only at 0, a number of type (2), and no maxima. The function j^3 (on { $-1 \leq j \leq 2$ }) has extrema only at −1 and 2.

Oresmes (in the 14th century) and Kepler (about 1615) noticed that, near an extremum, the altitude of a curve changes only little. About 1630, Fermat used this remark in an extensive study of extrema, for the sake of which he developed parts of the calculus of derivatives before Newton and Leibniz.

b. Extrema at Non-Interior Elements of Dom f.

THEOREM IV. *If* a *is an initial element of* Dom f *belonging to* Dom **D** f, *and if* f *has a relative maximum (a relative minimum) at* a, *then* **D** f a ≤ 0 (≥ 0). *If* b *is a terminal element of* Dom f *belonging to* Dom **D** f, *and* f *has a relative maximum (minimum) at* a, *then* **D** f a ≥ 0 (≤ 0).

Indeed, if a is an initial element of Dom f and **D** f a > 0, then, according to Theorem I^-, f b $> f$ a for every b in Dom f near a + 0 – thus, since a is initial, for every b in Dom f near a – and f has a proper relative minimum at a, and not a relative maximum. In other words, if f does possess a relative maximum at an initial element of Dom f belonging to Dom **D** f then necessarily **D** f a ≤ 0. The other three cases are proved similarly.

Theorem IV contains a necessary condition for a relative maximum at a. Whether, conversely, for a number a satisfying the condition, f actually possesses a relative maximum must, in each case, be investigated. Each of the three functions $-j^2$ (on $\{ O \leq j \}$), $-j^3$ (on $\{ O \leq j \}$), and j^2 (on $\{ O \leq j \}$) has the derivative 0 at 0 (and thus a derivative ≤ 0 at 0). Yet only the first two have a relative maximum at 0. The third has a relative minimum at 0.

In proving Theorem IV it has been shown that if a is an initial element of Dom f and if **D** f a > 0, then f has a relative minimum at a. Similarly, if **D** f a < 0, then f has a relative maximum at a. Hence the following sufficient conditions:

THEOREM V. *If* a *is an initial element of* Dom f *and* **D** f a < 0 (>0), *then* f *has a relative maximum (a relative minimum) at* a.

A corresponding statement is valid at a terminal element of Dom f.

The conditions in Theorem V are sufficient but not necessary. The functions $-j^2$ (on $\{ O \leq j \}$) and $-j^{1/2}$ have relative maxima at 0 although, at 0, the former has the derivative 0 (thus not < 0), the latter, no derivative.

A gap between necessary and sufficient conditions is typical for extremum problems. Necessary conditions are known that are not sufficient, and sufficient conditions that are not necessary.

c. The Role of the Second Derivative. While a systematic treatment of maxima and minima, based on a study of higher derivatives, will be found in Chapter X, the role of the second derivative $\mathbf{D}^2 f = \mathbf{D}\,\mathbf{D}\, f$ will be mentioned here.

Applied to $\mathbf{D}\, f$ (instead of to f), Theorem I$^+$ yields the following

Corollary. If $a < b$, then $\mathbf{D}\, f\, a < \mathbf{D}\, f\, b$, and if $b > a$, then $\mathbf{D}\, f\, b < \mathbf{D}\, f\, a$, provided $\mathbf{D}^2\, f\, a > 0$.

If $\mathbf{D}\, f\, a = 0$ and $\mathbf{D}^2\, f\, a > 0$, then, according to the Corollary,

$$\mathbf{D}\, f\, b > 0 \text{ for } b > a, \quad \text{and} \quad \mathbf{D}\, f\, b < 0 \text{ for } b < a.$$

It is intuitively clear from a study of polygons (see Exercise 5, p. 24) and will be proved in Chapter X that in this case f has a relative minimum at a. A corresponding statement holds for maxima. Hence

THEOREM VI. *For f to have a relative minimum (a relative maximum) at* c *it is sufficient that* $\mathbf{D}\, f\, c = 0$ and $\mathbf{D}^2\, f\, c > 0$ (that $\mathbf{D}\, f\, c = 0$ and $\mathbf{D}^2\, f\, c < 0$).

It will be noted that in this theorem c may be an interior or non-interior element. Combining the sufficient condition of Theorem VI with those of Theorem V, one has, of course, to distinguish initial and terminal elements. For instance,

sufficient for f to have a minimum at an initial element a is that either $\mathbf{D}\, f\, a > 0$ or $\mathbf{D}\, f\, a = 0$ and $\mathbf{D}^2\, f\, a > 0$;

sufficient for f to have a minimum at a terminal element b is that either $\mathbf{D}\, f\, b < 0$ or $\mathbf{D}\, f\, b = 0$ and $\mathbf{D}^2\, f\, b > 0$.

That the sufficient condition of Theorem VI is not necessary is demonstrated by the function j^4 which has a minimum at 0 even though $\mathbf{D}\, j^4\, 0 = 0$ and $\mathbf{D}^2\, j^4\, 0 = 0$.

EXERCISES

25. Determine the places of relative extrema and the relative extrema of the following functions: *(a)* $j^{-1} + 2j$; *(b)* $j^{-1} + j^2$; *(c)* $j^{-1} - 2j$; *(d)* $j^{-2} + 3j$; *(e)* $j^{2/3} + j$. Sketch the graphs of the functions.

26. Determine Ess $(j^3 - j)$, the essential domain of $j^3 - j$. This problem, whose solution is a prerequisite for the determination of $j//(j^3 - j)$, leads to a simple problem on maxima and minima.

27. Prove that if f has a relative maximum at c, then so do $f + 3$, $4f$, and f^2, whereas $-2f$ has a relative minimum at c. What can be said about $1/f$? What, if $f > O$, about $\sqrt{f}$? Utilize the last result to maximize $\sqrt{|\, j^2 + a\, j + b\,|}$.

28. Find the relative extrema of *(a)* $(j^3 - j)$ (on $\{-1 \le j \le 4\}$); *(b)* sin (on $\{-\pi \le j \le \pi/2\}$).

CHAPTER VII

THE APPLICATION OF CALCULUS TO SCIENCE

1. QUANTITIES

In the physical and social universe, man is surrounded by countless things with which numbers are somehow paired. A rod on which its length in feet is marked and a piece of merchandise to which its price in dollars is attached are examples. Any such pair consisting of a thing and a number (in this order) will hereinafter be referred to as a *quantity* – the first member in the pair as the *object,* the second member as the *value* of the quantity. Thus

$$quantity = (object,\ value).$$

If ρ_0 is a specific rod, then $(\rho_0, 1.5)$ is a quantity. If 1.5 is the length of ρ_0 in feet, then 18 is the length of ρ_0 in inches and $(\rho_0, 18)$ is another quantity. Similarly, if γ_0 denotes a specific instantaneous gas sample (that is, a gas sample at a definite instant), then $(\gamma_0, 70)$ and $(\gamma_0, 158)$ are two quantities. If 70 is the temperature of γ_0 in °C, then the value of the second quantity is the temperature of γ_0 in °F.

The values of all quantities studied in this book are pure numbers (such as 70) *and not so-called denominate numbers* (such as 70°C).

If it is claimed that the quantity $(\gamma_0, 70)$ is the temperature in °C of γ_0, then someone not knowing how to go about checking this statement would have to be instructed as to what instrument he should read (a thermometer and not, say, a manometer) and also what kind of thermometer (one that is calibrated in °C and not, say, in °F). In this way, the unit is, as it were, absorbed in the object of the quantity whose value is the pure number 70 (read on the proper scale). Of course, in view of the law connecting °C with °F, the reading on a °C scale might be replaced by one on a °F scale followed by a certain computation.

If 3 is the price in dollars of a piece of merchandise, μ_1, then, in general, 6 is the price in dollars of μ_1 and another piece μ_2 of the same kind. If 3 is the time in days required by a man, μ_1, for a certain painting job, then, in general, 1.5 is the time required by μ_1 and a similar man μ_2 to do the job jointly.

These and countless other laws connecting quantities are found by observation and experiment – not by logic or mathematics. If there were a regulation forcing joint workers to spend the first half workday on planning the work and the last half day on reporting about one another's working habits, then 2.5 days would be required by two men to do the work done by one man in 3 days.

The previously mentioned quantities (ρ_0, 1.5) and (ρ_0, 18) are connected by a mathematical law based on conventions concerning mensuration, and so are (γ_0, 70) and (γ_0, 158).

If one is told that the joint earnings in dollars of two brothers are 10, and that the earnings of the older, α, exceed those of the younger, β, by 2, then he can find the (not directly revealed) earnings of α and β as follows. If a and b are *ad hoc* designations of the earnings of α and β respectively, then the information about the quantities (α, a) and (β, b) yields the relations a + b = 10 and a – b = 2. According to pure algebra, if two numbers a and b are thus related, then a = 6 and b = 4. Hence the quantities are (α, 6) and (β, 4). Frequently used *ad hoc* designations for the values of unknown quantities are the letters x and y.

Are not the *values* of the two quantities all one is really interested in, so that any reference to α and β could be dispensed with? One is interested in the numbers 6 and 4 only inasmuch as they are the brothers' earnings in dollars. For instance, one wants to know which number represents the earnings of which brother. The problem, therefore, is not entirely one of pure algebra, which has nothing to do with men and earnings.

EXERCISES

1. Invent such stories about the brothers that algebra yields for either quantity (α, a) and (β, b): *(a)* two possible values; *(b)* infinitely many possible values; *(c)* no possible value (in which case the story cannot correspond to facts); *(d)* one value for a and two possible values for b.

If one is told that the square of the sum of the brother's earnings is 100, and that the square of the difference is 4, then algebra yields the possibility of negative earnings. Such possibilities are often called *extraneous solutions.* Note that these are solutions *incompatible with conditions* (in the example, inequalities) *that are not explicit parts of the specific information but are assumed to be valid in the physical or social world.*

2. Which of the following are quantities as defined in this section? *(a)* the number 5; *(b)* the constant function *5*; *(c)* (2, *log* 2); *(d)* (*log* 2, 2); *(e)* the line 3*j*; *(f)* (3, 3*j*); *(g)* (3*j*, 3); *(h)* the class of all pairs (x, *log* x) for any x > 0; *(i)* (j^2, 2*j*); *(j)* (j^2, 6).

2. CONSISTENT CLASSES OF QUANTITIES

Two quantities will be called *consistent* unless they have the same object and different values; in other words, (α, a) and (β, b) are consistent if and only if either $\alpha = \beta$ and a = b (in which case the quantities are called *equal*) or $\alpha \neq \beta$. For instance, (ρ_0, 1.5), (ρ_0, 3/2), and (γ_0, 70) are pairwise consistent; (ρ_0, 1.5) and (ρ_0, 18) are not consistent quantities although either is consistent with (γ_0, 70).

Of paramount importance for science are certain *consistent classes of quantities* (in the sequel sometimes abbreviated to c.c.q's), that is, *classes of mutually consistent quantities*. As a shorter synonym for c.c.q. we shall use Newton's term *fluent*.

As a first example, consider the class of all quantities whose objects are rods and whose values are the lengths in feet of the respective objects. For instance, (ρ_0, 1.5) does, whereas (ρ_0, 18) and (γ_0, 70) do not, belong to this class. If, for any rod ρ, the symbol $l_{ft}\rho$ denotes the length in feet of ρ, then the class just defined consists of all pairs (ρ, $l_{ft}\rho$) for any rod ρ. This class will be denoted by l_{ft} and will be referred to as *length in feet* (with regard to rods).

Similarly, if $l_{in}\rho$ denotes the length in inches of ρ, then the class of all pairs (ρ, $l_{in}\rho$) for any rod ρ is denoted by l_{in} and called the *length in inches*. The pair (ρ_0, 18) is an element of this class since $l_{in}\rho_0 = 18$.

Gas temperature in °C is the class of all pairs (γ, $t_C\gamma$) for any instantaneous gas sample γ, where $t_C\gamma$ denotes the temperature in °C of γ, and t_F is defined similarly.

The class of all first members in pairs belonging to a consistent class of quantities (that is, the class of all objects of those quantities) is called the *domain* of the c.c.q. – a term borrowed from the theory of functions. For instance, Dom l_{ft} = Dom l_{in} is the class of all rods, Dom t_C = Dom t_F is the class of all gas samples.

The *range* of a c.c.q. is the class of all numbers that are the second members in quantities belonging to the c.c.q. (that is, the class of all values of the quantities). For instance, 1.5 is called the value of l_{ft} for ρ_0 – *the* value for ρ_0, since it has been assumed that any rod has only one length in feet; (this is why l_{ft} is a c.c.q.). But 1.5 is also the value of l_{ft} for any rod that is congruent to ρ_0.

Consistent classes of quantities, such as l_{ft}, l_{in}, t_C, and t_F, whose domains consist of objects of scientific studies are what scientists call *variable quantities* – hereinafter often abbreviated to v.q.'s and always denoted by lower-case italics.

The methods for the determination of the values of a v.q. vary within its range. In defining t_C one uses mercury and alcohol thermometers, thermo-couples, etc.; and all instruments are so calibrated that, when applied to the same sample, they yield the same result. Moreover, by what Bridgman calls pencil-and-paper operations, the domain of t_C is gradually extended so as to include eventually even the sun a million years ago.

In reality, if ρ_0 is measured three times by the same observer, or even simultaneously by three observers, the results may well be 1.50, 1.49, and 1.51. In this case no number $l_{ft}\rho_0$ would be defined, nor would a class including the pairs

$$(\rho_0, 1.50), \qquad (\rho_0, 1.49), \qquad (\rho_0, 1.51)$$

be consistent. One might denote the average of all observed lengths of ρ_0 by $\overline{l}_{ft}\rho_0$ and define a v.q. $\overline{l}_{ft}$ or *average length in feet* as the class of all pairs $(\rho, \overline{l}_{ft}\rho)$ for any rod ρ. Moreover, one may define a v.q. $l_{ft}^{(0)}$ whose domain consists of the acts of measuring the particular rod ρ_0 and, if a is such an act, denote by $l_{ft}^{(0)}a$ the numerical result of the act a. An analogous v.q., *the observed length of ρ in feet,* might, of course, be defined for any rod ρ.

Since any two quantities $(\rho, l_{ft}\rho)$ and $(\gamma, t_C\gamma)$ are consistent, the class containing all quantities in the combined classes l_{ft} and t_C is consistent. But it is a class that is unlikely ever to be scientifically significant. What is, and what is not, of significance, practically important, or the like, are, of course, extra-mathematical questions. In what follows three very important types of c.c.q.'s will be mentioned.

a. Measurable Variable Quantities (according to Helmholtz, 1887). What scientists, psychologists, economists, and others approaching the study of a certain class of objects try first is the formulation of *equality criteria*. Two bodies are said to be *equal in temperature* if, upon contact, neither undergoes certain changes. Two rods are said to be *equal in length* (or congruent) if they can be so moved that both ends of the two rods coincide. Two bodies are said to be *equal in weight* if they are in equilibrium on a balance. Secondly, scientists try to formulate *order criteria* establishing, e.g., which of two elements that are not equal in temperature, in length, in weight, etc. is the warmer, the longer, the heavier, etc. These two criteria restrict the possibility of defining v.q.'s "temperature," "length," etc. Only such definitions are acceptable according to which the values of the temperature of bodies "equal in temperature," the lengths of rods "equal in length," etc. are equal numbers, and the temperature, the length, etc. of the warmer, the longer, etc. of two objects is the greater number.

In some domains there exists, thirdly, a scientific definition of *addition* of objects. For instance, weights are added by piling one on top of the other; rods, by using one as a straight extension of the other. Addition of equal objects is called multiplication. If ρ_0 and ρ_0' are congruent, a rod ρ'' that is congruent with ρ_0 extended by ρ_0' is said to be twice as long as ρ_0. If a straight chain consisting of three rods (each of which is congruent to a rod ρ^*) is congruent with ρ'', then ρ^* is said to be 2/3 times as long as ρ_0; and so on. In this way one can establish a scale of lengths and, similarly, a scale of weights. (In the domain of ordinary temperatures, no addition is defined. Temperature scales are somehow reduced to length scales.)

If now the length 1 is associated with a standard foot, then the v.q. l_{ft} whose value for any ρ is $l_{ft}\rho$ is determined. If with an inch the length 1 is associated, the v.q. l_{in} is determined, etc.

b. Kinematic Fluents. Let C be a specific car. If, with any act of observing the mileage gauge (or the speedometer), one pairs the result of that act, that is, the number read on the instrument, then the class of all pairs thus obtained is a fluent called the *distance in miles traveled by* C (or the *speed in mph of* C). While C is parked the speed assumes the value 0. A fluent whose range consists of only one number is said to be *constant*. Also the distance traveled by C is constant while C is parked.

If with each act of reading a clock (supplemented for longer periods by calendar readings) the numerical result of the act is paired, the class of all pairs obtained is a c.c.q. hereinafter referred to as the *historical time*. This fluent is observed, e.g., in a car. In contrast, by the time elapsed since the beginning of a process, briefly, the *elapsed time* (say, in seconds) is meant the time read on a timer (calibrated in seconds) that is set in motion when the process started.

Galileo introduced into science fluents pertaining to *classes* of kinematic processes. The *elapsed time in seconds* with regard to falling bodies is the class t of all pairs $(\tau, t\tau)$ for any act τ of reading a timer (calibrated in seconds) that is set in motion when an object is dropped. Here $t\tau$ denotes the result of the act τ, that is, the number read on the timer. If objects are dropped from the point 0 of a vertical scale, calibrated in feet, then the *distance traveled in feet* is the class σ of all pairs $(\sigma, s\sigma)$ for any act σ of reading the mark on the scale opposite a falling object, where $s\sigma$ denotes the number read as a result of σ.

c. Functions. A quantity is a pair of numbers if and only if its object is a number. Two pairs of numbers are consistent quantities unless their first members are equal and their second members unequal. Hence, *each function is a consistent class of pairs of numbers and each such class is a function.* In other words, *functions are the c.c.q.'s whose domains are classes of numbers.*

Some mathematicians propose to call all consistent classes of quantities "functions" — even scientific variable quantities such as temperature and time. While there is, of course, no *logical* reason against such a terminology, it should be noted that no scientific v.q. has a derivative. Only some of the c.c.q.'s such as *log* and j^3, whose domains consist of numbers, do. D *log* 2 is defined as the limit at 2 of the difference quotient $\frac{log - log\ 2}{j - 2}$. For l_{ft} or t_C neither difference quotients nor limits can be computed. Nor does there seem to exist a generalization of the traditional derivative that would yield useful definitions of $D\,l_{ft}\rho_0$

or $\mathbf{D}\, t_C\, \gamma$. The same is true for the integral: $\int_0^2 j^3$ is well defined, $\int_0^2 t_C$ is not. Hence, if all c.c.q.'s were called functions, one would still need a term for the functions (in the traditional, more restricted sense) which are the object of calculus. In this book, as is the usage in science, only the latter c.c.q.'s will be called functions, whereas length and temperature will be referred to as v.q.'s.

From other, and more basic, angles the question will be studied on p. 176 and p. 199.

In both terms "variable quantity" and "numerical variable" the word "variable" is used just as is the word "tangent" both in "tangent to a curve" and "tangent of an angle." Since the latter concepts are never confused, the reader should have no difficulty distinguishing between a variable quantity and a numerical variable. In the literature, unfortunately, both notions are frequently referred to by the single noun "variable." In absence of a clear definition of variable quantities, great confusion has indeed resulted from the ambiguous use of that noun. It is hoped that, after a careful study of Exercises 5 and 9 and of Section 7, the reader will be prepared to understand what the term "a variable" means in a given sentence in a classical or current book. He will discover that in many books this term, without explanation, changes its meaning from sentence to sentence.

EXERCISES

3. Let t be the historical time, in hours, s the distance in miles traveled by a specific car, v its instantaneous speed in mph. Plan a trip and describe the ranges of t, s, and v. What, if any, are the effects on t, s, and v *(a)* of state laws; *(b)* of specific curfews; *(c)* of parking the car.

4. Formulate scientific equality and order criteria for objects with *(a)* brightness, *(b)* loudness, *(c)* electrical resistance, and *(d)* for student's achievement. Give further examples of v.q.'s from life and from science.

5. State which of the following are consistent classes of quantities and give the reasons why the others are not: *(a)* the class of all numbers > 0; *(b)* the pair $(5, 5)$; *(c)* the class of all pairs of numbers; *(d)* the letter x in $x^2 + 1 \neq 0$ for any x; *(e)* $\mathbf{D}\, j^3$; *(f)* $\int_0^3 j^2$; *(g)* the class of all pairs $(l_{ft}\,\rho, \rho)$ for any rod ρ; *(h)* the class of all instantaneous gas samples; *(i)* the class of all pairs of rods; *(j)* the letter ρ in the statement: $l_{ft}\,\rho > 0$ for any rod ρ. Answer the same questions with regard to the classes

obtained by pairing *(k)* with any rod the negative of its length in feet (that is to say, with ρ_0 the number $-l_{ft}\,\rho_0$, with ρ_1 the number $-l_{ft}\,\rho_1$, and so on, for *any* rod); *(l)* with any instantaneous gas sample the logarithm of its volume in cu. ft.; *(m)* with any gas sample its volumes at various instants; *(n)* with any gas sample its volume at a definite instant; *(o)* with any throw of a die the number of points turning up at that throw.

6. What is the range of the last v.q. in Exercise 5? What is the range if the die is rolled *(a)* 4 times; *(b)* 7 times; one has to distinguish between the *potential* (possible) ranges before the games and the *actual* ranges (after the die has been rolled). How would one go about answering the following questions?

Let P_0 *ad hoc* denote the population (i.e. the class of all residents) of Chicago at present, P the population from now on. What in both cases are the ranges of the following v.q.'s? *(a)* the height in inches in the population? *(b)* the height to the nearest inch in the population?

7. Define $\int_0^2 j^3$ and $\mathbf{D}\, j^3\, 2$ and list the operations involved in these definitions that cannot be carried out if j^3 is replaced by l_{ft}.

8. Consider a plane as introduced on page 1. Any quantity whose object is a number, x, may be represented by a point whose projection on *O* is x and whose altitude above *O* is the value of the quantity. What is the representation of *(a)* consistent quantities; *(b)* classes of mutually consistent quantities of that kind?

3. FLUENT VARIABLES

Laws that are valid for any number or any function (belonging to certain classes) are usually expressed in terms of numerical or function variables – symbols accompanied by legends concerning their replacement by specific numbers or functions. Similarly, laws that are valid for any consistent class of quantities (or any fluent of a certain kind) are conveniently expressed in terms of what might be called *fluent variables* – symbols accompanied by legends concerning their replacement by specific c.c.q.'s. In what follows, the italics *u*, *v*, and *w* will be used for this purpose. Of course, function variables are special fluent variables, namely, variables whose scopes are classes of functions.

The following statements about fluent variables generalize remarks (p. 78) on function variables.

By Dom *u* (and Ran *u*) is meant the class of all objects that are the first (the second) member in a pair belonging to *u* for any variable quantity *u*. If here one replaces *u* by *log* or by *p* (for gas pressure), he obtains the definitions of Dom *log* and Dom *p*.

If a belongs to Dom u, then the value of u for a is denoted by ua; in other words and letters, if in a pair belonging to w the first member is β, then its second member is $w\beta$. Hence, w is the class of all pairs $(\beta, w\beta)$ for any β in Dom w.

Two c.c.q.'s, u and w, are *equal* if any pair belonging to either c.c.q. belongs also to the other. Clearly, if u and w are equal in this sense, then Dom u = Dom w and, for any element a of this (common) domain, $ua = wa$. Conversely, if two c.c.q.'s, u and w, have the same domain and assume equal values for any element of that domain, then they are equal, that is, the same class of pairs.

Two c.c.q.'s, u and w, may be called *disjoint*, in symbols $u \| w$, if no pair belongs to both classes. For instance, l_{ft} and the gas volume v are disjoint since their domains have no element in common. But l_{ft} and l_{in} are disjoint (if only objects of a length > 0 are considered) even though Dom l_{ft} = Dom l_{in}.

If each pair belonging to u also belongs to w, then u is said to be a *restriction* of w, and w an *extension* of u. For instance, let G_{100} denote the class of all instantaneous gas samples at a temperature of 100° C; in a formula,

$$G_{100} = \{\gamma \mid t_C\gamma = 100\}. \tag{1}$$

Just as the class $\{x \mid \sin x = 0\}$ has, without reference to the numerical variable x, been denoted by $\{\sin = O\}$, the class G_{100} might (without reference to the sample variable γ) be denoted by

$$G_{100} = \{t_C = \mathit{100}\}. \tag{1'}$$

Here, *100* is the constant v.q. of the value 100 whose domain is the class of all gas samples, that is, Dom t_C. To be quite precise, one would have to denote it by *100* (on Dom t_C). The class G_{100} consists of those gas samples for which t_C and *100* (on Dom t_C) assume the same value (that is, 100);

$$t_C \text{ (on } G_{100}) = \mathit{100} \text{ (on } G_{100}).$$

The restrictions p (on G_{100}) and v (on G_{100}) are the classes of all pairs $(\gamma, p\gamma)$ and $(\gamma, v\gamma)$ for any γ in G_{100}, that is, for any γ such that $t_C\gamma = 100$.

EXERCISE

9. Determine the nature of the letters *v* and v in the following statements:
(*a*) For the gas volume *v* in cu. ft., Ran *v* = P.
(*b*) If Ran *v* = P, then *v* || −*1* (on Dom P) for any variable quantity *v*.
(*c*) If v is any value of the gas volume *v*, then v > 0.
(*d*) If v is any value of the function *exp*, then v > 0.
(*e*) If Ran *v* = N, then *v* || *exp*, for any consistent class of quantities, *v*.

4. SUMS, PRODUCTS, AND FUNCTIONS OF CONSISTENT CLASSES OF QUANTITIES

The sum, $u + w$, of two c.c.q.'s u and w is defined as the class of all pairs $(a, ua + wa)$ for any element a belonging to both Dom u and Dom w. This definition is analogous to (and an extension of) that of the sum of two functions. Products can be defined similarly. For instance, by the product $p \cdot v$ of the pressure p and the volume v (in certain units) is meant the v.q. consisting of all pairs $(\gamma, p\gamma \cdot v\gamma)$ for any instantaneous gas sample γ.

Boyle's famous law concerning ideal gases can be expressed as follows: After units for p, v, and the temperature t have been chosen, there exists a number r such that

(2) $\quad p\gamma \cdot v\gamma = \mathrm{r}\, t\gamma$ for any instantaneous gas sample γ.

Here, γ is what on p. 174 was called a *sample variable* – a symbol that, in the formal part of (2), may be replaced by the designation of a specific gas sample, such as the oxygen γ_1 in a specific container at a specific instant. In the entire statement (2), γ may, without any change of the meaning, be replaced by any other letter, as in

(2′) $\quad p\delta \cdot v\delta = \mathrm{r}\, t\delta$ for any instantaneous gas sample δ.

If $\mathrm{r}t$ denotes the v.q. consisting of all pairs $(\gamma, \mathrm{r}\, t\gamma)$ for any sample γ, then (2) expresses the equality of the variable quantities $p \cdot v$ and $\mathrm{r}t$. Physicists express Boyle's Law in the formula

(2″) $$p \cdot v = \mathrm{r}t,$$

that is, without any reference to sample variables.

For samples of the temperature 100, it follows from (2) and (1) that

(3) $\quad p\gamma \cdot v\gamma \ = \ 100\ \mathrm{r}$ for any γ belonging to G_{100};

in words: $p(\text{on } G_{100}) \cdot v(\text{on } G_{100})$, the product of the restrictions of p and v to G_{100}, equals the constant v.q. whose domain is G_{100} and whose range consists of the single number 100 r. Consequently,

(4)
$$v(\text{on } G_{100}) = \frac{100 \text{ r}}{p(\text{on } G_{100})}.$$

Of course, (2) implies much more, namely, $p\gamma' \cdot v\gamma' \neq 100$ r for any sample γ' not in G_{100}; that is to say, outside of G_{100}, the v.q.'s v and 100 r/ p are what on page 174 was called disjoint. In a formula,

(5)
$$v = \frac{100 \text{ r}}{p} \text{ just on } \{ t = 100 \}$$

expressing that $v\gamma = 100$ r/ $p\gamma$ if and only if γ belongs to $G_{100} = \{ t = 100 \}$; or

(5′)
$$\{ t = 100 \} = \left\{ v = \frac{100 \text{ r}}{p} \right\} = G_{100}.$$

An operation to which, of all c.c.q.'s, only functions lend themselves (just as only functions lend themselves to derivation and integration) is *substitution.* Into the function *log* one can substitute functions as well as scientific v.q.'s, such as gas pressure in atmospheres. Let p be the class of all pairs $(\gamma, p\gamma)$ for any instantaneous gas sample γ, where $p\gamma$ denotes the pressure in atmospheres of γ. Then *log* p may be defined as the class of all pairs $(\gamma, \textit{log } p\gamma)$ for any gas sample γ. Similarly, if v is the gas volume in a certain unit, $j^{-1}v$ or $1/v$ is the class of all pairs $(\gamma, j^{-1}v\gamma) = (\gamma, 1/v\gamma)$ for any gas sample γ. To obtain values of *log* p and $j^{-1}v$ for γ, one has to evaluate the function *log* for $p\gamma$ and the function j^{-1} for $v\gamma$.

More generally, for any c.c.q. u and any function f,

$f(u)$ – briefly, fu – is the class of all pairs $(a, f(ua))$ for any a in Dom u for which $u\,a$ belongs to Dom f.

Thus fu is a c.c.q. whose domain is a subclass of Dom u.

On the other hand, try to define pv or p *log* or, in fact, to substitute any c.c.q. into p. The result will be the empty class of pairs (thus insignificant). Indeed, values of p are defined only for gas samples, that is to say, p can be evaluated only for something like the oxygen γ_1 in a specific container at a specific instant – not for values of v or of *log*, which are numbers. There is no pressure of *log* 2 (= .69...), wherefore p cannot be evaluated for *log* 2.

There is a logarithm of the pressure; but there is no pressure of the logarithm.

EXERCISES

10. Into what function must the variable quantity $v(\text{on } G_{100})$ be substituted to yield $p(\text{on } G_{100})$? What are the effects on the number r in (4) if *(a)* the unit of pressure is doubled; *(b)* the unit of volume is doubled; *(c)* both units are doubled; *(d)* one unit is doubled, the other halved? If for certain units, r is found to be 30, by what changes of those units could one achieve that r = .01 in (4) and hence $v(\text{on } G_{100}) = 1/p(\text{on } G_{100})$?

11. Let t be the time in years elapsed since the observation of a piece of 5 grams of radium. Define as a class of pairs the v.q. obtained by substituting t into the function $5\ exp(-4.33 \cdot 10^{-4} j)$. When (that is, for what value of t) is the value of that v.q. equal to 2.5?

5. FUNCTIONAL CONNECTIONS BETWEEN CONSISTENT CLASSES OF QUANTITIES

The v.q.'s p and v in (4) have the same domain, G_{100}. Related v.q.'s with unequal domains are

1. the *observed pressure* p^* in atmospheres; that is, the class of all pairs $(\mu, p^* \mu)$ for any act μ of reading a manometer calibrated in atmospheres, where $p^* \mu$ denotes the numerical result of the act μ, i.e., the number read on the scale under observation;

2. the *observed volume* v^* in cu.ft., the class of all pairs $(\beta, v^* \beta)$ for any volumetric act β.

Since the elements of Dom v^* are volumetric acts, whereas the domain of any function of p^* consists of manometric acts (see p.176), v^* cannot be strictly equal to any function of p^*. To establish a connection between v^* and p^* (say, for gas samples at the temperature 100), one first must somehow pair the domains of the two v.q.'s. The traditional pairing follows what we will call *Galileo's Principle:* one pairs *simultaneous acts of observation directed to the same object.* Let $\Gamma(p^*, v^*)$, briefly Γ, denote the class of all Galileo pairs with respect to p^* and v^*, that is, the class of all pairs of simultaneous acts of manometric and volumetric observation of the same gas sample. Then the analogue of (4) reads

If (μ, β) belongs to Γ, then $v^* \beta = 100\ \mathrm{r}/p^* \mu$.

Three other ways of expressing this law follow.

If (μ, β) belongs to Γ, then the pair of numerical results $(p^* \mu, v^* \beta)$ belongs to the function $100\ \mathrm{r}\ j^{-1}$.

If $\Gamma \mu$ denotes the volumetric act that is simultaneous with μ and directed to the same sample, then $v^* \Gamma \mu = \dfrac{100\ \mathrm{r}}{p^* \mu}$, for any act μ in Dom p^*.

Without any reference to an act variable, the last statement may be abbreviated to

$$v^* \Gamma = \frac{100\ \mathrm{r}}{p^*}; \tag{6}$$

in words, v^* is equal to $\dfrac{100\ \mathrm{r}}{p^*}$, relative to the Galileo pairing of Dom v^* with Dom p^*.

It was in dealing with his kinematic v.q.'s s and t (see p. 171) that Galileo introduced his principle of pairing. He considered the class $\Gamma(\tau, s)$ (*ad hoc,* again Γ), of all pairs (τ, σ) such that, if τ is an act of reading a timer, set in motion when an object is released, σ is the simultaneous act of reading the scale mark opposite that same object. If the timers are calibrated in seconds, and the scales in feet, then Galileo's immortal discovery may be formulated in any of the following forms:

If (τ, σ) belongs to Γ, then $s\sigma = 16(t\tau)^2$; and $(t\tau, s\sigma)$ belongs to the function $16\, j^2$;

$s\Gamma\tau = 16(t\tau)^2$ for any τ in Dom t, where $(\tau, \Gamma\tau)$ belongs to Γ;

(7) $$s\Gamma = 16\, t^2 .$$

Galileo's pairing has become second nature to physicists. It is tacitly understood in laws such as Galileo's or Boyle's, which, consequently, in physics are written without any reference (not only to act variables but even) to the specific pairing of the domains of the v.q.'s involved, that is, in the forms

(7*) $$s = 16\, t^2 \text{ and } v^* = \frac{100\ r}{p^*}.$$

Formulas (4) and (5) connect v.q.'s with the same domain. According to (3), they are based on the pairing of each gas sample with itself, in other words, on the identity pairing or the class I of all pairs (γ, γ). With reference to I, one would write (5) in the form

(5′) $$vI = \frac{100\ r}{p} \text{ just on } \{\, t = 100 \,\}.$$

Similarly, mathematicians connecting functions compare their values for the same number. For instance, if they say that j^6 is the square of j^3, they mean that $j^6 x = j^2 (j^3 x)$ for any x – of course, the same x on both sides; that is to say, mathematicians take the identity pairing for granted or, in other words, a class of pairs (x, x). But such a class is the identity function j or a restriction of j. Indeed, the connection of j^6 with j^3 can also be expressed as follows:

If (x, y) belongs to j, then $j^6 y = j^2 (j^3 x)$; and $(j^3 x, j^6 y)$ belongs to j^2.

$j^6 (j\, x) = j^2 (j^3 x)$ for any x, briefly, $j^6 j = j^2 j^3$, which means simply $j^6 = j^2 j^3$.

If the value of j^6 for, say, $j^2 x = x^2$ were compared with the value of j^3 for x, then j^6 would be the 4th power of j^3:

$j^6 (j^2 x) = j^4 (j^3 x)$ for any x, which simply means $j^6 j^2 = j^4 j^3$.

Or consider the functions *sin* and -2 *sin*. Relative to the identity pairing, -2 *sin* is obtained by substituting *sin* into the function $-2j$. If the value of -2 *sin* at $x + \pi$ were compared with the value of *sin* at x, then -2 *sin* would be obtained by substituting *sin* into $2j$, since

$$-2 \sin(x + \pi) = 2 \sin x \text{ for any x;}$$

in other words, relative to the pairing $j + \pi$, the function -2 *sin* is twice (and not minus twice) the function *sin*.

In general, *the function connecting two (non-constant) consistent classes of quantities depends upon the underlying pairing of the domains.* As an example in the realm of scientific v.q.'s, consider two pendulums of the same length that are simultaneously released. Let a be the class of all pairs $(\alpha, a\alpha)$ for any act α of observing the angle between the first pendulum and a vertical line, where $a\alpha$ is the numerical result of α. Let b be the v.q. defined similarly with regard to the second pendulum. Let t be the time (in some unit) elapsed since the release of the two pendulums, i.e., the class of all pairs $(\tau, t\tau)$ for any act τ of reading a timer. Suppose that, relative to the pairing of simultaneous acts of angle and time observations

$$a = \sin t \quad \text{and} \quad b = -2 \sin t.$$

What function connects b with a? The answer depends entirely upon the pairing of the domains of a and b. If simultaneous acts are paired, then $b = -2a$. But $b = 2a$ if acts are paired that are π units of time apart.

EXERCISES

12. The mass of what at some instant was 5 grams of radium is connected to the time in years elapsed since that instant by the function in Exercise 11. Why is the value asked for in that exercise called the half-life of 5 g of radium? When are 5 g reduced to 1.25 g? What follows from this reasoning concerning the function connecting a mass of originally 2.5 g of radium with the time in years since then?

13. Suppose a car has been driven 15,400 miles. Find an example of a pair of numbers, a and b, such that on a long straight road (without intersections or traffic) the car could be driven from 9 A.M. on for 2½ hours according to the following condition: The distance in miles, s, driven altogether by the car is connected with the time by the function $aj + b$; that is to say, $s = at + b$, where t denotes *(a)* the historical time in hours; *(b)* the time in hours elapsed since the beginning of the trip.

14. By what function is the distance in feet traveled by a falling object related to the historical time in seconds if the object is dropped 9 seconds after midnight from a height of 256 ft.?

15. Give further examples of *(a)* two functions, *(b)* two scientific v.q.'s, *(c)* two v.q.'s from life, which are connected by different functions according to different pairings of their domains. Give an example of two c.c.q.'s such that the second is paired with the first by one and the same function no matter how the domains are paired.

6. IS w A FUNCTION OF u?

The following statements, which are valid for any two consistent classes of quantities, will be expressed in terms of fluent variables.

If u and w are any two c.c.q.'s, is $w = 1/u$? From what was said in the last section it is clear that this question is answerable only relative to a given pairing of Dom w with Dom u.

For any two domains that happen to be identical, one common (though by no means cogent) principle of pairing is the identity pairing consisting of all pairs (α, α) for any α in Dom u. If, however, Dom u and Dom w are disjoint, then not only is no principle of pairing "traditional" or "natural," but no common principle that is applicable to every pair of domains can even be *formulated.*

Let Π be a class of pairs associating with any element α of Dom u an element $\Pi\alpha$ of Dom w. Then w is said to be the reciprocal of u relative to Π, or to be equal to $j^{-1}u$ relative to Π, if and only if

(8) $$w\,\Pi\,\alpha = j^{-1}\,u\,\alpha \text{ for any } \alpha \text{ in Dom } u \text{ – briefly, } w\,\Pi = \frac{1}{u}$$

The fluent variables u and w and the pairing variable Π in the preceding definition may be replaced by specific c.c.q.'s and pairings; for instance, by p^*, v^*, and Γ. In this case (8) implies, in agreement with (6), that $v^*\Gamma = \frac{1}{p^*}$ if and only if the units are so chosen that $r = .01$. If u, w, Π are replaced by *cos, sec,* and j, then (8) yields *sec* $j = 1/$*cos* (that is, *sec* $= 1/$*cos*) in agreement with the fact that *sec* is the reciprocal of *cos* (relative to the tacitly understood identity pairing).

After this preparation, one can readily answer the following question, which, besides u, w, and Π, involves a function variable. Is w a function of u relative to Π?

The answer is affirmative if and only if there is a function f such that

$$w\Pi\alpha = f(u\,\alpha) \text{ for any } \alpha \text{ in Dom } u \text{ – briefly, } w\,\Pi = fu.$$

This condition can be restated as follows: w is a function of u relative to Π if and only if there is a function f such that, for any pair (α, β) in Π, the corresponding pair of numerical values $(u\,\alpha, w\beta)$ belongs to f; in other words, such that the class of all pairs $(u\,\alpha, w\beta)$, for any pair (α, β) in Π, is a subclass of f. Obviously this condition holds if and only if the class of all pairs $(u\,\alpha, w\beta)$ is itself a function (clearly, a restriction of f). Conversely, if that class is a function, say g, then one readily shows that $w\,\Pi = g(u)$ – briefly, gu.

Thus w is a function of u relative to Π *if and only if the class of all pairs* $(u\,\alpha, w\beta)$ *for any* (α, β) *in* Π, *is a function.*

But that class of pairs of numbers is a function if and only if it does not contain two pairs of numbers with equal first, and unequal second, members. This in turn, is equivalent to the following condition: If (α, β)

and (α', β') are any two pairs in Π such that $u\alpha = u\alpha'$, then $w\beta = w\beta'$. In words: *w is a function of u relative to* Π *if and only if to any two elements of* Dom *u for which u assumes equal values, there correspond by* Π *two elements in* Dom *w for which w assumes equal values.*

Example 1. Is the gas volume *v* a function of the gas pressure *p*? (Both domains consist of the instantaneous gas samples, and the identity pairing is tacitly understood.) The answer is negative. Indeed, there are gas samples such that

$p\gamma = p\gamma'$ and yet $v\gamma \neq v\gamma'$ —samples with different temperatures.

Confining the question to gas samples of one and the same temperature, however, Boyle not only discovered that the answer is affirmative but succeeded in identifying the function connecting, say, v (on G_{100}) and p (on G_{100}), namely, with $100\, \mathrm{r}\, j^{-1}$.

Example 2. The weight in pounds, *w*, in the population of New York clearly is not a function of the height *h* in inches with the same domain. If, however, for any number h, one denotes by $\overline{w}_h$ the average weight in pounds of all residents of New York having a height of h inches (to the nearest half inch), then the class of all pairs $(h, \overline{w}_h)$ clearly is a function. But it is unlikely that one will succeed in identifying the connecting function with a specific elementary function.

Statisticians in such cases select from a somehow chosen class of functions (e.g., the linear or the cubic or the exponential functions) one that, according to certain principles, is closer to the class of all pairs $(h, \overline{w}_h)$ than any other member of that class. It is called the *regression function* of *w* on *h*.

Example 3. Consider a large horizontal sheet of coarse sand paper and a set of marbles, all of the same size and material. One at a time, these marbles may be pushed and then released (each marble many times). Choose units and let $v\,\mu$ and $s\,\mu$ denote the initial velocity and the distance traveled by the marble before coming to a standstill in any such experiment μ. Experience shows that, relative to the identity pairing, *s* is a function of *v* and that the connecting function is *continuous* in the following sense: slight changes in the value of *v* result in slight changes of *s*. But it may be very difficult or impossible to identify the connecting function with a specific elementary function.

Beginners dislike statements such as that in Example 3. They are disappointed if functions connecting scientific v.q.'s remain unidentified with elementary functions. They overlook the importance of the mere statement that one scientific v.q. pertaining to a large class of phenomena is connected with another such v.q. by a function that is continuous in the sense of Example 3. In view of the generality of his kinematic v.q.'s, Galileo might have founded modern science even if he had only discovered that *s* is a continuous function of *t* without identifying that function with $16\,j^2$. Also the statement in Example 3 expresses a general, if not very important, law of nature. This fact will be put in relief by contrasting *s* and *v* therein with physical v.q.'s connected by a non-continuous function (Example 4) or not pertaining to large classes of phenomena (Example 5) or not connected by any function (Example 6).

Example 4. The following experiment has all the essential features of tossing a coin – head and tail being replaced by 0 and 1. Consider a vertical half-plane bounded at the bottom by a horizontal line, L. Let ρ be a homogeneous slender rod, 2 ft. long, marked "0" on one side, and "1" on the other. Lift the center of ρ to the altitude a ft. ($a > 1$) and, in the half-plane, bring ρ into a vertical position with the mark 0 on the right side. Then, within the plane, spin ρ about its center counterclockwise with an angular velocity of half a revolution per second, and drop ρ.

t seconds after release, the center of ρ has traveled $16\,t^2$ ft. The initial angular velocity of ρ remains constant until ρ touches L. For the sake of simplicity, make the following assumptions:

(1) If ρ, when first touching L, is in a vertical position (which happens 1, 2, 3, ... seconds after release), then ρ will remain in this unstable equilibrium and the outcome of the tossing will be said to be V(ertical).

(2) If ρ first touches L after having traveled between 0 and 1 or between 2 and 3 or between 4 and 5 seconds, etc., then it will so fall that 0 turns up. If the first contact with L occurs either between 1 and 2 or between 3 and 4 or between 5 and 6 seconds, etc., then 1 turns up.

It is easy to see that, t seconds after release, the altitude of the lowest point of ρ above L is $a - 16\,t^2 - |\cos(\pi t)|$. Hence, if the first contact of ρ with L occurs t_1 seconds after release, then t_1 is the smallest number such that

$$a - 16\,t^2 - |\cos(\pi t)| = 0.$$

Now determine the integer k for which

$$\frac{\sqrt{a-1}}{4} - 1 \leq k < \frac{\sqrt{a-1}}{4}, \text{ that is, } 16\,k^2 + 1 < a \leq 16\,(k+1)^2 + 1.$$

Clearly, if $t \leq k$, then $a - 16\,t^2 - |\cos(\pi t)| \geq a - 16\,k^2 - 1$, which is > 0. Hence the first contact between ρ and L does not occur before k seconds after release.

On the other hand, if $t = k + 1$, then $a - 16\,(k+1)^2 - |\cos(\pi(k+1))|$ equals $a - 16\,(k+1)^2 - 1$, which is ≤ 0. Thus the first contact occurs not later than $k + 1$ seconds after release.

Since, therefore, the first contact occurs between k and k + 1 seconds after release, 0 or 1 turns up according to whether k is even or odd.

Let the v.q. r be the result of tossing ρ, that is, the class of all pairs (ϵ, $r\,\epsilon$) for any experiment ϵ, where $r\,\epsilon$ is 0 or 1 according to whether "0" or "1" turns up. If the v.q. a is the initial altitude of the center, i.e., the class of all pairs (ϵ, $a\,\epsilon$) for any ϵ, then, according to what has been said, relative to the identity pairing, r is connected with a by the function whose range is $\{0, 1\}$ and which assumes the values

0: between 1 and 17; 65 and 145: etc.

1: between 17 and 65; 145 and 257; etc.

The function is not defined for 17, 65, 145, ... (corresponding to the outcome V). But clearly, as near to these numbers as one pleases, there are both altitudes for which the outcome is "0" and altitudes for which it is "1." In fact, if one cannot measure altitudes accurately (which indeed one cannot) and he lifts the center to what appears to be 17 or 65 feet, then *the outcome is unpredictable.* It is hoped that the study of this "chance experiment" will make the reader better appreciate the law in Example 3 in spite of its limitations.

Example 5. Every scientific v.q. assuming at no instant more than one value clearly is a function of the historical time: the mileage and the speed of a specific car; the temperature of a room and that of a patient; the price of a G.E. share on the New York Stock Exchange; etc. It would, however, be next to impossible to identify any of the connecting functions (except for deliberately regulated v.q.'s as in Exercise 13). Nor would such an identification, based on the past record, be of interest even if it were possible. Predictions of stock prices based on such a function would be disastrous.

The futility of such functions would be due to the lack of generality of the v.q.'s connected with the historical time. Galileo connected s with t, *in the sense of time elapsed* rather than historical time, *for any falling object.* Analogous would be connections of: the distance traveled by a particular man's car with the time elapsed after leaving his home, *for any trip to his office;* the temperature of a patient with the time elapsed since the first symptoms, *for any case of undulant fever;* and the like. But these v.q.'s are not functions of the time elapsed; only averages (in the sense of Example 2) are. Thus in some such cases one creates v.q.'s of some generality that can be connected with the time elapsed by functions, if not by identifiable functions. But even identifiable functions connecting non-general v.q.'s are inferior in significance to the function in Example 3.

Example 6. t_P, the temperature of a patient is not in general a function of t_R, the temperature of his room (pairing by simultaneity being tacitly understood). In a special case, the *actual* class of pairs of simultaneous values may consist of only

$$(68, 99), \quad (69, 97.6), \quad (70, 99).$$

But the function character of the connection would be regarded as a coincidence, since *potentially* the class includes the fourth pair (70, 102).

Simultaneously with the same value of t_R, one may well observe values of t_P that are *appreciably* different – a remark that is important since, if the differences lay below the limits of accuracy, one might well decide to disregard them and to *create* a function. It should further be noted that averages in the sense of Example 2 would be medically insignificant. Nor can it be claimed that the situation is comparable to that of the gas temperature in relation to gas volume. The former is a function of the latter for gases of one and the same pressure. But t_P is not a function of t_R even while the patient's blood pressure has one and the same value. Nor is there any known group of v.q.'s such that, while their values remain the same, t_P would be a function of t_R.

Interestingly enough, just the fact that t_P is *not* a function of t_R or of any other known v.q. accounts for the medical importance of a patient's temperature as a symptom in its own right.

So-called many-valued functions (p. 70), which serve no useful purpose in pure calculus, are, in full generality utterly useless in applications. If they were admitted, then every c.c.q. would be a function of every other c.c.q.

Note that, as the result of incomplete studies of two scientific variable quantities, u and w, (e.g., on the basis of past records), one may apodictically claim that, relative to a given pairing, w is *not* a function of u, but never that it *is*. A definite positive decision presupposes a complete survey of all pairs of corresponding values $(u\,\alpha,\ w\,\Pi\,\alpha)$, which is possible if either that class of pairs is finite (as are the classes in Exercise 15, p. 72) or constructed according to a mathematical law, as the class of all (x, x^3) for any x.

EXERCISES

16. Is the time between midnight and noon observed on a clock C_1 a function of the time on a clock C_2, regardless of whether either clock keeps the "correct" time? What if C_1 but not C_2, or C_2 but not C_1, stops? What if C_1 and C_2 stop?

17. Characterize those physical v.q.'s of which the historical time is a function. Give examples of functions that are, and that are not, of this type. Think of Exercise 11.

18. Give examples of pairs of physical v.q.'s with the same domain and of functions such that, relative to the identity pairing

(a) either; *(b)* neither; *(c)* exactly one

is a function of the other. Then do the same for v.q.'s in the field of economics.

19. If u and w are measurable v.q.'s with the units α and β such that relative to some pairing, $w = fu$, how are u' and w' (the same objects measured in units α' and β') connected, assuming that $\alpha/\alpha' = m$ and $\beta/\beta' = n$? In the general statement, replace f by *(a)* j^k; *(b)* exp; *(c)* sin.

20. Compare in detail the statements that any physical v.q. is a function of the historical time and that a general physical v.q. pertaining to a class of processes is a function of some time elapsed. What about retrieving missed opportunities in the two cases?

7. VARIABLE QUANTITIES IN GEOMETRY AND KINEMATICS

In this section, only the Euclidean plane will be studied. But there is no difficulty in extending the results to the Euclidean space.

a. Three Types of Geometry. A careful distinction will be maintained between three theories each of which is known under the name of plane geometry, even though they are based on totally different foundations:

(1) *Physical Geometry,* dealing with the shapes, positions, etc. of objects in planes such as this page of paper, a blackboard, or a large (but, on account of the curvature of the earth, not too large) prairie. On a blackboard, dots of chalk are called points; streaks drawn by means of a straightedge, lines; curves drawn by means of a compass, circles; etc.

(2) *Postulational Geometry,* dealing with three classes of undefined elements. The elements of the first class are called *points;* those of second, *lines;* while those of the third class are (unordered) pairs each consisting of an element of the first, and an element of the second, class. A point and a line are called *incident* (and either one is said to be *on* the other) if and only if they are members of a pair belonging to

the third class. A few simple assumptions about points, lines, and incidence are postulated; e.g., that any two distinct points are on exactly one line (called the line *joining* the two points); that for any line λ, on any point σ that is not on λ, there is exactly one line not intersecting λ (called the line on σ *parallel* to λ). Postulational geometry is the system of statements that can be derived from the postulates by purely logical reasoning.

(3) *Pure Analytic Geometry,* dealing with ordered pairs of numbers, called points; certain classes of such pairs, called lines; certain classes, called circles; etc. Thus pure analytic geometry, in contrast to postulational geometry, starts with definitions of points and lines; but, in contrast to physical geometry (where points are defined as chalk dots or the like), with definitions in purely mathematical terms (points are pairs of numbers, etc.)

Notwithstanding their different foundations, these three theories exhibit a far reaching parallelism because the assumptions in postulational geometry and the definitions in pure analytics are so formulated as to reflect observable facts in physical planes. This parallelism justifies the reference to each of these theories as "plane geometry." Yet only a careful distinction between the various foundations (which, unfortunately, is not maintained in most current textbooks) makes it possible for a reader to understand clearly what, at any stage, the author is talking about.

b. Variable Quantities in Physical Planes. In physical geometry, one measures straight segments like rods (p. 170) and one defines a variable quantity l, the *length* in a chosen linear unit. Moreover, by slightly more complicated procedures, one measures squares, rectangles, and some other figures, and defines a variable a, the *area* in a chosen square unit. If the latter unit is a square with sides of length 1, then

$$a\ \rho = l^2\ \rho \text{ for any square } \rho ,$$

where $l\ \rho$ denotes the length of side of ρ .

After geographers had described the earth by longitude and latitude, Descartes and Fermat around 1635 introduced a similar method into the geometry of physical planes. They selected what is now called a *Cartesian frame of reference* consisting essentially of three points not lying on one and the same line: a point o, called the *origin;* a point ϵ_1, called the *1st unit point;* and a point ϵ_2, called the *2nd unit point.* (Usually, one selects three points that form an isosceles triangle with a right angle at o.) The line joining o to ϵ_1 (to ϵ_2) is called the 1st (the 2nd) *axis of reference.*

Such a frame is not inherent in a physical plane. It is a man-made device for a description of the plane. To achieve a simple description, one will adapt the frame to the object under consideration. In modern cities with square blocks, a busy downtown intersection is chosen as the origin, and the intersections one block East and one block North of the origin serve as the unit points. On a blackboard, the frame of reference consists of three chalk dots, usually so selected that the axes of reference are parallel to the sides on the blackboard. – An important forerunner of Descartes and Fermat in the 14th century was the great French schoolman Oresmes.

After selecting a frame, one pairs with any point σ in the plane the distance from the 2nd axis to σ and calls this number the *abscissa of* σ. It will be here denoted by $x\,\sigma$. In particular,

(9) $$x\,o = 0, \quad x\,\epsilon_1 = 1, \quad x\,\epsilon_2 = 0.$$

Thus a variable quantity x is defined, the *abscissa* (relative to the chosen frame). Its domain is the class of all points in the physical plane, say a blackboard. Its range is the class of all numbers that are the abscissas of points – an interval depending on the frame and the size of the blackboard.

In a similar way, one defines a v.q., called the *ordinate* and denoted by y, as the class of all pairs $(\sigma, y\,\sigma)$ for any point σ in the plane. In particular,

(9′) $$y\,o = 0, \quad y\,\epsilon_1 = 0, \quad y\,\epsilon_2 = 1.$$

It can be shown that the class of all points σ such that

$$2\,x\,\sigma + 3\,y\,\sigma = 5$$

is a fragment of a straight line (e.g., on a blackboard, a chalk streak that can be drawn by means of a straightedge). It may be denoted by $\lambda_{2,3,5}$. The points on that line include the point σ' for which $x\,\sigma' = 1$ and $y\,\sigma' = 1$; and σ'' for which $x\,\sigma'' = 4$ and $y\,\sigma'' = -1$. Thus $\lambda_{2,3,5}$ is the class $\{\sigma \mid 2\,x\,\sigma + 3\,y\,\sigma = 5\}$.

Here, σ is what might be called a *point variable*, that is, a symbol that may be replaced by the designation of any specific point in the plane; e.g., by σ' and σ'', which satisfy the condition and lie on $\lambda_{2,3,5}$, or by o, ϵ_1, and ϵ_2, which do not satisfy the condition and consequently do not lie on $\lambda_{2,3,5}$. Moreover, σ may, without any change of the meaning, be replaced by any other Greek letter, as in the definition of $\lambda_{2,3,5}$ as the class of all points α such that $2\,x\,\alpha + 3\,y\,\alpha = 5$. Without any reference to point variables:

$$\lambda_{2,3,5} = \{2x + 3y = 5\},$$

where 5 is the constant v.q. of the value 5 whose domain is the class of all points. Clearly, for any number $k \neq 0$,

$$\{2k\,x + 3k\,y = k\,5\} = \{2x + 3y = 5\} \text{ or } \lambda_{2,3,5} = \lambda_{2k,3k,5k}.$$

More generally, for any three numbers a, b, c (of which not both a and b are 0) and any $k \neq 0$, there is a line

$$\lambda_{a,b,c} = \{a\,x + b\,y = c\} = \{ka\,x + kb\,y = k\,c\} = \lambda_{ka,kb,kc}.$$

If the distance in the physical plane is measured in the ordinary way (e.g., in a city, along the bee line), then the class of all points σ such that $x^2\sigma + y^2\sigma = 1$ is a circle C about *o* of radius 1 – briefly,

$$C = \{x^2 + y^2 = 1\}.$$

Similarly, $\{y = x^2\}$ and $\{x = y^2\}$ are parabolas through *o* whose axes are the 2nd and the 1st axis of reference, respectively. More generally, for any function *f*, the class of all points σ such that $y\,\sigma = f(x\,\sigma)$ – briefly, $\{y = fx\}$ – is a curve intersecting no line parallel to the 2nd axis in more than one point.

c. Polar Coordinates in a Physical Plane. Another description of the plane is based on what is called a *polar frame of reference.* Also this frame consists of three points: a point *o* called *pole,* and two points, α and β, equidistant from *o* and called the *radial* and the *angular unit point,* respectively. The ray issuing from *o* and passing through α is called the *polar axis;* the circle about *o* passing through α and β, the *unit circle.* On the axis one establishes a linear scale with the marks 0 for *o*, and 1 for α; on the unit circle, a circular scale, with the marks 0 for α, and 1 for β. Usually, the circular scale is calibrated in radians (the 2πth part of the circle) or in degrees (the 360th part).

Two v.q.'s, denoted by *r* and *a* and called the *polar coordinates* relative to the frame, are defined as follows: For any point σ, draw the circle about *o* through σ and call the scale mark of its intersection with the axis the *radius vector* of σ or $r\,\sigma$. For any point $\sigma \neq o$, draw the line joining σ with *o* to its nearest intersection with the unit circle and call the scale mark of that point on the circle the *angle* of σ or $a\,\sigma$. The domain of the v.q. *r* is the entire plane, that of *a* is the plane with the exception of the pole. Clearly,

$$r\,o = 0; \quad r\,\alpha = 1, \quad a\,\alpha = 0; \quad r\,\beta = 1, \quad a\,\beta = 1.$$

Suppose that a Cartesian frame of reference is chosen in such a way that the origin coincides with the pole, ϵ_1 with α, while ϵ_2 is the point on the unit circle with the scale mark $\pi/2$ or 90 according to the calibration. Then the resulting Cartesian coordinates are connected with the polar coordinates as follows:

$x\,\sigma = r\,\sigma \cdot \cos a\,\sigma$ and $y\,\sigma = r\,\sigma \cdot \sin a\,\sigma$ for any point $\sigma \neq o$. or

(10) $$x = r \cdot \cos a \quad \text{and} \quad y = r \cdot \sin a\,.$$

d. Motions in a Physical Plane. If an observer studies a physical plane during a period of time, he can historico-geographically locate each event in that plane by a pair (τ, σ) consisting of an instant τ and a place σ. If he observes the motion M of a particle in that plane, the result is a class of such events: the changing positions of the particle. According to classical physics, at any instant τ, the particle occupies a definite position σ_τ and, for kinematical purposes, M is fully described by the class of all pairs (τ, σ_τ) for any instant τ belonging to the period of observation.

Suppose that the observer has a clock associating with any instant τ the historical time $t\,\tau$ of the instant τ in some unit; and that he has chosen a Cartesian frame whereby with any point σ an abscissa $x\,\sigma$ and an ordinate $y\,\sigma$ are associated. Then the class of all pairs of numbers $(t\,\tau, x\,\sigma_\tau)$, for any instant τ during the period of observation, is a function, which may be denoted by g; that is to say, $x\,\sigma_\tau = g(t\,\tau)$ – briefly, $x = gt$. (Substitution! If one prefers, he may write $x = g(t)$.) Similarly, $y = ht$. Just as $\lambda_{2,\,3,\,5}$ is the class of all points σ such that $2x\,\sigma + 3y\;\sigma = 5$ or $\{\, 2x + 3y = 5 \,\}$, the motion M is the class of all (τ, σ) such that

$$x\,\sigma = g(t\,\tau) \text{ and } y\,\sigma = h(t\,\tau) \text{ or}$$

(11) $$M = \{\, x = gt \text{ and } y = ht \,\}.$$

The class of all positions occupied by the moving particle is called the *trajectory* of M. If, and only if, the trajectory consists of a single point, then the particle is said to be at rest, and both g and h are constant. If and only if the trajectory is parallel to the 2nd axis, then g is constant; if and only if it is a part of the line $\lambda_{1,\,-1,\,0} = \{\, x - y = O \,\}$, then $g = h$. For instance, the trajectory from eternity to eternity of the motion $\{\, x = t$ and $y = t \,\}$, for which $g = h = j$, is the entire line $\{\, x - y = O \,\}$. The same is true for the motion $\{\, x = t^3$ and $y = t^3 \,\}$, for which $g = h = j^3$. Yet the last two motions are different, since at any instant (except when t assumes the values -1, 0, 1) the positions of the two moving particles are different. The motion $\{\, x = \cos t$ and $y = \cos t \,\}$ is a (so-called harmonic) oscillation along the segment of length $2\sqrt{2}$ on the line $\{\, x - y = O. \,\}$.

The plane in which M occurs will be called *plane of the motion.* If in a different plane, called *plane of timetables,* a Cartesian frame is chosen relative to which the abscissa is denoted by t and the ordinate by z, then the curves $\{ z = gt \}$ and $\{ z = ht \}$ are called the graphical timetables of the abscissa and the ordinate in the motion M, respectively. (In this setup, the unit of length is also used to represent the unit of time.)

Suppose the physical plane is perpendicular to the surface of the earth and the 2nd axis is vertical. If g denotes the gravitational acceleration and a and b are two numbers, then

$$\left\{ x = a\,t \text{ and } y = b\,t - \frac{g}{2}\,t^2 \right\} \tag{12}$$

is the motion of a particle ejected from the origin when t assumes the value 0 and under the influence of gravity. The trajectory is a part of the parabola $\left\{ y = \frac{b}{a}\,x - \frac{g}{2a^2}\,x^2 \right\}$, obtained by substituting $t = \frac{1}{a}\,x$ from the first relation (12) into the second.

The trajectory of the motion $\{ x = \cos t \text{ and } y = \sin 2t \}$ (where $\sin 2t$ stands for $\sin (2t)$) is found by establishing a relation between x and y as follows: $x^2 = \cos^2 t$ and $y^2 = 4 \sin^2 t \cdot \cos^2 t$ imply $y^2 = 4\,(1 - x^2) \cdot x^2$. Of course, only a part of the curve $\{ y^2 = 4\,(1 - x^2) \cdot x^2 \}$ may be traversed during a certain period. E.g., if the period extends between the values 0 and 1 of t, then only an arc between the points (1, 0) and $(\cos 1, \sin 2)$ is traversed. On the other hand, the trajectory of $\{ x = \cos 4t \text{ and } y = \sin 4t \}$, as the value of t increases from 0 to 3π, is the full circle $\{ x^2 + y^2 = 1 \}$, each point being traversed 6 times.

e. Descartes' Method in a Postulational Plane. Also in a postulational plane Descartes' method can be introduced by selecting, as a frame of reference, three "non-collinear" elements from the first of the three basic classes, i.e., "points" o, ϵ_1, and ϵ_2 for which there is no "line" (i.e., element of the second basic class) "on" which the three points would be (in the sense of the third basic class). Two v.q.'s, called the abscissa x and the ordinate y relative to that frame (each having the class of all points as its domain) are introduced as follows: First define their values for o, ϵ_1, and ϵ_2 by (9) and (9'). Then

call	the line	and, for any point σ of it, set
λ_{010}	joining o and ϵ_1	$y\sigma = 0$
λ_{100}	joining o and ϵ_2	$x\sigma = 0$
λ_{101}	through ϵ_1 and parallel to λ_{100}	$x\sigma = 1$
λ_{011}	through ϵ_2 and parallel to λ_{010}	$y\sigma = 1$

E.g., for the point σ' of intersection of λ_{101} and λ_{011}: $x\sigma' = 1$, $y\sigma' = 1$.

Then call

$\lambda_{1,1,1}$ the line joining ϵ_1 and ϵ_2;
$\lambda_{1,1,2}$ the line through σ' and parallel to $\lambda_{1,1,1}$;
$\lambda_{1,-1,0}$ the line joining o and σ'.

For the points

σ_2 where $\lambda_{1,1,2}$ and $\lambda_{0,1,0}$ intersect, set $x\sigma_2 = 2$ and $y\sigma_2 = 0$;
σ^* where $\lambda_{1,1,1}$ and $\lambda_{1,-1,0}$ intersect, set $x\sigma^* = \frac{1}{2}$ and $y\sigma^* = \frac{1}{2}$.

Proceeding in this way, gradually define the values of x and y for many points. Moreover, it is easy to show that

$$\lambda_{1,1,2} = \{x + y = 2\} \quad \lambda_{1,-1,0} = \{x - y = O\}; \text{ etc.}$$

Since the postulates about the undefined points and lines reflect also facts about printed points and lines on this sheet of paper, the reader can follow the preceding constructions (which in postulational geometry are entirely based on the assumptions about undefined elements) in the Figure 33 on the following page.

f. Variable Quantities in Pure Analytic Geometry. This topic will be studied in Chapter XI. But the *numerical variables* in terms of which general statements are formulated in pure analytic geometry may well be considered at this point.

The pure analytic plane is the class of all pairs of numbers. There are, in particular, points (0, 0), (1, 0), and (0, 1); but it would not make much sense to call them a frame of reference, since nothing is referred to them. Lines are defined as certain classes of pairs of numbers; e.g.,

(13) the class of all pairs (x, y) such that $2x + 3y = 5$

will be called the line $L_{2,3,5}$. In (13), x and y are numerical variables

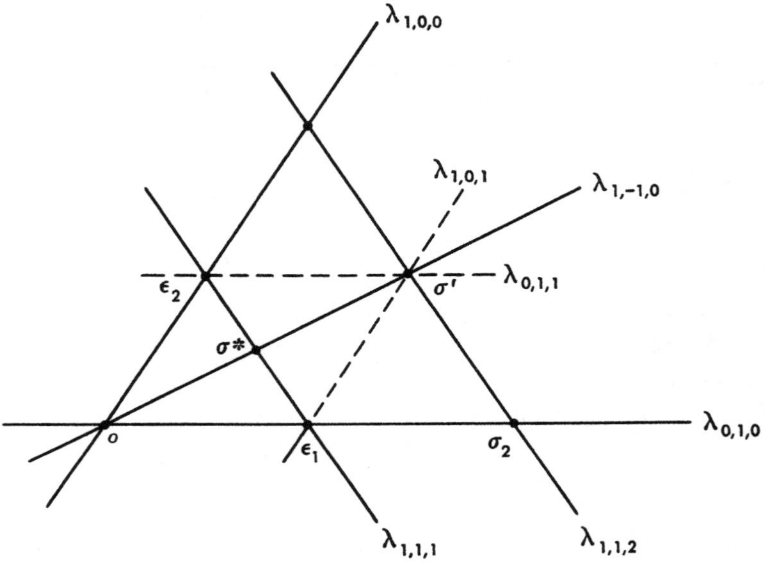

Fig. 33

used conjunctively. If they are replaced by 1 and 1 or by 4 and −1, the condition $2x + 3y = 5$ is satisfied, wherefore the points (1, 1) and (4, −1) are said to be on $L_{2,3,5}$. If x and y are replaced by 1 and 0 or by 0 and 0, the condition is not satisfied, wherefore the points (1, 0) and (0, 0) are said not to be on $L_{2,3,5}$. Being numerical variables, x and y may moreover, without any change of the meaning, be replaced by any two non-identical letters such as a and b, or b and x, or y and x; that is to say, the following definitions of $L_{2,3,5}$ are equivalent with (13):

(13′)
the class of all (a, b) such that $2a + 3b = 5$;
the class of all (b, x) such that $2b + 3x = 5$;
the class of all (y, x) such that $2y + 3x = 5$.

There is a far-reaching parallelism between physical, postulational, and pure analytic geometry (see p. 186). But there is, of course, no parallelism whatever between numerical variables and variable quantities. This is most strikingly illustrated by the fact that (13) and (13′) are the same line, whereas in the physical and postulational plane $\{2x + 3y = 5\}$ and $\{2y + 3x = 5\}$ are non-identical lines. *Numerical variables are interchangeable, variable quantities are not.*

In a variant of pure analytic geometry (which is analogous to the algebra of polynomial forms), straight lines are defined as triples of numbers rather than as classes of pairs of numbers. For instance, the line (13) = (13′) would be defined as the triple [2, 3, 5]. Clearly, [2k, 3k, 5k] is the same line for any $k \neq 0$. Any ordered triple of numbers [a, b, c] in which not both a and b are 0 is a line; and [a′, b′, c′] is defined to be equal to the former if and only if there is a number k such that $a = ka'$, $b = kb'$, $c = kc'$. The point (x, y) is said to be on the line [a, b, c] if and only if $ax + by = c$.

The basic role of arithmetical definitions in pure analytic geometry was stressed by E. Study in the first quarter of this century. The relations between mathematical and physical geometry were clarified by Helmholtz, Poincaré, and Einstein.

EXERCISES

21. Find the trajectory of *(a)* $\{x = a\,t + b$ and $y = c\,t + d\}$ for any four numbers a, b, c, d; *(b)* $\{x = t$ and $y = ht\}$; *(c)* $\{x = gt$ and $y = t\}$; *(d)* $\{x = t^2$ and $y = t^2\}$.

22. For the motion (12) determine *(a)* the highest point and the time when it is traversed; *(b)* the instant and position at which the particle hits the line $\{y = c\}$.

23. For any three numbers a, b, c, determine the trajectory of $\{x = a \cos c\,t$ and $y = b \sin c\,t\}$ and show that it is independent of c. Then determine the smallest figure enclosing all preceding trajectories, for any a and b. Finally discuss the motion $\{x = \cos t$ and $y = \sin (t + c\,t)\}$.

24. If in the physical plane a polar frame is related to the Cartesian frame by (10), determine the motions of the preceding examples relative to the polar frame, that is, in the form $\{r = gt$ and $a = ht\}$.

25. In a polar plane, draw the curves $\{r = 3\}$, $\{a = 1\}$, $\{r = a\}$, $\{r = 1/a\}$, $\{r = \log a\}$, $\{r = \cos a\}$. According to the definition of polar coordinates in the text, the range of *r* is the class of all numbers ≥ 0. Hence $\{r = a\}$ and $\{r = \log a\}$ can be drawn only for points where *a* assumes values ≥ 0 and ≥ 1, respectively, and $\{a = 1\}$ is a ray.

26. Since on a circular scale (in radians) ...

$$\ldots a - 2\pi = a = a + 2\pi = a + 4\pi = \ldots,$$

the angle $a\,\sigma$ (unless it is confined to an interval such as between 0 and 2π) has infinitely many values for any point σ. Another convention is to associate with the point σ for which $r\,\sigma = r$ and $a\,\sigma = a$ also the polar coordinates $-r$ and $a + \pi$. Then the curves $\{r = a\}$ etc. can be drawn also for points with negative angles and $\{a = 1\}$ is a line. Draw the so-called roses $\{r = \sin 3a\}$ and $\{r = \sin 4a\}$ and explain why the former has 3, and the latter 8 (and not 4), leaves.

★27. If, in a city with square blocks (and without one-way streets), the distance between two points is measured by the length of a car trip from one to the other (rather than in the bee line), draw on a city map all points at the distance 4 from the origin and determine the equation in city coordinates (see p. 187) of this "circle."

28. Draw the graphical timetables of the abscissa and the ordinate in the two motions in Exercise 23 for $a = \frac{1}{2}$, $b = 2$, $c = \pi$. For the first motion, draw also the timetable of $\sqrt{x^2 + y^2}$, which equals the radius vector r in (10).

29. Show that for any number a and any function h, the graphical timetable of y in the motion $\{x = t + a \text{ and } y = ht\}$ is congruent to the trajectory in the plane of the motion, whereas for $\{x = t^2 \text{ and } y = ht\}$ this is not the case. How are the time table of y and the trajectory related in case of the motion $\{x = 2t \text{ and } y = ht\}$?

8. REMARKS CONCERNING THE TRADITIONAL NOTATION IN APPLIED MATHEMATICS AND IN SCIENCE

(This Section may be omitted in a first study of the book.)

Remark 1. Let s denote the position of a certain harmonic oscillator. The fact that the variable quantity s is a function of the time t is frequently expressed by writing first $s = s(t)$ and then (after observation has revealed that the function is, say, the cosine) $s = s(t) = \cos t$. But, in this formula, s has two totally different meanings: on the left side of the equality sign, s stands for a physical variable quantity, on the right side, for a function. If s is the distance in feet traveled, v the velocity in ft./sec. of a falling object, and t the time in seconds after its release, then in the formula

$$s = s(t) = s(v) = 16\,t^2 = \frac{1}{64}\,v^2$$

s has three different meanings: of a variable quantity and of two (unequal) functions. Such formulas, when taken literally (and in what other way should one take formulas?), are inconsistent beyond repair and ought to be simply abandoned.

That a certain rod ρ_0 has a length in inches would not be expressed by writing first $\rho_0 = \rho_0$ in. and then (after mensuration has revealed that the number is, say, 18) $\rho_0 = \rho_0$ in. $= 18$ in. (Here, ρ_0 would have two different meanings.) Rather one would write

first: $\rho_0 = a$ in., for some number a;

then: $\rho_0 = a$ in. $= 18$ in. or $\rho_0 = a$ in. and $a = 18$.

Similarly, one would describe the oscillator by writing

first: $s = f(t)$, for some function f;
then: $s = f(t) = \cos t$ or $s = f(t)$ and $f = \cos$.

In these formulas, the physical and mathematical elements are clearly separated: the two objects (ρ_0 and an inch) and the two variable quantities (s and t), on the one hand, and the connecting number ($a = 18$) and function ($f = \cos$), on the other. Of course, such a complete separation is possible only where the connecting function possesses a symbol. The falling object might be described by

$$s = f(t) = g(v) = 16\ t^2 = \frac{1}{64} v^2;$$

but, without a symbol for the identity function, one cannot say that $f = 16\ j^2$ and $g = \frac{1}{64}\ j^2$, and therefore faces asymmetries even in the realm of applications.

Similarly, when Gulliver bought four arrows from the islanders whose deficient arithmetical vocabulary has been mentioned on p. 100, he received the following bill: "4 arrows... 6s." But when he bought two arrows, the bill read "arrow-arrow... III." Naturally, they would not write "II arrows."

Remark 2. Scientists shun the introduction of tentative function symbols, such as f and g. They would much rather write $s(t)$ and $s(v)$ --- undoubtedly because the latter symbols are *self-explanatory*. In thermodynamics, where on one page physicists often study ten or more variable quantities and functions connecting them, the introduction of tentative function symbols would presuppose an entirely arbitrary labelling (e.g., by calling the functions $f_1, f_2, \ldots$) and it would be difficult to remember what, say, f_{13} designates. Self-explanatory symbols for functions are a practical necessity.

However, one would not want to identify such self-explanatory symbols with designations of functions that refer to x, for instance, $s(t)$ with $\cos x$ or $s(v)$ with $\frac{1}{64} x^2$. On the other hand, just as one writes

$$\rho_0 \text{ in in.} = 18 \text{ or } \rho_0 / \text{in.} = 18,$$

one may write

$s(t) = \cos$ for the oscillator;
$s(t) = 16\ j^2$ and $s(v) = \frac{1}{64}\ j^2$ for a falling object.

In this way the self-explanatory symbols for functions connecting variable quantities can be used with perfect consistency.

By writing $s//t$ instead of $s(t)$, in analogy to $\rho_0/\text{in.}$, one brings out the following parallelism between functions connecting two variable quantities and numbers connecting two objects:

ρ_0/in. = 18 and in./ft. = $\frac{1}{12}$ imply ρ_0/ft. = $18 \cdot \frac{1}{12} = 1.5$ (multiplication!);

$s//t = 16j^2$ and $t//v = \frac{1}{32}j$ imply $s//v = 16j^2\left(\frac{1}{32}j\right) = \frac{1}{64}j^2$ (substitution!).

More generally, in terms of object variables and fluent variables:

$\gamma/\beta = \mathrm{n}$ and $\beta/a = \mathrm{m}$ imply $\gamma/a = \mathrm{n} \cdot \mathrm{m}$; $\beta/\gamma = 1/\mathrm{n}$; $\beta/\beta = 1$, for any three objects a, β, γ and any two numbers m and n ($\neq 0$);

$w//v = g$ and $v//u = f$ imply $w//u = gf$; $v//w = j//g$; $v//v = j$, for any three consistent classes of quantities u, v, w and any two functions f and g.

Without numerical and function variables:

$$(\gamma/\beta) \cdot (\beta/a) = (\gamma/a), \quad \beta/\gamma = 1/(\gamma/\beta), \quad \beta/\beta = 1;$$

$$(w//v)(v//u) = (w//u), \quad v//w = j//(w//v), \quad v//v = j.$$

Remark 3. Numbers followed by units, such as 18 inches, often are referred to as *denominate numbers*. Similarly, functions followed by variable quantities, such as cos t, might be called *denominate functions*. Arithmetic deals exclusively with what, in contrast to denominate numbers, are sometimes called *pure numbers*, such as 18; and calculus, with what might be called *pure functions*, such as cos.

The formulas ρ_0/ in. = 18 and $s//t = \cos$ or $s(t) = \cos$ are free of denominate entities. The first equates the pure number connecting a certain rod ρ_0 with an inch and the pure number 18; the second, the pure function connecting s with t and the pure function cos. The absence of denominate entities in these formulas, far from jeopardizing their applicability, simplifies the formulation of rules (such as those stated at the end of Remark 2) concerning the application of mathematics to the real world.

Remark 4. Let a_1 be a unit object for the consistent class of quantities, u; that is to say, assume $u\,a_1 = 1$. If a new c.c.q., u', is defined by choosing a unit m times as large, then $u'a_1 = \frac{1}{\mathrm{m}}$ and $u' = \frac{1}{\mathrm{m}}u$.

Similarly, assume $w' = \frac{1}{\mathrm{n}}w$.

Now suppose that $w = f(u)$ relative to a certain pairing Π. How is w' connected with u' relative to Π? Since $u = \mathrm{m}\,u'$ and $w = \mathrm{n}\,w'$, one finds $w' = \frac{1}{\mathrm{n}} f(\mathrm{m}\,u')$. In pure form:

$$u' = \frac{1}{\mathrm{m}}u, \; w' = \frac{1}{\mathrm{n}}w, \text{ and } w//u = f \text{ imply } w'//u' = \frac{1}{\mathrm{n}} f(\mathrm{m}\,j).$$

In particular, if $f = j^k$, then $w'//u' = \frac{\mathrm{m}^k}{\mathrm{n}} j^k$; in denominate form: if $w = u^k$, the $w' = \frac{\mathrm{m}^k}{\mathrm{n}}(u')^k$.

Remark 5. If, relative to a certain pairing, w is a function of u, then u is sometimes called *independent*, w *dependent*. The terms seem to have been coined for numerical variables (see p. 64) and then applied to variable quantities. If p and v are connected by the law $p \cdot v = 1$, then p or v is independent according to whether $v = 1/p$ or $p = 1/v$ is being considered. Consequently, if, relative to a Galileo pairing, either of two physical variable quantities is a function of the other, then the question "which v.q. is independent?" can be asked only *relative to a specific formula* and is tantamount to the question *"which v.q. in this formula is being represented as a function of the other v.q.?"* Obviously, therefore, the terms "dependent" and "independent" variable quantities may well be dispensed with.

Remark 6. Many books give the following definition. "A variable w is a function of a variable u if and only if, whenever u assumes equal values, so does w." In the literature, this sentence is either not preceded by any definition of the term "variable" or it is based on a definition in the sense of numerical variable.

On p. 180, the following statement has been proved from the explicit definitions of consistent classes of quantities and of functions: w is a function of u relative to the pairing Π if and only if, to any two elements of Dom u for which u assumes equal values, there correspond by Π two elements in Dom w for which w assumes equal values. In this statement (which, relative to a certain pairing Π, gives a precise meaning to the definition quoted at the beginning of Remark 6) u and w stand for *consistent classes of quantities*, each of which has a domain and assumes a value for each element of that domain; in other words, they are *fluent variables*.

But the quoted definition cannot be stated for *numerical variables*. One cannot say "a symbol that stands for any element of a certain class of numbers is a function of another symbol that stands for any element of a certain class of numbers if and only if, whenever the latter assumes equal values, so does the former." A numerical variable may be replaced by the designation of any element of its scope, but it has no domain for each element of which it would "assume a value." Nor is it clear what the word "whenever" should mean with regard to numerical variables, These difficulties are enhanced by the fact that, traditionally, the letters x and y (in italics) are used (1) for numerical variables, (2) for the specific variable quantities, abscissa and ordinate, in analytic geometry, (3) for fluent variables. In this book, the letters u, v, w are used for the latter, x and y are reserved for abscissa and ordinate, and x and y in roman type are used as numerical variables. The concept "y is a function of x" will herein remain undefined.

Remark 7. The comparison of the self-explanatory symbols for functions connecting consistent classes of quantities with the traditional symbols for values of functions, e.g., of

(14) $$s(t) \text{ with } \cos x \text{ and } f(x)$$

strongly suggests analogies between

(I) s and cos or f;

(II) t and x.

It seems that s plays the role of a function, and t that of a numerical variable or, which is tantamount, that the numerical variable x and the "independent" variable quantity *t* play the same role – appearances that are accentuated by the traditional use of italics for both numerical variables and variable quantities.

Neither of these analogies is valid.

(I) *s* is a physical variable quantity and not a function as is the cosine. What in (14) is comparable to the cosine function is $s(t)$ and not *s*. In fact, for certain oscillators, $s(t)$ is not only comparable to the cosine function but, according to Remark 2, equal to *cos*.

(II). *t* likewise is a physical variable quantity and not a numerical variable. The function connecting *s* with *t* depends upon *t* as much as upon *s*. For a falling object, *s* is connected with *t* by the function $16\,j^2$ and with *v* by a different function (namely $\frac{1}{64}\,j^2$), just as *v* is connected with *t* by a function different from $s(t)$.

The profound difference between *t* and a numerical variable is strikingly illustrated by the following example in which, in line with the typographical convention adopted in this book, roman letters will be used for numerical variables – a convention that strongly accentuates the contrast between an "independent" variable quantity and a numerical variable. The example is based on the fact that $s(t) = 16\,j^2 = 16 \cdot j^2$, wherefore $\frac{s(t)}{j^2}$ is the constant function *16*, whereas $\frac{s(v)}{j^2}$ is the constant function of value $\frac{1}{64}$. It reads:

$$\frac{s(t)}{j^2} > 1 \quad \text{or} \quad \frac{s//t}{j^2} > 1, \text{ where } t \text{ may not be replaced by } v,$$

in contrast to

$$\cos \mathrm{x} \geq -1 \text{ for any x, where x may be replaced by any letter, as in}$$
$$\cos \mathrm{y} \geq -1 \text{ for any y}.$$

The last line involves the function *cos*, the number -1, and the numerical variables x and y; the preceding line, the specific variable quantities *s*, *t*, and *v*, and the specific functions j^2 and *1*. This preceding line is free of variables of any kind. So is the formula $\cos \geq -1$.

Remark 8. (14) not only simulates non-existing parallelisms but also conceals analogies that exist. The numerical variable x that appears in the formula

(15) $$\sec \mathrm{x} = 1/\cos \mathrm{x} \text{ for any } \mathrm{x} \neq \pi/2,\ 3\pi/2, \ldots$$

does possess an analogue in Boyle's Law concerning volume and pressure of gas at a certain temperature. But this analogue is not the "independent" variable quantity *p* in $v = 1/p$ (proper units being chosen). To bring out the parallelism one has to formulate the law in the form:

(16) $v\gamma = 1/p\gamma$ for any instantaneous gas sample γ at the proper temperature.

Here, the analogue of the numerical variable x in (15) clearly is the sample variable γ in (16). Without any change of the meaning, γ may be replaced by, say, δ, just as x may be replaced by a.

Similarly, in the analytic treatment of a physical or postulational plane, one finds

$$y\sigma = 1/x\sigma \text{ for any point } \sigma \text{ on a certain hyperbola, H,}$$

where the point variable σ, and not the variable quantity x (the abscissa), is the analogue of the numerical variable, x, in (15).

Summarizing, one can formulate

Laws of	Physics	Geometry	Calculus
using variables	$v\gamma = 1/p\gamma$ for any gas sample γ	$y\delta = 1/x\delta$ for any point δ	$\sec x = 1/\cos x$ for any $x \neq \pi/2, 3\pi/2, \ldots$
without variables	$v = 1/p$ for gas at 100°C	$y = 1/x$ on H	$\sec = 1/\cos$ on $U - \{\pi/2, 3\pi/2, \ldots\}$

The traditional formulations of laws in physics and geometry do not make use of variables; those of the laws of calculus do. Hence, in the formulations emphasized in this book, the laws of calculus, such as $\sec = 1/\cos$, or $\mathbf{D}\ \sin = \cos$, are in complete parallelism with a large class of scientific laws, such as $v = 1/p$ or $s = 16\ t^2$. The laws connect consistent classes of quantities by means of functions: the scientific and geometric laws connect scientific and geometric variable quantities; the mathematical laws connect functions.

Very general laws establish functional connections between any two c.c.q.'s (or any two c.c.q.'s of a certain kind) and are therefore expressed in terms of fluent variables in science and geometry, and of function variables in mathematics.

C.c.q.'s other than functions (e.g., the pressure p or the volume v) are unfit to serve as connectors since no c.c.q. can be substituted into them. This explains why c.c.q.'s such as p and v (pertaining to the physical world) belong to restricted chapters of knowledge, whereas c.c.q.'s such as *log* and j^{-1}(creations of the human mind) are not only studied in mathematics but used throughout science. Denying the significance of the difference between functions and other c.c.q.'s is denying the significance of the role of mathematics as a universal tool.

It is almost symbolic for the history of calculus that the term *function*, coined by Leibniz, is in general use, whereas the term *fluent*, introduced by Newton, is half forgotten and, which is worse, the idea behind it has been neglected. Under the name *variable quantities*, fluents have been often mistaken for numerical variables. Just as Newton seems to have confined the term fluent to c.c.q.'s "with fluxions" (i.e., connected with the time or with other c.c.q.'s by functions having derivatives), Leibniz studied only differentiable functions. Later, however, Leibniz' notion was made precise and more general, whereas Newton's idea does not seem to have been used in a strictly deductive theory before its revival in this book in the concept *consistent class of quantities*.

EXERCISES

30. Call r, s, and v the radius, the surface, and the volume of a sphere. Express all six functional relations between pairs of these three variable quantities *(a)* in denominate, *(b)* in pure, form. Verify the laws mentioned at the end of Remark 2. What pairing of the domains of r, s, and v is tacitly understood? What convention concerning the units for length, area and volume is tacitly made in the classical formulas such as $s = 4\pi r^2$. Apply the Remark 4 to the case where r is measured in inches, and s in square feet.

31. Express the connections between j^3, j^5, and j^9 in denominate and in pure form.

⋆32. Let Π be the pairing of Dom v with Dom u relative to which $v // u = f$ and P the pairing of Dom w with Dom v relative to which $w // v = g$. Describe the pairing of Dom w with Dom u with regard to which $w // u = gf$. With regard to what pairings is $u // v = j // f$ and $v // v = j$?

33. Replace the three laws mentioned in the table at the end of Remark 8 by three other examples from physics, geometry, and calculus, and add an example from another branch of science.

9. APPROXIMATE FUNCTIONAL CONNECTIONS

If, in a physical plane, length can be measured only within .001 in., then one cannot distinguish the length of the diagonal of a unit square from 1.415 in., even though from certain postulates and in analytic geometry the diagonal can be proved to have the irrational length $\sqrt{2}$.

Similarly, a function f connecting one physical variable quantity with another cannot, on the basis of observation, be distinguished from any function between $f + c$ and $f - c$ for some small number c. The class of all these functions includes, for instance, for any n

(17) $f + c \sin(nj)$ assuming for any x the value $fx + c \sin(nx)$.

On a blackboard, every curve has a breadth.

Nor can observation prove that a function connecting two physical v.q.'s is continuous, except within certain limits of accuracy. Suppose, for instance, that the positions of an oscillating particle can be determined only within .02 linear units. Then, if s appears to be connected with the time t by the cosine function, no observation can rule out the possibility that the connecting function assumes the values $\cos x - .01$ for $x > 0$ and $\cos x + .01$ for $x \leq 0$. The latter function, which jumps from .99 to 1.01, is discontinuous at 0.

Not even the question as to whether or not another variable quantity, w, is a function of s can be answered strictly. An affirmative answer would presuppose the certainty that two values of s to which unequal values of w correspond are unequal. If those two values of s differ by more than .02, then they are certainly unequal. But if they differ by less than .02, then no observation can decide whether or not they are equal.

On the other hand, the physical and social universe abounds in variable quantities connected with others by certainly discontinuous functions; for instance,

(1) the outcome of tossing a coin, with the altitude to which the coin was raised (see p. 183);

(2) the postage of a letter, with its weight;

(3) the volume of water spilled from a conical cup by dipping a spherical ball, with the radius of this ball. (By submerging a ball whose radius is ever so slightly smaller than the radius of the rim of the cone one spills more than twice the water spilled by dipping (as far as possible) a ball whose radius is ever so slightly larger than that of the rim.)

(4) the velocity of a particle, with the time; traditional mechanics assumes abrupt changes at the instant of impacts; highspeed photography reveals very rapid but more or less gradual changes;

(5) the acceleration, a, in ft./sec.^{2} of a falling object of mass m, with the time (a jumps from 0 to 32 at the instant of release, but is connected by the function m j with the acting force, which undergoes a similar abrupt change);

(6) quantum physics is based on the assumption of abrupt changes of energy in multiples of an elementary amount.

"Approximation mathematics" in contrast to "Precision mathematics" was stressed by the German mathematician F. Klein at the turn of the century.

EXERCISES

34. Determine the function in example (3); then that in example (1) in terms of the function that assumes the value 0 between 0 and 1, between 2 and 3, . . . ; and the value 1 between 1 and 2, between 3 and 4, . . . (This function may, *ad hoc*, be denoted by f.) Then determine the function connecting the fare in a taxicab with the distance traveled.

35. Assuming that the motion of a falling object is observed during 3 seconds and that, at any instant, the distance traveled can be determined within .001 ft., find functions that are compatible with Galileo's Law $s = 16\,t^2$ and are of the form
(a) $(16 + c)\,j^2$; *(b)* $16\,j^2 + b$; *(c)* $(16 + c)(j^2 + b)$; *(d)* $16\,j^{2+c}$.
Then answer the questions (a) - (d) under the assumption that the motion is observed during 4 seconds.

36. Let q be a v.q. whose values can be determined within 1%. Assume that after an observation of d days, q is reported to be a constant function of the time, more precisely, to be connected with t by the constant function of value c. Give examples of exponential, trigonometric, and polynomial functions that are not excluded by the observations.

10. INTEGRALS AND THE CUMULATION OF ONE VARIABLE QUANTITY WITH REGARD TO ANOTHER

Consider a particle whose mass is 1 gram. If, during an interval of t seconds, it is acted upon by a constant force of p dynes, then the particle is said to receive an *impulse* of $p \cdot t$ dyne-seconds. If it moves s cm. in the direction of the force, the latter is said to do *work* of $p \cdot s$ ergs. If one moves the particle s cm. against the force, he is said to do work of $p \cdot s$ ergs.

Now suppose the particle is acted upon by a force that may change from instant to instant and from place to place. What impulse does the particle receive during a certain interval of time and what work is done by moving it from one place to another?

Just as in the case of the area under a curve, two points of view are possible (see p. 140):

(I) that the meaning of the terms "impulse" and "work" is understood even if the force is non-constant, and that the only problem lies in the determination of those numbers;

(II) that, with regard to non-constant forces, impulse and work first have to be defined and then, if possible, numerically determined.

Example 1 (Impulse). The force is a variable quantity, p: the class of all pairs $(\pi, p\pi)$ for any act π of reading a force-meter calibrated in dynes, say a spring balance, where $p\pi$ denotes the number read as a result of the act π. The time in seconds, t, after the beginning of the observation of the moving particle is the class of all pairs $(\tau, t\tau)$ for any act τ of clock reading. If τ_0 is the initial act, then $t\tau_0 = 0$. Let Γ be the pairing of Dom p with Dom t by simultaneity. Thus, if $\Gamma\tau$ denotes the act of reading the force meter performed simultaneously with the act τ, then

$$(18) \qquad p = g(t), \text{ that is, } p(\Gamma\tau) = g(t\tau) \text{ for any } \tau \text{ in Dom } t.$$

For instance, $p = \sin t$ in the case of some elastic forces.

Let T be a sequence of acts of clock reading

$$T:\ \tau_0, \tau_1, \ldots, \tau_{n-1}, \tau_n = \tau,$$

with the results

$$t_0 < t_1 < \ldots < t_{n-1} < t_n = t \quad (\text{where } t_i = t\tau_i)$$

The results of the simultaneous acts

$$\Gamma\tau_0, \Gamma\tau_1, \ldots, \Gamma\tau_{n-1},$$

of reading the force meter may be denoted by

$$p_0, p_1, \ldots, p_{n-1} \quad (\text{where } p_i = p(\Gamma\tau_i))$$

It is easy to determine the impulse that the particle would receive from τ_0 to τ if, between any two consecutive observations, the force were constant (namely, equal to what it actually is at the beginning of that short interval). This impulse is

(19) $$p_0 \cdot (t_1 - t_0) + p_1 \cdot (t_2 - t_1) + \ldots + p_{n-1} \cdot (t - t_{n-1})$$

and may be called *the approximate impulse computed by means of* T. According to (18), $p_i = gt_i$ or, if parentheses are used,

$$p_0 = g(t_0),\ p_1 = g(t_1), \ldots, p_{n-1} = g(t_{n-1}).$$

Hence the approximate impulse (19) is equal to

(20) $$g(t_0) \cdot (t_1 - t_0) + g(t_1) \cdot (t_2 - t_1) + \ldots + g(t_{n-1}) \cdot (t - t_{n-1}).$$

If all n differences $t_{i+1} - t_i$ are small (which presupposes that n is large), then the product sum (20) is close to $\int_{t_0}^{t} g$, and so is the approximate impulse (19), since it is equal to (20).

From here on, the reasoning is slightly different according to which of the two views is held.

Adopting view (I) one knows (or, at any rate, one believes that he knows) what the impulse due to the force p is. One denotes this number by $\int_{\tau_0}^{\tau} p\, dt$, a symbol reminiscent of the product sums (19) and one considers (20) as an approximation to that precise value of the impulse. Since, for sequences of sufficiently many clock readings in sufficiently rapid succession, the sum (20), which is equal to (19), is as close as one pleases to $\int_{t_0}^{t} g$, one considers the latter number as the precise value of the impulse received from τ_0 to τ; that is to say,

$$\text{(21)} \qquad \text{if } p = g(t) \text{ and } t\tau = \mathrm{t}, \text{ then } \int_{\tau_0}^{\tau} p \, \mathrm{d}t = \int_{\mathrm{t}_0}^{\mathrm{t}} g .$$

For instance, for an elastic force,

$$\text{if } p = \sin t, \text{ then } \int_{\tau_0}^{\tau} p \, \mathrm{d}t = \int_{\mathrm{t}_0}^{\mathrm{t}} \sin = [\cos]_{\mathrm{t}_0}^{\mathrm{t}} = \cos \mathrm{t} - \cos \mathrm{t}_0 .$$

Gravity is a constant force of 981 dynes, and hence is connected with t by the constant function *981*. According to (21), the impulse in dyne-seconds received during the first 3 seconds is

$$\int_{\tau_0}^{\tau} p \, \mathrm{d}t = \int_0^3 981 = 981 \cdot 3,$$

which is in agreement with the remark at the beginning of this section.

Adopting the point of view (II) one considers the number $\int_{\mathrm{t}_0}^{\mathrm{t}} g$ as a *definition* of $\int_{\tau_0}^{\tau} p \, \mathrm{d}t$ and, as in case (I), the product sums (19) and (20) as approximations to that number.

Values of the c.c.q.'s p and t have been denoted by p and t with subscripts. In many examples, specific values of a c.c.q. or numerical variables whose scope is the range of a c.c.q. will be indicated by the same letter as the c.c.q. but in roman type. These letters are, as it were, self-explanatory numerical variables.

Just as $p = g(t)$ is an abbreviation for $p\Gamma = g(t)$, the pairing Γ being tacitly understood, $\int_{\tau_0}^{\tau} p \, \mathrm{d}t$ is an abbreviation for $\int_{\tau_0}^{\tau} p\Gamma \, \mathrm{d}t$ – a symbol for the impulse that contains an explicit reference to the pairing by simultaneity.

Example 2 *(Work).* Suppose the particle moves along a straight line calibrated in centimeters, and s denotes its position, i.e., the class of all pairs $(\sigma; s\sigma)$ for any act σ of observing the scale mark opposite the particle. Suppose further that, relative to Γ, the force p acting on the particle is connected with its position s by the function f. Thus briefly, $p = f(s)$. Consider a sequence of acts of mark reading

$$\Sigma: \sigma_0, \sigma_1, \ldots \sigma_{m-1}, \sigma_m = \sigma$$

with the results

$$0 = \mathrm{s}_0 < \mathrm{s}_1 < \ldots < \mathrm{s}_{m-1} < \mathrm{s}_m = \mathrm{s} \quad (\mathrm{s}_i = s\sigma_i).$$

Let p'_i denote the result $p(\Gamma\sigma_i)$ of the act of reading the force meter simultaneously with σ_i, then that approximate work equals

$$p'_0 \cdot (s_1 - s_0) + p'_1 \cdot (s_2 - s_1) + \ldots + p'_{n-1} \cdot (s - s_{n-1}) =$$

$$f(0) \cdot (s_1 - 0) + f(s_1) \cdot (s_2 - s_1) + \ldots + f(s_{n-1}) \cdot (s - s_{n-1}).$$

The work done by the force, which may be denoted by $\int_{\sigma_0}^{\sigma} p\, \mathrm{d}s$, equals (or is defined as) $\int_0^{s} f$.

For instance, for an elastic force,

if $p = -s$, that is $-j(s)$ and $s\sigma = s$, then

$$\int_{\sigma_0}^{\sigma} p\, \mathrm{d}s = \int_0^{s} (-j) = \left[-\frac{1}{2}j^2\right]_0^{s} = -\frac{1}{2}s^2 .$$

For gravity, p is connected with s (as it is with t) by the constant function 981. Hence the work in ergs done by moving the particle $\frac{1}{2} \cdot 981 \cdot 3^2 = 4414.5$ cm. (which is the distance traveled during the first 3 seconds) is

$$\int_{\sigma_0}^{\sigma} p\, \mathrm{d}s = \int_0^{4414.5} 981 = 981 \cdot 4414.5,$$

which is in agreement with the remark at the beginning of this section.

Chapter VI contains the definition of the integral from a to b of a function such as *sin* or *-j*. For consistent classes of quantities other than functions, the integral has not been defined; and symbols such as $\int_{\pi_0}^{\pi} p$ and $\int_0^{P} p$ will remain undefined throughout this book. However, one can integrate the function connecting one variable quantity with another, for instance, p with t or p with s. In fact, integrations of such functions represent the scientific applications of integral calculus.

If $p = \sin t$, then the impulse $\int_{\tau_0}^{\tau} p\, \mathrm{d}t$ obviously can also be written in the form $\int_{\tau_0}^{\tau} \sin t\, \mathrm{d}t$. Consequently, according to (8),

$$\text{if } t\tau = \mathrm{t}, \text{ then } \int_{\tau_0}^{\tau} \sin t\, \mathrm{d}t = \int_0^{\mathrm{t}} \sin .$$

It should be noted that, here, t is not a numerical variable; t is a specific variable quantity, just as *sin* is a specific function. It should also be noted that the impulse is from τ_0 to τ, whereas the integral is from 0 to t. The symbol $\int_{\tau_0}^{\tau} sin$ remains undefined just as does $\int_0^p p$.

The impulse $\int_{\tau_0}^{\tau} p\, \mathrm{d}t$ and the work $\int_{\sigma_0}^{\sigma} p\, \mathrm{d}s$ will be called *cumulations* of the force with regard to the time t and the position s, respectively, more specifically, *the* cumulations between τ_0 and τ, and between σ_0 and σ.

More generally, let u and w be any two consistent classes of quantities: u, the class of all pairs (a, ua) for any a in Dom u; and w, the class of all pairs $(\beta, w\beta)$ for any β in Dom w. Suppose that, relative to the pairing Π of Dom w with Dom u, the c.c.q. w is connected with u by a continuous function f:

$$w = f(u) \text{ or, more precisely, } w\Pi = f(u);$$

that is to say, $w(\Pi a) = f(ua)$ for any a in Dom u.

Consider a sequence

$$A\colon\ a_0, a_1, \dots, a_{k-1}, a_k = a$$

of elements of Dom u such that the corresponding values ua_i, which will be denoted by u_i, satisfy the inequalities

$$u_0 < u_1 < \dots < u_{k-1} < u_k = u.$$

If the corresponding values $w(\Pi a_i)$ are denoted by w_i, then

$$(22)\qquad w_0 \cdot (u_1 - u_0) + w_1 \cdot (u_2 - u_1) + \dots + w_{k-1} \cdot (u - u_{k-1})$$

will be called *the approximate cumulation of w with respect to u, computed by means of* A. Clearly, the number (22) equals

$$f(u_0) \cdot (u_1 - u_0) + f(u_1) \cdot (u_2 - u_1) + \dots + f(u_{k-1}) \cdot (u_k - u_{k-1})$$

and thus is an approximation to $\int_{u_0}^{u} f$. As in Examples 1 and 2, one may define the cumulation of w with respect to u from a_0 to a as

$$(23)\qquad \int_{a_0}^{a} w\, \mathrm{d}u = \int_{u_0}^{u} f \text{ for any two c.c.q.'s } u \text{ and } w, \text{ if } f \text{ is continuous.}$$

One thus arrives at the following

BASIC SCHEME FOR THE APPLICATION OF INTEGRALS TO CONSISTENT CLASSES OF QUANTITIES. If $w = f(u)$ then $\int_{a_0}^{a} w\, \mathrm{d}u = \int_{ua_0}^{ua} f$

more precisely, that is, with reference to the pairing Π:

$$\text{if } w\Pi = f(u), \text{ then } \int_{a_0}^{a} w\Pi \, \mathrm{d}u = \int_{ua_0}^{ua} f.$$

Example 3. If, in (23), w is replaced by the constant v.q. *1* (of value 1), then the product sum (22) collapses into $\mathrm{u} - \mathrm{u}_0 = ua - ua_0$ for any A.

Hence $\int_{a_0}^{a} 1 \, \mathrm{d}u = ua - ua_0$ and, more generally,

$$\int_{a_0}^{a} \mathrm{c} \, \mathrm{d}u = \mathrm{c} \cdot (ua - ua_0).$$

If, in (23), u is replaced by c, then in (22) the second factors of all terms equal 0 and

$$\int_{a_0}^{a} w \, \mathrm{d}\mathrm{c} = 0.$$

Example 4. If $w = u$ and Π is the identity pairing, then the function connecting w with u is j and

$$\int_{a_0}^{a} u \, \mathrm{d}u = \int_{\mathrm{u}_0}^{\mathrm{u}} j = \left[\tfrac{1}{2} j^2\right]_{\mathrm{u}_0}^{\mathrm{u}} = \tfrac{1}{2}(\mathrm{u}^2 - \mathrm{u}_0^2) = \tfrac{1}{2}(u^2 a - u^2 a_0).$$

Example 5. For any u and w, if A is so chosen that

$$\mathrm{u}_1 - \mathrm{u}_0 = \mathrm{u}_2 - \mathrm{u}_1 = \ldots = \mathrm{u} - \mathrm{u}_{k-1} = \frac{1}{\mathrm{k}}(\mathrm{u} - \mathrm{u}_0),$$

then (22) reduces to

$$\frac{1}{\mathrm{k}}(\mathrm{w}_0 + \mathrm{w}_1 + \ldots + \mathrm{w}_{k-1}) \cdot (\mathrm{u} - \mathrm{u}_0).$$

This number is the average of k values of w (namely, of the values for the elements corresponding to $a_0, a_1, \ldots, a_{k-1}$) multiplied by $\mathrm{u} - \mathrm{u}_0$, which is the rise of u from a_0 to a. Thus $\int_{a_0}^{a} w \, \mathrm{d}u$ is the product of this rise and the limit of $\frac{1}{\mathrm{k}}(\mathrm{w}_0 + \mathrm{w}_1 + \ldots + \mathrm{w}_{k-1})$ as $\mathrm{k} \to \infty$. It is customary to refer to this limit as the *mean* of w for all elements corresponding to elements a' of Dom u for which ua' is between u_0 and u. (Occasionally, the limit is called the *average* of all those values of w; but usually the word "average" is applied only to finite classes of numbers.) Denoting the mean by $M_{\mathrm{u}_0}^{\mathrm{u}} w$ or $M_{ua_0}^{ua} w$, one can write

$$\int_{a_0}^{a} w \, \mathrm{d}u = M_{ua_0}^{ua} w \cdot (ua - ua_0).$$

Example 6 (Areas). Replace, in (23), *u* by the abscissa x, and *w* by the ordinate *y*, of the points on the curve *f* in Fig. 1, page 1. Then

$$y = f(x);\ \text{that is to say, } y\sigma = f(x\sigma) \text{ for any point } \sigma \text{ on the curve } f.$$

If α and β denote the points on *f* for which $x\alpha = 2$ and $x\beta = 4.5$, then the product sum (22) is the area under a step line approximating *f*. Hence

$$\int_{\alpha}^{\beta} y\,dx = \int_{2}^{4.5} f$$

is the area of the shaded domain in Fig. 1.

If, along a curve between the points γ and δ, the abscissa x is connected with the ordinate *y* by a function *g*, then the reader can easily determine the geometric meaning of $\int_{\gamma}^{\delta} x\,dy = \int_{y\gamma}^{y\delta} g.$

Example 7 (Volume). Rotating the shaded domain in Fig. 1 about the line *O* one obtains a solid of revolution. Its volume is close to (or defined as the limit of) the volumes of the "cylinder sums" obtained by rotating about the line *O* step lines approximating the curve *f*. The reader can easily prove that this volume is

$$\int_{\alpha}^{\beta} \pi y^2\,dx = \int_{a}^{b} \pi f^2 \quad \text{or} \quad \pi \int_{a}^{b} f^2,$$

where a = 2 and b = 4.5.

Clearly, $\int_{a}^{b} \pi f^2$ is also the area from a to b under the curve πf^2. The equality of this area in square units and the volume in cubic units of the solid of revolution is, of course, as unparadoxical as is the equality of the length in linear units of a segment and the area in square units of a domain (see p. 9).

Example 8 (Moments). A mass m at a distance r from a point is said to have the *moment* $m \cdot r$ about this point and the *2nd order moment* (or moment of inertia) $m \cdot r^2$. For a system of k masses (m_1 at the distance r_1, ..., m_k at the distance r_k) one calls the number $m_1 \cdot r_1 + \ldots + m_k \cdot r_k$ the moment, and $m_1 \cdot r_1^2 + \ldots + m_k \cdot r_k^2$ the 2nd order moment, about the point. Now consider a thin rod ρ of inhomogeneous material placed along a linear scale beginning at the point α and ending at β. Let *m* denote the class of all pairs $(\sigma, m\sigma)$ for any point σ between α and β, where $m\sigma$ denotes the mass of the portion of ρ between α and σ. (The variable quantity *m*, thus defined, might be called the *initial mass* of ρ.) A sequence

$$\Sigma:\ \alpha = \sigma_0, \sigma_1, \ldots, \sigma_{k-1}, \sigma_k = \beta \ \text{ such that}$$

$$0 = m\sigma_0 < m\sigma_1 < \ldots < m\sigma_{k-1} < m\beta = m$$

breaks ρ into k splinters. The i-th splinter has the mass $m\sigma_{i+1} - m\sigma_i$.

Let x be the class of all pairs $(\sigma, x\,\sigma)$, where $x\,\sigma$ is the scale mark of σ. The cumulation of x with respect to m computed by means of Σ equals the moment about 0 of a system of k masses of which the ith equals the mass of the ith splinter and is concentrated at σ_i. This product sum is an approximation to the moment of ρ about 0, denoted by $\int_\alpha^\beta x\,d m$. If x is connected with m by the continuous function f, then

$$\int_\alpha^\beta x\,d m = \int_0^m f \text{ and, similarly, } \int_\alpha^\beta x^2\,d m = \int_0^m f^2.$$

What about the cumulation of m with respect to x? The approximation computed by means of Σ equals (if the k splinters are equally long) the product of the length of the bar and an average of masses of initial segments of ρ. This number is not particularly significant. Nor is $\int_\alpha^\beta m\,d x$.

Example 9. One may replace u and w in (23) by functions, (i.e., c.c.q.'s whose domains consist of numbers), e.g., by j^3 and j^5. Thus, in view of $j^3\,a = a^3$ and $j^3\,b = b^3$,

$$j^5 = j^{5/3}\,j^3 \text{ implies } \int_a^b j^5\,d j^3 = \int_{a^3}^{b^3} j^{5/3}, \text{ which equals } \frac{3}{8}(b^8 - a^8).$$

More generally,

$$\text{if} \qquad h = fg, \text{ then } \int_a^b h\,d g = \int_{g\,a}^{g\,b} f.$$

In particular, if $g = j$, then $f = fj$ implies

$$\int_a^b f\,d j = \int_a^b f;$$

that is to say, between any two numbers a and b, the cumulation of a function f with respect to j is the integral of f.

If h is continuous and g is an increasing (though not necessarily continuous) function, then it can be proved that the product sums

$$h\left(\tfrac{0}{k}\right)\cdot\left(g\left(\tfrac{1}{k}\right)-g\left(\tfrac{0}{k}\right)\right)+h\left(\tfrac{1}{k}\right)\cdot\left(g\left(\tfrac{2}{k}\right)-g\left(\tfrac{1}{k}\right)\right)+\ldots+h\left(\tfrac{k-1}{k}\right)\cdot\left(g\left(\tfrac{k}{k}\right)-g\left(\tfrac{k-1}{k}\right)\right)$$

have a limit as $k\to\infty$. This limit $\int_0^1 h\,d\,g$ is called the *Stieltjes integral* of h with respect to g from 0 to 1. It is named after the Dutch mathematician who discovered this important concept toward the end of the 19th century. One can prove that $\int_0^1 h\ d\ g$ is also the limit of the product sums

$$h(x_0^*)\cdot\left(g\left(\tfrac{1}{k}\right)-g\left(\tfrac{0}{k}\right)\right)+h(x_1^*)\cdot\left(g\left(\tfrac{2}{k}\right)-g\left(\tfrac{1}{k}\right)\right)+\ldots+h(x_{k-1}^*)\cdot\left(g\left(\tfrac{k}{k}\right)-g\left(\tfrac{k-1}{k}\right)\right),$$

no matter how x_i^* is chosen between $\frac{i}{k}$ and $\frac{i+1}{k}$; and that one may even take intervals of unequal length into consideration.

If g has a derivative $\mathbf{D}g$, then, even if h is not a function of g, it can be shown that

$$\int_a^b h\ \mathrm{d}g = \int_a^b (h\cdot\mathbf{D}g)$$

But even if g has no derivative and, in fact, is not continuous, $\int_a^b h\ \mathrm{d}g$ can be shown to exist provided h is continuous and g is monotonic (i.e., increasing or decreasing) or the sum of an increasing and a decreasing function. More generally, the cumulation of one c.c.q., w, with respect to another u may exist without w being a function of u. But these situations go beyond the ordinary calculus.

Since the basic scheme for the application of integrals is valid for any element a of Dom u, one may define a consistent class of quantities, called *the cumulation of w with respect to u from* a_0 *on*. It has the same domain as u and is the class

$$\int_{a_0} w\ \mathrm{d}u \text{ of all pairs } \left(a, \int_{a_0}^{a} w\ \mathrm{d}u\right) \text{ for any } a \text{ in Dom } u.$$

According to the basic scheme, $w = f(u)$ implies that, if $u a_0 = u_0$, then $\int_{a_0} w\ \mathrm{d}u$ is connected with u by the function $\int_{u_0} f$.

For instance, if the force p is connected with the time t by the function *sin*, then the impulse i_{τ_0} from τ_0 on is connected with t by the function $\int_{\tau_0} sin = -\cos + \cos t_0$; in other words,

if $\quad p = \sin t$ and $t\tau_0 = t_0$, then $i_{\tau_0} = -\cos t + \cos t_0$.

Similarly, if p is connected with s by $-j$, then the work $w\sigma_0$ from σ_0 on is connected with s by $\int_{s_0} -j = -\frac{1}{2}(j^2 - s_0^2)$; in other words,

if $\quad p = -s$ and $s\sigma_0 = s_0$, then $w_{\sigma_0} = -\frac{1}{2}(s^2 - s_0^2)$.

EXERCISES

37. Determine the volume of the barrel-shaped solid obtained by revolving about the line O the portion between -3 and 3 of the semi-ellipse $\frac{1}{2}\sqrt{16 - j^2}$. Then determine the volume of the entire ellipsoid. (All that is needed is the integration of a polynomial, whereas the determination of the length of the ellipse would lead to an elliptic integral, see p. 56). In 1615, Kepler determined the approximate volume of a barrel by *(a)* measuring the perimeters at various heights; *(b)* computing the areas of the corresponding circular cross sections; *(c)* computing the combined volume of the cylinders between those cross sections.

38. By integration, determine the volume of a conical frustrum and of the section from a to b of the sphere obtained by rotating about O the semicircle $\sqrt{r^2 - j^2}$. Find the volumes of the solids obtained by rotating the portion from 0 to b of the parabola j^2 *(a)* about the line O; *(b)* about the line perpendicular to O through the point 0. Then solve the same problem for the portions between 0 and c of the parabola $j^{1/2}$ and under the curve $\sqrt{j^3 + j + 1}$. (Note that the area under last curve is given by an elliptic integral.)

39. In a plane with a polar frame, consider the curves along which the radius vector r is connected with the angle a by the functions j, $\cos$, j^2, and f; that is to say, along which $r = a$, $r = \cos a$, $r = a^2$, and $r = f(a)$, respectively. Find the area of the triangular domain bounded by (1) the portion of the curve from the polar axis to the angle a; (2) the segments from the pole to the end points of this portion. Hint. Break up the domain into n sectors between the angles $\frac{k}{n}$ a and $\frac{k+1}{n}$ a(k = 1, . . . , n), and replace each of them by a sector of a circle with the angle $\frac{1}{n}$ a and the radius equal to the radius vector of the curve at $\frac{k}{n}$ a.

40. Consider a thin bar. Let m be the initial mass (p. 208). Suppose that $x = f(m)$. At which point should the entire mass of the bar be concentrated to have the same moment about the point 0 as has the bar? Then answer the same question for the 2nd order moment. In particular, answer the second question for a homogeneous bar; that is to say, if m and $x - a$ are proportional.

41. If x and y are the Cartesian coordinates along a curve that intersects no line parallel to either axis in more than one point, prove that, if α and β are the terminal points of the curve,

$$\int_{\alpha}^{\beta} y\,dx + \int_{\alpha}^{\beta} x\,dy = [x \cdot y]_{\alpha}^{\beta} = x\beta \cdot y\beta - x\alpha \cdot y\alpha.$$

Hint. Approximate the domain under the curve by that under a step line, and the domain to the left of the curve by a sum of horizontal rectangles that interlock with the steps under the step line.

Prove that

$$\int_a^b (g_1 + g_2)\,\mathrm{d}h = \int_a^b g_1\,\mathrm{d}h + \int_a^b g_2\,\mathrm{d}h; \quad \int_a^b g\,\mathrm{d}(h_1 + h_2) =$$

$$\int_a^b g\,\mathrm{d}h_1 + \int_a^b g\,\mathrm{d}h_2; \quad \int_a^b (\mathrm{p}g)\,\mathrm{d}(\mathrm{q}h) = \mathrm{pq}\int_a^b g\,\mathrm{d}h;$$

$$\int_a^b g\,\mathrm{d}h + \int_b^c g\,\mathrm{d}h = \int_a^c g\,\mathrm{d}h.$$

42. Find *(a)* $\int_{-1}^{2} j\,\mathrm{d}\,abs$; *(b)* $\int_{-1}^{2} j^3\,\mathrm{d}j^2$; *(c)* $\int_{-a}^{b} j^m\,\mathrm{d}j^n$.

Verify that the Stieltjes integral *(c)* equals $\int_{-a}^{b} (j^m \cdot \mathbf{D}j^n)$.

Hint. In *(a)*, use $\int_{-1}^{2} = \int_{-1}^{0} + \int_{0}^{2}$.

43. Let *s* be the distance traveled by a moving point, *t* the time. What is the meaning of the cumulation $\int_{T_0}^{T} s\,\mathrm{d}t$? What are its values for a falling body and for an oscillator satisfying $s = \cos t$?

11. DERIVATIVES AND THE RATE OF CHANGE OF ONE VARIABLE QUANTITY WITH RESPECT TO ANOTHER

Example 1 (Speed). A driver who traveled s miles in t hours is said to have made an average speed of s/t m.p.h.

Thus the average speed is not defined as an average of speeds (which would be in conformity with the common use of the word "average"). It is defined as the ratio of the distance traveled to the duration of the motion. Yet, s/t turns out to be intimately related to an average of speeds. Suppose, for instance, a driver starts at 1 p.m., accelerates almost instantaneously to a speed of 25 m.p.h. and travels with this constant speed until 1^{30}. At that instant he accelerates to a speed of 40 m.p.h. with which he travels until 2 p.m. Then he decelerates to 28 m.p.h. and travels with this constant speed until 2^{30}, when he almost instantaneously stops. He traveled 12.5, 20, and 14 miles during the three half-hour periods, thus altogether 46.5 miles in 1.5 hours. His average speed in m.p.h. was 46.5/1.5 = 31. Also the average of the three speeds, 25, 40, and 28, is 31. (On the other hand, the average of the initial and terminal speed is 0 and has nothing to do with the average speed.)

What is the speed of a moving car at a certain instant? As with regard to the slope of a curve at a certain point, two attitudes are possible (see p. 131): (I) that all one has to do is to determine the speed numerically; (II) that the term "instantaneous speed" should be first *defined,* since it can be taken for granted only if the motion is uniform (i.e., if in any two periods of equal length equal distances are traveled). In this special case, the average speed (which is the same in any interval) may be called the speed at any instant.

One may define the speed v as the class of all pairs $(\phi, v\phi)$ for any act ϕ of reading the speedometer, just as t and s are the classes of pairs $(\tau, t\tau)$ and $(\sigma, s\sigma)$. In this case, the question arises how speedometers should be constructed and what gauges should be called speedometers.

Let $t_0 = t\tau_0$ and $t_1 = t\tau_1$ be results of reading the clock, and s_0 and s_1 the simultaneously obtained results of reading the mileage gauge.

Then

$$\frac{s_1 - s_0}{t_1 - t_0} \tag{11}$$

is the average speed between the instants t_0 and t_1. If, relative to the pairing by simultaneity (which here and in Examples 2 and 3 will be tacitly understood), s is connected with t by the function f, then (11) equals

$$\frac{f(t_1) - f(t_0)}{t_1 - t_0}. \tag{12}$$

The limit of the difference quotients (12), which is $\mathbf{D}ft_0$, may be considered as the definition of the value of the speed at the instant t_0. (This is the number read on a speedometer as the result of an act performed simultaneously with τ_0.)

The speed, as a result of operating on s and t, is denoted by $\frac{ds}{dt}$, a symbol that is reminiscent of the difference quotients (11). Hence the following theorem:

if $$s = f(t), \text{ then } \frac{ds}{dt} = \mathbf{D}f(t) \text{ or } v = \mathbf{D}f(t)\,.$$

Suppose, for instance, a car that has traveled 12,000 miles is started at 2 p.m. in such a way that

$$s = \begin{cases} 12{,}000 + 400\,(t-2)^2 & \text{between 2 p.m. and } 2^{03} \\ 12{,}001 + 40\,(t - 2^{03}) & \text{between } 2^{03} \text{ and 3 p.m.} \end{cases}$$

Then

$$v = \begin{cases} 800\,(t-2) & \text{in the first 3 minutes} \\ 40 & \text{thereafter.} \end{cases}$$

Since there are only 26 letters while the number of c.c.q.'s is unlimited, different *ad hoc* uses of the same letter are inevitable. In Example 2, the letter v will not designate velocity.

Example 2 (Time Rate of Expansion). Let v denote the volume of an expanding sphere, and let v_0 and v_1 be the volumetric results obtained at the instants t_0 and t_1, respectively. Then

$$\frac{v_1 - v_0}{t_1 - t_0}$$

is called the average time rate of change of the volume between the instants t_0 and t_1. If $v = g(t)$, then that difference quotient equals

$$\frac{g(t_1) - g(t_0)}{t_1 - t_0}.$$

Its limit, as t_1 approaches t_0, which is $\mathbf{D}\, g\, t_0$, is called an instantaneous rate of change of v, namely, the rate of expansion at the instant t_0. The variable quantity $\frac{d\,v}{d\,t}$ is connected with t by the function $\mathbf{D}\, g$. For instance, if $f = exp$, then $\mathbf{D}\, f = exp$; hence if $v = exp\ t$, then $\frac{d\,v}{d\,t} = exp\ t$; that is to say, if the volume at the instant t is e^t, then its rate of change at that instant is e^t.

Example 3 (Rate of Radial Expansion). Let r be the radius of the sphere just mentioned. If (r_0, v_0) and (r_1, v_1) are pairs of results of simultaneous length and volumetric observations, then

$$\frac{v_1 - v_0}{r_1 - r_0}$$

is called the average rate of change of v with regard to r between r_0 and r_1. If v is measured in the cube of the linear unit, then $v = \frac{4}{3}\,\pi\, r^3$; that is to say, v is connected with r by $\frac{4}{3}\,\pi\, j^3$. Consequently, $\frac{d\,v}{d\,r}$ is connected with r by the function $\mathbf{D}\left(\frac{4}{3}\,\pi\, j^3\right) = 4\,\pi\, j^2$; that is to say, $\frac{d\,v}{d\,r} = 4\,\pi\, r^2$.

Chapter VI contains the definitions of the derivatives of functions such as exp and $\frac{4}{3}\,\pi\, j^2$. The derivative of consistent classes of quantities other than functions has not been defined. Symbols such as $\mathbf{D}\, v$, $\mathbf{D}\, s$, $\mathbf{D}\, t$ and $\mathbf{D}\, (sin\ t)$

remain undefined throughout this book, just as **D** 3 and the derivatives of a sphere or a chair. What one may derive is the function connecting one variable quantity with another, for instance, v with t or v with r.

More generally, let the consistent class of quantities w be connected with the c.c.q. u by the function f relative to the pairing Π of Dom w with Dom u. If (u_0, w_0) and (u_1, w_1) are pairs of values of u and w for corresponding elements of Dom u and Dom w, then

(13) $$\frac{w_1 - w_0}{u_1 - u_0}, \text{ which equals } \frac{f(u_1) - f(u_0)}{u_1 - u_0},$$

is called the average rate of change of w with regard to u between u_0 and u_1. The limit of (13) as u_1 approaches u_0, that is, $\mathbf{D}\, f\, u_0$, may be considered as the definition of a value of a consistent class of quantities, $\frac{\mathbf{d}\, w}{\mathbf{d}\, u}$, whose domain is a subclass of the domain of u; more specifically, $\mathbf{D} f u_0$ is the value of $\frac{\mathbf{d}\, w}{\mathbf{d}\, u}$ for a_0, if $u_0 = u\, a_0$. Hence the following

BASIC SCHEME FOR THE APPLICATION OF DERIVATIVES TO CONSISTENT CLASSES OF QUANTITIES. *If* $w = f(u)$, *then* $\frac{\mathbf{d}\, w}{\mathbf{d}\, u} = \mathbf{D}\, f(u)$.

$\frac{\mathbf{d}\, w}{\mathbf{d}\, u}$ is called the rate of change of w with regard to u.

Here, Dom w is paired with Dom u by Π, and Dom $\frac{\mathbf{d}\, w}{\mathbf{d}\, u}$ with Dom u by the identity pairing. For instance,

(14) $$\text{If } w = \sin u, \text{ then } \frac{\mathbf{d}\, w}{\mathbf{d}\, u} = \cos u.$$

Replacing w by $\sin u$ one may write

(15) $$\frac{\mathbf{d} \sin u}{\mathbf{d}\, u} = \cos u \text{ for any consistent class of quantities, } u.$$

Here, u is a c.c.q. variable and not a numerical variable as is x in $\mathbf{D} \sin x = \cos x$ for any x such as 1 or π. One cannot in (14) replace u by π and w by 0, even though $0 = \sin \pi$. But one may replace u and w by specific variable quantities; for instance, u by the time, and s by the position of a harmonic oscillator. Indeed,

$$\text{if } s = \sin t, \text{ then } \frac{\mathbf{d}\, s}{\mathbf{d}\, t} = \cos t \text{ or } \frac{\mathbf{d} \sin t}{\mathbf{d}\, t} = \cos t.$$

Example 4. If w is any constant variable quantity, and u is nonconstant, then all difference quotients (13) are equal to 0, and $\frac{\mathbf{d}\, c}{\mathbf{d}\, u} = O$. On the other hand, for obvious reasons, $\frac{\mathbf{d}\, w}{\mathbf{d}\, c}$ remains undefined.

If $w = u$ and Π is the identity pairing, then all quotients (13) have the value 1, wherefore $\frac{\mathrm{d}\,u}{\mathrm{d}\,u} = 1$. Here, *1* and *0* are the constant c.c.q.'s having the same domain as *u*, while their ranges consist of the single numbers 1 and 0, respectively. The last formula is also a consequence of the basic scheme, since *u* is connected with *u* by *j*.

Example 5 (Slope). If *u* and *w* are replaced by the abscissa and ordinate along the curve *f*, then $\frac{\mathrm{d}\,y}{\mathrm{d}\,x} = \mathbf{D}\, f\,(x)$ is the slope of the curve. For instance, if $y = \sin x$, then $\frac{\mathrm{d}\,y}{\mathrm{d}\,x} = \cos x$. The reader can easily determine the geometric meaning of $\frac{\mathrm{d}\,x}{\mathrm{d}\,y}$ along a curve in a plane with a Cartesian frame.

Example 6 (Acceleration). Consider a motion for which $s = f\,(t)$ and hence $v = \mathbf{D}\, f\,(t)$. The quotient

$$\frac{v_1 - v_0}{t_1 - t_0} = \frac{\mathbf{D}\, f\, t_1 - \mathbf{D}\, f\, t_0}{t_1 - t_0}$$

is called the average acceleration between t_0 and t_1. The limit $\mathbf{D}^2 f\, t_0$ may be considered as a definition of the acceleration at the instant t_0. Hence

if $$s = f\,(t), \text{ then } \frac{\mathrm{d}\,v}{\mathrm{d}\,t} = \mathbf{D}^2\, f\,(t)\,.$$

Instead of $\frac{\mathrm{d}\,v}{\mathrm{d}\,t}$ or $\frac{\mathrm{d}}{\mathrm{d}\,t}\left(\frac{\mathrm{d}\,s}{\mathrm{d}\,t}\right)$ one writes also $\frac{\mathrm{d}^2\,s}{\mathrm{d}\,t^2}$ and calls the acceleration the *second rate of change* of s with *t*.

Example 7 (Functions). The consistent classes of quantities in the Basic Scheme may be replaced by functions; for instance, *u* by j^2, and *w* by j^8. Then $f = j^4$ and

$$\frac{\mathrm{d}\,j^8}{\mathrm{d}\,j^2} = \mathbf{D}\, j^4\,(j^2) = 4\,j^3\,(j^2) = 4\,j^6\,.$$

If *u* is replaced by *j*, and *w* by *f*, then (since $fj = f$ and $\mathbf{D}\,fj = \mathbf{D}\,f$)

$$\frac{\mathrm{d}\,f}{\mathrm{d}\,j} = \mathbf{D}\,f.$$

In words: the rate of change of a function with regard to *j* is the derivative of that function. In particular,

$$\frac{\mathrm{d}\,\sin}{\mathrm{d}\,j} = \mathbf{D}\,\sin = \cos.$$

Both **D** and $\frac{\mathrm{d}}{\mathrm{d}}$ are called *operators*. But they are different.

D associates a function with a function; for instance, *cos* with *sin*;

$\frac{\mathrm{d}}{\mathrm{d}}$ associates a fluent with two fluents; for instance, v with s and t; and *cos* with *sin* and j.

The operator **D** may be considered as a restriction of the operator $\frac{\mathrm{d}}{\mathrm{d}}$ to functions with j in the denominator; $\mathbf{D} = \frac{\mathrm{d}}{\mathrm{d}\,j}$. Similarly, if the a-integral of f were denoted by $\mathbf{S}_a f$, rather than $\int_a f$, then $\mathbf{S}_a = \int_a \mathrm{d}\,j$, that is, $\mathbf{S}_a$ is the restriction to functions and to an initial number of the operator $\int \mathrm{d}$ (with j as the second function).

If v denotes spherical volume, then, in the symbols $\frac{\mathrm{d}\,v}{\mathrm{d}\,t}$ and $\frac{\mathrm{d}\,v}{\mathrm{d}\,r}$, the letters t and r are as important as is v. Similarly, *sin* and j are equally important in $\frac{\mathrm{d}\,sin}{\mathrm{d}\,j}$, But the numerical variable x in **D** *sin* x = *cos* x is not as important as are the functions *sin* and *cos*.

The preceding ideas are summarized in the following table concerning the connection of sine and cosine in the calculus of derivatives.

This connection can be

expressed in terms of the operator:	**D**	$\frac{\mathrm{d}}{\mathrm{d}}$
using variables:	**D** *sin* x = *cos* x for any x	$\frac{\mathrm{d}\,sin}{\mathrm{d}\,j}\,x = cos\ x$
	D *sin* a = *cos* a for any a	$\frac{\mathrm{d}\,sin}{\mathrm{d}\,j}\,a = cos\ a$
	. . . .	. . . . . .
without variables:	**D** *sin* = *cos*	$\frac{\mathrm{d}\,sin}{\mathrm{d}\,j} = cos$
and be applied (1) to specific fluents: (a) the time t		
using variables:	(**D** *sin*) $t\ r = cos\ t\ r$ *for any* r	$\frac{\mathrm{d}\,sin}{\mathrm{d}\,t}\,r = cos\ t\ r$
without variables:	(**D** *sin*) $t = cos\ t$	$\frac{\mathrm{d}\,sin\ t}{\mathrm{d}\,t} = cos\ t$

(continued)

(b) the abscissa x along the sine curve		
using point variables	(**D** *sin*) x σ = *cos* x σ for any σ	$\frac{d\ sin\, x}{d\ s}\sigma = cos\, x\, \sigma$
without variables:	(**D** *sin*) x = *cos* x	$\frac{d\ sin\ x}{d\ x} = cos\ x$
(2) to fluent variables		
using element variables:	(**D** *sin*) *u* a = *cos* *u* a for any a in Dom *u*	$\frac{d\ sin\ u}{d\ u} a = cos\, u\, a$
without element variables:	**D** *sin* *u* = *cos* *u* for any c.c.q., *u*	$\frac{d\ sin\ u}{d\ u} = cos\ u$

To simplify the reading, parentheses have been introduced in (**D** *sin*) *t*. But they are unnecessary since **D** *(sin t)* is an undefined symbol. On the other hand, d *sin* being undefined, $\frac{d\ sin\ t}{d\ t}$ cannot mean anything but $\frac{d\ (sin\ t)}{d\ t}$.

In the literature, the theorem **D** *sin* = *cos* is frequently expressed in the formula

$$\frac{d \sin x}{dx} = \cos x$$

and the (italicized) *x* is interpreted variously in its several appearances

(a) as a numerical variable, which is replaced, e.g., by 1 or π;
and/or (b) the identity function *j*;
and/or (c) a specific variable quantity, namely the abscissa along a sine curve;
and/or (d) as a fluent variable, which is replaced, e.g., by the time *t* or the abscissa *x*.

The foregoing analysis should enable the reader of traditional texts to decide in which of these basically different meanings the letter is used at a given place.

Newton called time rates of variable quantities *"fluxions"* of *"fluents"* and denoted them, as early as 1665, by "pricked" letters. To this day, in mechanics, time rates are often denoted by dots; e.g., the velocity by $v = \dot{s}$ and the acceleration by $\dot{v} = \ddot{s}$. Outside of mechanics, Newton's dots have been superseded by Lagrange's primes and, of course, by Leibniz' and Cauchy's symbols (see p. 153 f.).

EXERCISES

44. If, for an oscillator, $s = -3\ sin\ t$, find the average speed and average acceleration between any two instants, and the instantaneous speed and acceleration. Are the latter v.q.'s functions of *s*, and if so, what functions of *s*?

45. For an expanding sphere, find the rate of change (a) of the volume *v* with respect to the surface *s*; (b) of *s* with respect to *v*. For

what value of the radius, r, are the values of $\frac{d\,v}{d\,r}$ and $\frac{d\,s}{d\,r}$ equal? Can one say at what instant these values are equal? Can one answer the last question if one is told that $r = \exp t$? Assuming that $r = 1 + t^2$, find $\frac{d\,v}{d\,t}$ and $\frac{d\,s}{d\,t}$ or (in Newton's notation) $\dot{v}$ and $\dot{s}$. Are the values of $\dot{v}$ and $\dot{s}$ equal at the instant at which $\frac{d\,v}{d\,r}$ and $\frac{d\,s}{d\,r}$ assume equal values?

46. The curvature of a semi-circle of radius r in the position of a cup (a cap) is defined as the number 1/r (the number −1/r). The curvature of a semi-circle passing through the three points of a curve f above a − h, a, and a + h is called *approximate curvature* of f at a. The *curvature* of f at a is the limit (if it exists) of that approximate curvature as h approaches 0. For the parabola c j^2, prove that the curvature at 0 equals $\mathbf{D}^2\,(c\,j^2)\,0$, whereas the curvature at 1 is not equal to $\mathbf{D}^2\,(c\,j^2)\,1$. What is the curvature of j^3 at 0?

★47. Show that, for any curve f, for which $\mathbf{D}^2 f$ is continuous, the curvature equals $\mathbf{D}^2\,f\,a$ at any a for which $\mathbf{D}\,f\,a = 0$. More generally, the curvature of f is $\frac{\mathbf{D}^2\,f\,a}{[1 + (\mathbf{D}\,f\,a)^2]^{3/2}}$ at any a. Thus the variable quantity $\frac{d^2\,y}{d\,x^2}$ along a curve is not very significant. What is the meaning of the sign of the values of this variable quantity? (See Exercise 6, p. 24).

48. The federal income tax t is connected with the taxable income i by a function defined as follows

if	$i \leq 2000$	$2000 \leq i \leq 4000$	$4000 \leq i \leq 6000$	$6000 \leq i \leq 8000$ ...
then $t=$	$.2\,i$	$400 + .22\,(i - 2000)$	$840 + .26\,(i - 4000)$	$1,360 + .30\,(i - 6000)$...

Show that, for $i \leq 8000$, the v.q. t is a continuous function of i having a discontinuous derivative.

49. Determine $\frac{d\,j^3}{d\,j^2}$, $\frac{d\,j}{d\,g}$, $\frac{d\,j}{d\,abs}$, $\frac{d\,j^2}{d\,abs}$.

★50. Prove that the average speed defined by (11) equals the mean of the speeds between t_0 and t_1. Hint. From $v = \mathbf{D}\,f\,(t)$ prove, according to the Basic Scheme for the application of integrals that

$$\int_{\tau_0}^{\tau_1} v\,d\,t = \int_{t_0}^{t_1} \mathbf{D}\,f.$$

Applying the Second Reciprocity Law of Calculus prove that the integral on the right side equals $s_1 - s_0$. The cumulation on the left side may, according to p. 207, be replaced by the product of $t_1 - t_0$ and the mean of v between t_0 and t_1.

12. WHAT IS THE SIGNIFICANCE OF CALCULUS IN SCIENCE?

Important problems in applying calculus arise from the fact that functional connections between scientific variable quantities, established on the basis of observations, are only approximate.

a. The Operational Significance of Integrals. Slight changes of a curve affect the area under it only slightly (see p. 38). Hence the approximate knowledge of a curve yields conclusions about the approximate area under it. Similarly, from an approximate connection of the force with the time t or the position s, one can infer an approximate connection of the impulse with t or of the work with s.

From the mere information that the values of the acceleration in ft./sec.^{2} are between $32 + c$ and $32 - c$, one can conclude that

$$(32 - c)\, t < v < (32 + c)\, t$$

and that the average speeds during short intervals are between $32 + c$ and $32 - c$.

b. The Operational Insignificance of Derivatives. The slope of a curve may, if the latter is altered ever so slightly, change completely (see p. 37). Hence the approximate knowledge of a curve does not yield any conclusion whatever about its slope at a point. Similarly, from an approximate connection of the speed with the time, one cannot draw any inferences whatever concerning the connection of the acceleration with the time.

For instance, if $31.9\, t < v < 32.1\, t$, then, simultaneously with the values 1 and 1.01 of t, one may observe the following values of v:

(a) 31.9 and 32.4, thus an average acceleration of $.5/.01 = 50$;

(b) 32.1 and 32.22, thus an average acceleration of $.12/.01 = 12$.

In fact, the inequality for v is compatible with

$$v = 32 - .1 \cos (10^6\, t) \text{ and } \frac{dv}{dt} = 10^5 \sin (10^6\, t).$$

c. How Does Calculus Connect the Laws of Falling Bodies? Many mathematicians and physicists have claimed that

$s = 16\, t^2$ implies $v = 32\, t$ by virtue of differential calculus;

$v = 32\, t$ implies $s = 16\, t^2$ by virtue of integral calculus.

Considering the two formulas as laws of observation, however, one can admit only the second of these implications.

What is here under discussion is the logico-mathematical relation between the two physical laws. That either law is valid can be verified by experiments, e.g., on a simple machine constructed by Atwood (1784) – essentially an easily running pulley over which a thread is thrown. On either end, a weight of 1 lb. is attached to the thread. If not disturbed, the two weights are at rest. Now put an extra ounce on top of one of the two pounds. At that instant, due to gravity acting on this ounce, the 33 oz. start an accelerated motion: 17 oz. move downward, while the pound on the other side is moving upward. Since the force acting on 1 oz. has to move 33 oz., the normal acceleration of 32 ft./sec.^{2} is reduced in the ratio 33: 1, that is, to approximately 1 ft./sec^2.
If, t seconds after the start of the motion, the extra ounce is lifted, then, due to inertia, the two pounds continue moving – but uniformly, with a constant speed equal to their speed at the instant when the ounce was lifted. Hence by the simple process of measuring that uniform speed (i.e., by observing the number of ft. travelled during one second) one determines the instantaneous speed of the accelerated motion t seconds after its start. Repeated observations yield, within the limits of accuracy, $v = t$. Similarly, by repeated observations of the distance travelled along a vertical scale, one finds, within the limits of accuracy, $s = \frac{1}{2} t^2$. These results are in agreement with Galileo's $v = 32\ t$ and $s = 16\ t^2$ for his v and s.

Having verified $v = 32\ t$ within the limits of observation, one can, in presence of the law of inertia, prove that s must be approximately $16\ t^2$. One thus can predict the latter law without actual observation. But having verified $s = 16\ t^2$ (no matter how small the possible inaccuracy) even under assumptions about inertia one cannot prove that $v = 32\ t$. All that can be predicted are the approximate means of v during various periods. Values of v (even approximate values) can be ascertained only by other observations. Other observations indeed establish $v = 32\ t$ and thereby identify the directly observable variable quantity v with the rate of change of s with respect to t.

$s = 16\ t^2$ within limits of accuracy would be compatible, e.g., with the observation that, 1 second after the start of the motion, the speed is 96 or 0 or −32 rather than 32; in other words, that in Atwood's experiment, when the extra ounce is lifted, the two pounds move three times as fast as they actually do or stop or even reverse their motion. Thus the preceding analysis of the application of calculus to the variable quantities of classical mechanics approaches the latter somewhat to statistical mechanics and to quantum mechanics.

There are other cases where an observable variable quantity u can be identified with the rate of change of a second fluent with respect to a third. But each such identification presupposes independent observations of u. The only consistent classes of quantities whose rates of change are amenable to proofs are functions. Newton's program of applying the results concerning the rates of functions to the determination of the fluxions of physical fluents – one of the most fertile ideas ever conceived – cannot be carried out operationally.

Some c.c.q.'s that used to be considered as *fluents* with *fluxions* seem to be rather *salients* with *saltus* (e.g., the energy, see p. 201). Others, especially the resultants of large numbers of salients, are what might be called *tremblents* to whose connections with the time one can apply integral but not differential calculus; e.g., the position of a smoke particle in a Brownian motion ascribed to many irregular impacts of gas molecules. The ink curves drawn by self-registering barometers and thermometers in meteorological stations (i.e., the graphs of the functions connecting atmospheric pressure and temperature with the time) reminded Boltzmann of curves without tangents (see p. 134).

13. OPTIMAL DIFFERENCES FOR THE APPROXIMATE DETERMINATION OF SLOPES

The tangent to the parabola j^2 at .75 is the line $-.75^2 + 1.5\,j$. Joining the point of j^2 above .75 to its neighbor point above a, one obtains the line $-.75\,a + (.75 + a)\,j$. If a = .8, the line is $-.6 + 1.55\,j$; if a = .76, it is $-.57 + 1.51\,j$ – an even better approximation to the tangent.

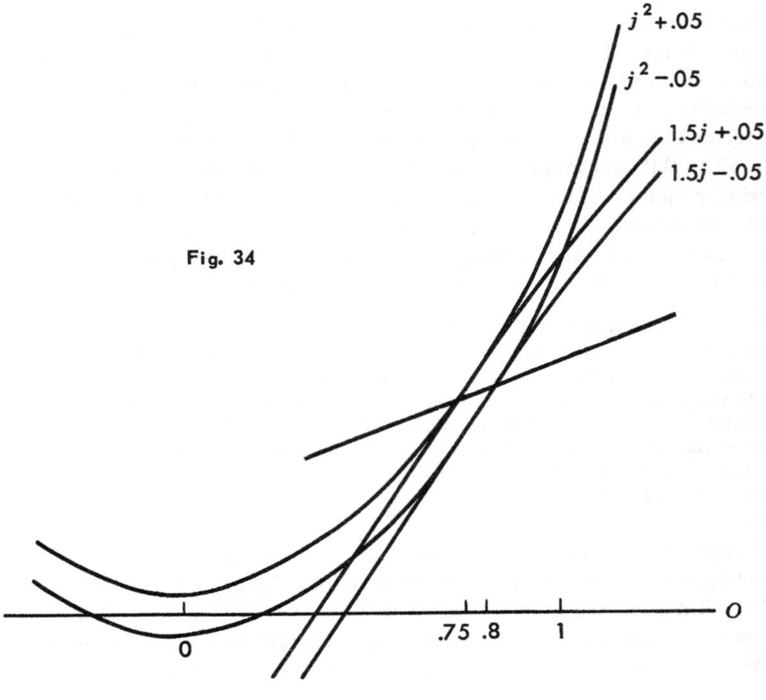

Fig. 34

Now suppose that the parabola is given only within d linear units; in other words, that the parabolic strip between $j^2 + d/2$ and $j^2 - d/2$ is given. The straight strip of slope 1.5 between $-.75^2 + d/2 + 1.5\,j$ and $-.75^2 - d/2 + 1.5\,j$ may be said to be tangential to the parabolic strip (see Fig. 34).

What points above .75 and a (within the parabolic strip) should one join to obtain a good approximation to that tangential strip? Clearly, the best choice are the two points on the parabola itself, i.e., the points at the altitudes $.75^2$ and a^2. The worst choice would be one point on the upper rim, and one point on the lower rim of the strip, that is,

	either the altitudes	or the altitudes
above .75:	$.75^2 - d/2$	$.75^2 + d/2$
above a:	$a^2 + d/2$	$a^2 - d/2$
yielding the slope:	$\dfrac{(a^2 + d/2) - (.75^2 - d/2)}{a - .75}$	$\dfrac{(a^2 - d/2) - (.75^2 + d/2)}{a - .75}$
	$= .75 + a + \dfrac{d}{a - .75}$	$= .75 + a - \dfrac{d}{a - .75}$.

Suppose, for instance, that d = .01; that is to say, that the strip between $j^2 + .005$ and $j^2 - .005$ is given.

First, set a = .8. Joining the point of altitude $.75^2 - .005 = .5575$ above .75 to the point of altitude $.8^2 + .005 = .645$ above .8, one obtains the slope .0875/ .05 = 1.75. In Fig. 34, the line joining the highest point above .75 to the lowest point above .8 has the slope .47.

Next, set a = .76. Joining the point of altitude .5575 to the point of altitude $.76^2 + .005 = .5826$ above .76, one obtains the slope .0251/.01 = 2.51, which differs from the slope 1.5 of the tangential strip much more than does 1.75.

Thus the choice of closer neighbor points within the strip may result in considerably worse approximations to the slope. In fact, it is clear that, near any point above .75, the parabolic strip contains points which, when joined to the former, yield the slope 1000, and others which yield the slope −100.

If one cannot precisely determine the altitude of a point within the strip, then it is important to make sure that even the worst choice does not result in unnecessary inaccuracies of the approximate slope. The question therefore arises: For which point, a, near .75 has the steepest line joining a point above a to a point above .75 a slope as close as possible to 1.5? In other words, for which number a is

$$\left| .75 + a + \frac{.005}{a - .75} - 1.5 \right|$$

as small as it can be? The solutions of this simple minimum problem are .85 and .65. For a = .85, the steepest chord has the slope 1.7, which is indeed closer to 1.5 than is 1.75, obtained by means of .8. Thus .85 – .75 = .1 might be called the *optimal difference for the approximate determination of the slope.* Both smaller differences (as .8 – .75) and larger differences (as .9 – .75) result in less favorable approximations if the points within the strip are poorly chosen.

More generally, in the strip between $j^2 + d/2$ and $j^2 - d/2$ joining the highest point above c to the lowest point above b results in the slope $b + c + d/(c - b)$. The reader should verify that, for given b, this slope is as close as possible to the slope, 2b, of the tangential strip at b if

$$c = b + \sqrt{d} \quad \text{or} \quad c = b - \sqrt{d},$$

in which cases the slope differs from 2b by $2\sqrt{d}$.

To approximate the slope of the tangent to an inaccurately given curve at a, *one should join a point above* a *with a point above a neighbor number that is neither too far from* a *nor too close to* a.

Similarly, if $s = t^2$ within d linear units, then to find the approximate speed at a certain instant t_0, one should determine the average speed between t_0 and $t_0 + \sqrt{d}$ or $t_0 - \sqrt{d}$.

EXERCISE

51. For six positive numbers a, b, c, and d, d′, d″, determine the best approximation to the slope at x of the parabolic strip between

$$(a + d)\, j^2 + (b + d')\, j + c + d'' \text{ and } (a - d)\, j^2 + (b - d')\, j + c - d''.$$

CHAPTER VIII

THE CALCULUS OF DERIVATIVES

In this branch of calculus, the derivatives of functions that are obtained from simpler functions by addition, multiplication, substitution, etc. are determined by, as it were, algebraic procedures. They are expressed in terms of those simpler functions and the derivatives of the latter. In particular, one can in this way determine the derivatives of all elementary functions and prove that also these derivatives are elementary.

1. CONVENTIONS CONCERNING THE REACH OF THE SYMBOL D

Unambiguous statements concerning derivatives presuppose a clarification of the use of the symbol **D**.

Basic Convention. *Only the immediately following function is within the reach of* **D**; that is to say,

	stands for	and not for	For instance,	
$\mathbf{D} f + g$	$(\mathbf{D} f) + g$	$\mathbf{D}(f + g)$	$\mathbf{D} j^4 + j^2 = 4j^3 + j^2$;	$\mathbf{D}(j^4 + j^2) = 4j^3 + 2j$
$\mathbf{D} f \cdot g$	$(\mathbf{D} f) \cdot g$	$\mathbf{D}(f \cdot g)$	$\mathbf{D} j^4 \cdot j^2 = 4j^5$;	$\mathbf{D}(j^4 \cdot j^2) = 6j^5$
$\mathbf{D} fg$	$(\mathbf{D} f)g$	$\mathbf{D}(fg)$	$\mathbf{D} j^4 j^2 = 4j^6$;	$\mathbf{D}(j^4 j^2) = 8j^7$

The derivative of a number remains undefined. $\mathbf{D} j^2 3$ cannot mean $\mathbf{D}\, 9$. It cannot mean anything but the value of $\mathbf{D} j^2$ for 3, which is 6.

EXERCISE

1. Give examples of two functions f and g such that

(a) $\mathbf{D}(f \cdot g) \neq \mathbf{D} f \cdot \mathbf{D} g$;

(★b) $\mathbf{D}(f \cdot g) = \mathbf{D} f \cdot \mathbf{D} g$ (such an example is much harder to find);

(c) $\mathbf{D}(fg) \neq \mathbf{D} f \mathbf{D} g$ and $\mathbf{D}(fg) \neq \mathbf{D} fg$ and $\mathbf{D}(fg) \neq f \mathbf{D} g$, that is to say, an example in which $\mathbf{D}(fg)$ cannot be obtained by substituting $\mathbf{D} g$ into $\mathbf{D} f$, nor by substituting g into $\mathbf{D} f$, nor by substituting $\mathbf{D} g$ into f.

(d) $\mathbf{D}(fg) = \mathbf{D} f g$.

2. THE SUM RULE

For any two functions f and g and any number a for which f as well as g has a derivative, $\mathbf{D}(f+g)\,a$ is obtained by the Three-Step Rule as follows.

Step I. Write down the difference quotient of $f+g$ at a:

$$\frac{(f+g)-(f+g)\,a}{j-a}\text{, that is, }\frac{(f+g)-(fa+ga)}{j-a}.$$

Step II. Obtain a handier form of the difference quotient by regrouping the terms in the numerator:

$$\frac{(f+g)-(fa+ga)}{j-a}=\frac{(f-fa)+(g-ga)}{j-a}=\frac{f-fa}{j-a}+\frac{g-ga}{j-a}.$$

Step III. Compute the limit at a of the transformed difference quotient. Since, in the new form, it is the sum of two quotients, its limit (by the Law about the Limits of Sums, p. 115) equals the sum of the limits of the two quotients, that is,

$$\lim_{a}\frac{f-fa}{j-a}+\lim_{a}\frac{g-ga}{j-a}.$$

Now the limit at a of $\frac{f-fa}{j-a}$ is $\mathbf{D}f\,a$ and that of $\frac{g-ga}{j-a}$ is $\mathbf{D}g\,a$.

Hence

(1) $\mathbf{D}(f+g)\,a=\mathbf{D}f\,a+\mathbf{D}g\,a$ for any a in Dom $\mathbf{D}f$ and Dom $\mathbf{D}g$.

Some beginners ask: Why is the limit at a of $\frac{f-fa}{j-a}$ equal to $\mathbf{D}f\,a$? Answer: $\mathbf{D}f\,a$, by the very definition of this symbol, is the limit at a of $\frac{f-fa}{j-a}$. Hence the limit at a of $\frac{f-fa}{j-a}$ is $\mathbf{D}f\,a$. Others ask: Why all this reasoning? Is it not self-evident that the derivative of a sum of two functions equals the sum of their derivatives? Answer: Certainly it is not always true that the derivative of the product of two functions is equal to the product of their derivatives (see Exercise 1a).

If one knows about two curves f and g that, at 4, their slopes are 5 and -3, respectively, then, even if he knows nothing else about the curves (not even their altitudes at 4), he can affirm that the curve $f+g$ (obtained by superposition of f and g) has the slope $5+(-3)=2$ at 4.

Since, for any a, the value at a of $\mathbf{D}(f+g)$ equals the sum of the values at a of $\mathbf{D}f$ and $\mathbf{D}g$, the definition of the sum of two functions applied to the derivatives yields the following

SUM RULE. $\mathbf{D}(f+g)=\mathbf{D}f+\mathbf{D}g$ (on Dom $\mathbf{D}f\cdot$ Dom $\mathbf{D}g$), where the legend indicates that the formula is valid for any number belonging to both Dom **D** f and Dom **D** g.

The slope curve of a curve obtained by the superposition of two curves can be found by the superposition of the two slope curves. Also one might say: If $s = f + g$, then **D** s = **D** f + **D** g. By induction, the Sum Rule can be extended to any finite number of terms:

$$\mathbf{D}\,(f_1 + f_2 + \dots + f_n) = \mathbf{D}\,f_1 + \mathbf{D}\,f_2 + \dots + \mathbf{D}\,f_n\,.$$

EXERCISES

2. Find **D** $(f + g)$ for
(a) $f = j$ and $g = \mathit{sin}$; *(b)* $f = j$ and $g = \mathit{exp}$; *(c)* $f = j$ and $g = j^3$.
Then draw the graphs of $j + \mathit{sin}$, $j + \mathit{exp}$, and $j + j^3$ and determine their slope curves graphically. Where, if anywhere, are the slopes of the three curves 0, 1, -1?

Neither the function *abs* nor the function $-$*abs* has a derivative at 0. What about the sum of these functions? Show that if f has a derivative at a, whereas g has not, then $f + g$ has no derivative at a.

3. Let v stand for a positive number. Consider right prisms ("boxes") of volume v with square bases. *(a)* For what height of the prism and what side of the base is the surface of the prism as small as it can be? (Hint. Express first the height, and then the surface, in terms of the side of the base. Then find the number minimizing the function connecting the surface with the side. Finally determine the corresponding value of the height.) *(b)* Answer the same question for the surface minus the area of the top square (the "box without cover"). Then answer corresponding questions for a circular cylinder of volume v, with and without its top circle. Questions of this type are of great practical importance since the answers indicate how to build boxes or cans of a given volume with a minimum expenditure of paper or tin. Finally, let s stand for a positive number and consider all plane domains of area s consisting of four rectangles and a square that can be folded into a box without cover. For what side of the square and what height of the rectangles is the volume of the box as large as it can be? Compare the answer with that to *(b)*.

4. Find *(a)* $\mathbf{D}^k\,(f_1 + f_2 + \dots + f_n)$, the k–th derivative of the sum, for any two positive integers k and n; *(b)* **D** $(f - g)$. In *(a)*, one can use the Sum Rule; in *(b)*, only the method by which it was proved.

★5. Consider the function $f_a = \dfrac{a}{j - 1} + \dfrac{1}{j^2}$ for any positive number a, and determine **D** f_a and $\mathbf{D}^2 f_a$. There is exactly one number a such that

$\mathbf{D}\, f_a\, x = \mathbf{D}^2\, f_a\, x = 0$ for one and the same number x. Find this a and this x. A problem of this kind arises if the critical temperature for a van der Waals gas is to be determined. For such a gas there are three numbers r, b, and c such that pressure, volume, and temperature are related by $(p + c/v^2) \cdot (v - b) = r\, t$. The critical temperature is that value of t for which $\dfrac{d\,p}{d\,v} = \dfrac{d^2\,p}{d^2\,v} = 0$. Is there any critical temperature in an "ideal" gas satisfying Boyle's Law (that is, b = c = 0)?

3. THE PRODUCT RULE AND ITS CONSEQUENCES

For any two functions f and g and any number a for which f as well as g has a derivative, $\mathbf{D}\,(f \cdot g)$ a is obtained by the Three-Step Rule as follows.

Step I. Write down the difference quotient of $f \cdot g$ at a:

$$\frac{f \cdot g - (f \cdot g)\, a}{j - a}, \text{ that is, } \frac{f \cdot g \;-\; (fa \cdot ga)}{j - a}$$

Step II. In the numerator, subtract and add one and the same term, namely, $fa \cdot g$ (the product of the constant function fa and the function g), and transform the resulting expression into the sum of two products:

$$\frac{f \cdot g - fa \cdot ga}{j - a} = \frac{f \cdot g - fa \cdot g + fa \cdot g - fa \cdot ga}{j - a} = \frac{f - fa}{j - a} \cdot g + fa \cdot \frac{g - ga}{j - a}.$$

(From the point of view of algebra, the introduction of two terms which cancel one another complicates the expression. But the computation of the limit at a is easier in the new form.)

Step III. Compute the limit at a of the transformed difference quotient. Since, in the new form, it is the sum of two products, its limit is the sum of the limits of the two products

$$\lim_{a} \left(\frac{f - fa}{j - a} \cdot g\right) + \lim_{a} \left(fa \cdot \frac{g - ga}{j - a}\right).$$

By the Law about the Limits of Products (p. 115), the limit of either of these products equals the product of the limits of the factors, so that one obtains

$$\lim_{a} \frac{f - fa}{j - a} \cdot \lim_{a}\, g + \lim_{a}\, fa \cdot \lim_{a} \frac{g - ga}{j - a}.$$

The *first* limit is $\mathbf{D}\, f$ a. The *second* limit is g a, since from the fact that g has a derivative at a it follows that g is continuous at a, and therefore $\lim_{a} g = g\, a$. The *third* limit is $f\, a$ since fa is a constant function with the value $f\, a$ and therefore has everywhere the limit $f\, a$. The *fourth* limit is $\mathbf{D}\, g\, a$. Thus altogether

(2) $$\mathbf{D}\,(f \cdot g)\, a = \mathbf{D}\, f\, a \cdot g\, a + f\, a \cdot \mathbf{D}\, g\, a.$$

Example 1. Find $\mathbf{D}\,(exp \cdot sin)\, 1$. Since $\mathbf{D}\, exp = exp$ and $\mathbf{D}\, sin = cos$; $\mathbf{D}\,(exp \cdot sin)\, 1 = exp\, 1 \cdot sin\, 1 + exp\, 1 \cdot cos\, 1 = e \cdot (sin\, 1 + cos\, 1) =$

$$2.718\ldots \cdot (.841 + .540).$$

Example 2. The derivative of $(j^4 - j^2 + 2) \cdot (j^2 - 3j + 3)$ at 0 is $0 \cdot 3 + 2 \cdot (-3) = -6$; at 1, it is $2 \cdot 1 + 2 \cdot (-1) = 0$. To obtain these results, one need not first determine an elaborate expression for the derivative of the product and then evaluate this derivative for 0 and for 1. All one has to find are the values for 0 and 1 of the factors and their derivatives.

Example 3. Suppose that all one knows about two functions, f and g, is that $f\, 7 = 2$, $\mathbf{D}\, f\, 7 = -3$, $g\, 7 = 1$, $\mathbf{D}\, g\, 7 = 4$. Then by virtue of (2), he may conclude that $\mathbf{D}\,(f \cdot g)\, 7 = -3 \cdot 1 + 2 \cdot 4 = 5$.

Example 4. If a and b are the altitude and the base of an expanding cylinder, then, for the volume v of the cylinder: $v = a \cdot b$. Suppose that the functions connecting a and b to the time are unknown and that only the following information is available: at some instant

(1) a and b assume the values 3 and 4,

(2) their instantaneous rates of change are $-.1$ and $.2$, respectively.

At that instant, according to (2), the volume v (which assumes the value 12) changes with the rate $-.1 \cdot .4 + .2 \cdot 3 = .2$.

Just as (1) yields the Sum Rule, (2) yields the

PRODUCT RULE. $\mathbf{D}\,(f \cdot g) = \mathbf{D}\, f \cdot g + f \cdot \mathbf{D}\, g$ on Dom $\mathbf{D}\, f$ · Dom $\mathbf{D}\, g$.

For instance, $\mathbf{D}\,(j^3 \cdot j^5) = 3j^2 \,.\, j^5 + j^3 \cdot 5j^4 = 8j^7$, which confirms the previous result about $\mathbf{D}\, j^8$. Also the derivative of $exp \cdot sin$ which, by the Product Rule, is $exp \cdot sin + exp \cdot cos$, might be obtained directly by computing the limit of difference quotients of $exp \cdot sin$. But in so doing one would have to retrace the steps by which the general Product Rule was demonstrated (including the addition and subtraction of $exp\, a \cdot sin$ in the numerator of the quotient). The proof of the Product Rule obviates the necessity of computing the limit of difference quotients of particular products, such as $exp \cdot sin$.

According to Boyle, pressure, volume, and temperature (in proper units) of an "ideal" gas are related by the equality $p \cdot v = T$. Clearly, p, v, and T are functions of the time t and, by the Product Rule, at any instant

(3) $\dot{T} = \dot{p} \cdot v + p \cdot \dot{v}$ in Newton's notation; $\frac{dT}{dt} = \frac{dp}{dt} \cdot v + p \cdot \frac{dv}{dt}$

according to Leibniz.

If, for a process, the values at one and the same instant of four of the five variable quantities $\dot{T}, p, v, \dot{p}, \dot{v}$ are known, then the value of the fifth at that instant can be computed from (3). A process is called *isothermic* if in its course the temperature does not change. In this case $\dot{T} = O$. If at some instant the values of three of the quantities $p, v, \dot{p}, \dot{v}$ are known, that of the fourth can be computed.

If for a process the functions connecting some of the five variable quantities in (3) with the time are known, then other functions can be determined. For instance, if p and v are linear functions of the time, say, $p = 3 - .1\,t$ and $v = 4 + .2\,t$, then $\dot{p}$ and $\dot{v}$ are constant ($-.1$ and $.2$, respectively), and $\dot{T}$ is a linear function of the time,

$$\dot{T} = -.1\,v + .2\,p = .2 - .04\,t\,.$$

When does $\dot{T}$ assume the value 0? Clearly, when t assumes the value 5. From that instant on, the temperature of the gas decreases. When t assumes the value 6, the value of $\dot{T}$ is $-.04$.

Since the derivative of any constant function c is O, the Product Rule yields the

CONSTANT FACTOR RULE $\mathbf{D}(c \cdot g) = c \cdot \mathbf{D}g$ or $\mathbf{D}(c\,g) = c\,\mathbf{D}g$.

If q is the quotient of two functions, $q = f_1/f_2$, then $f_1 = q \cdot f_2$. According to the Product Rule,

$$\mathbf{D}f_1 = \mathbf{D}q \cdot f_2 + q \cdot \mathbf{D}f_2, \text{ whence } \mathbf{D}q = \frac{\mathbf{D}f_1 - q \cdot \mathbf{D}f_2}{f_2}\,.$$

If, in the numerator, q is replaced by f_1/f_2, and both numerator and denominator are multiplied by f_2, the result is the following

QUOTIENT RULE $\mathbf{D}\frac{f_1}{f_2} = \frac{f_2 \cdot \mathbf{D}f_1 - f_1 \cdot \mathbf{D}f_2}{f_2^2}$.

What is the derivative of the product of three functions? Clearly, $f_1 \cdot f_2 \cdot f_3 = (f_1 \cdot f_2) \cdot f_3$. Hence

$$\mathbf{D}(f_1 \cdot f_2 \cdot f_3) = \mathbf{D}(f_1 \cdot f_2) \cdot f_3 + (f_1 \cdot f_2) \cdot \mathbf{D}f_3$$

$$= (\mathbf{D}f_1 \cdot f_2 + f_1 \cdot \mathbf{D}f_2) \cdot f_3 + (f_1 \cdot f_2) \cdot \mathbf{D}f_3$$

$$= \mathbf{D}f_1 \cdot f_2 \cdot f_3 + f_1 \cdot \mathbf{D}f_2 \cdot f_3 + f_1 \cdot f_2 \cdot \mathbf{D}f_3\,.$$

For instance,

$$\mathbf{D}(j^2 \cdot \sin \cdot \exp) = 2j \cdot \sin \cdot \exp + j^2 \cdot \cos \cdot \exp + j^2 \cdot \sin \cdot \exp.$$

By induction, this reasoning yields the following

GENERAL PRODUCT RULE.

$$\mathbf{D}(f_1 \cdot f_2 \cdot \ldots \cdot f_n)$$
$$= \mathbf{D}f_1 \cdot f_2 \cdot \ldots \cdot f_n + f_1 \cdot \mathbf{D}f_2 \cdot \ldots \cdot f_n + \ldots + f_1 \cdot f_2 \cdot \ldots \cdot \mathbf{D}f_n .$$

If each of the n factors is equal to f, then each term on the right side is equal to $f^{n-1} \cdot \mathbf{D}f$. Hence, for any positive integer n, the

POWER RULE. $\mathbf{D}f^n = n f^{n-1} \cdot \mathbf{D}f$.

If f is replaced by j, then, in view of $\mathbf{D}\,j = 1$, the Power Rule yields $\mathbf{D}\,j^n = n\,j^{n-1}$; if f is replaced by *sin*, then $\mathbf{D}\,sin = cos$,

$$\mathbf{D}\,sin^n = n\,sin^{n-1} \cdot cos.$$

EXERCISES

6. Find the derivatives of the following functions: (a) $j^n \cdot sin$; (b) $exp\,(n\,j) = exp^n$; (c) $\frac{sin}{j}$; (d) $csc = \frac{1}{sin}$; (e) $j^{-n} = \frac{1}{j^n}$; (f) $\frac{1}{exp}$, *sinh* and *cosh* (p. 125); (g) $tan = \frac{sin}{cos}$ and $cot = \frac{cos}{sin}$ (use $\mathbf{D}\,cos = -sin$); (h) $\frac{sin}{exp}$ (whose graph shows damped oscillations). (i) For every f, express $\mathbf{D}\,(1/f)$ in terms of f and $\mathbf{D}f$.

7. Determine the relative extrema of (a) $j^n \cdot exp$, for $n > 0$; (b) *sin/exp* (assuming for any x the value $e^{-x}\,sin\,x$); (c) $j^n \cdot sin$ for $n > 0$.

8. Using the equalities $sin\,(2j) = 2\,sin \cdot cos$ and $cos\,(2j) = cos^2 - sin^2$, find $\mathbf{D}[sin\,(2j)]$ and $\mathbf{D}\,[cos\,(2j)]$. Then find $\mathbf{D}\,(cos^2 + sin^2)$ and explain the result.

9. Show that the derivative of any polynomial is a polynomial; of any rational function, a rational function. Find

$$\mathbf{D}\,\frac{a \cdot j + b}{c \cdot j + d} \quad \text{for any } a, b, c, d.$$

10. Let p and q be two positive integers. Find $\mathbf{D}\,f^{p/q}$. Hint: Set $f^{p/q} = g$. Then $f^p = g^q$. Hence $\mathbf{D}\,f^p = \mathbf{D}\,g^q$. Apply the Power Rule for integer exponents, and express g in terms of f. As a result, the Power Rule above will be found to be valid if $n = p/q$. Then verify that the rule also holds for negative rational numbers n.

11. Using the result of Exercise 11, find the derivatives of the following functions: *(a)* $\sqrt{4-j^2}$; *(b)* $-\sqrt{4-j^2}$; *(c)* $\sqrt{f^2+g^2}$, for any two functions f and g; *(d)* $(a^2-j^2)^{-1/2}$ for any a; *(e)* $(a^2+j^2)^{n/2}$ for any a and any positive or negative integer n.

12. Find the ratio of the height to the diameter of a cylinder of the largest possible volume that can be inscribed into a sphere of given radius r. In 1615, Kepler was amazed to find a ratio different from 1:1 (that is, a non-square cross section). Hint. Call x the radius, and h the height of a cylinder inscribed into a sphere of radius r. Find the functions connecting (1) h with x; (2) the volume of the cylinder with x. Maximize the latter function using the Product Rule and essentially the result of Exercise 11 *(a)*. Then note how much simpler it would have been to maximize the square of the volume of the cylinder. The application of the rules in Exercise 27, p. 166 often bring about great simplifications. What are the cylinders of *smallest* possible volume that can be inscribed into a sphere of radius r? One can guess the answer or obtain it from the theory of minima at non-interior elements of the domain (see p. 165) if he takes into account the non-negativity of all quantities involved.

13. Suppose that a function f (having a derivative $\mathbf{D}\,f$) is implicitly defined by the condition

(a) $j^2 - 4 + f^2 = O$.

Since, according to (a) the functions $j^2 - 4 + f^2$ and O are equal, so are their derivatives; that is to say, $2j + 2f \cdot \mathbf{D}\,f = O$. From this relation between f and $\mathbf{D}\,f$, the latter can be expressed in terms of the former:

(b) $\mathbf{D}\,f = -j/f$.

Show that $f_1 = \sqrt{4-j^2}$ and $f_2 = -\sqrt{4-j^2}$ are (explicitly defined) functions satisfying (a). Any other function satisfying (a) is not continuous and has a graph made up of parts of the two semicircles f_1 and f_2. Using the results of Exercise 12 *(a)*, *(b)*, show that f_1 and f_2 indeed satisfy (b); that is to say: $\mathbf{D}\,f_1 = -j/f_1$ and $\mathbf{D}\,f_2 = -j/f_2$.

In the same way treat the functions implicitly defined by:

(a) $f^2 - 2j \cdot f + 1 = O$; *(b)* $f^2 - 2j \cdot f - 1 = O$. Find also explicit definitions for those functions. The last task would be difficult if

(c) $j^3 - 3j \cdot f + f^3 = O$. The functions defined by (d) $f^5 - 2j \cdot f + j^7 = O$ and *(e)* $f + \exp f - j = O$ are non-elementary; that in *(e)* is $j//(j + \exp)$. Yet, even in problems *(c)*, *(d)*, *(e)*, $\mathbf{D}\,f$ can easily be expressed in terms of f.

14. Find $\mathbf{D}^2 (f \cdot g)$. In analogy to the binomial formula, $j^n (a + b) = \sum_{k=0}^{k=n} \binom{n}{k} j^{n-k} a \cdot j^k b$, prove by induction

$$\mathbf{D}^n(f \cdot g) = \sum_{k=0}^{k=n} \binom{n}{k} \mathbf{D}^{n-k} f \cdot \mathbf{D}^k g, \text{ where } \mathbf{D}^0 h = h \text{ and } \mathbf{D}^1 h = \mathbf{D}\, h.$$

Here, $\binom{n}{k} = \dfrac{n!}{k!\,(n-k)!}$.

15. If the radius of a sphere is an exponential function of the time, $r = \exp t$, and v and s denote the volume and the surface of the sphere, find $\dfrac{dv}{dt}, \dfrac{ds}{dt}, \dfrac{dv}{dr}, \dfrac{dv}{ds}$.

16. A conical funnel of height h and radius r at the top is partly filled with water. Let a be the altitude of the water level above the tube of the funnel, and v the volume of the water in the cone. Establish the connection, at any instant, between a, $\dfrac{da}{dt}$, and $\dfrac{dv}{dt}$ (the rate with which the water leaves the funnel). Show that, at any instant, each of these three numbers can be determined from the other two.

4. THE SUBSTITUTION RULE

$\mathbf{D}\, g^n = n\, g^{n-1} \cdot \mathbf{D}\, g$, by the Power Rule. Now g^n and $n\, g^{n-1}$ are the results of substituting g into j^n and $n\, j^{n-1}$ $(= \mathbf{D}\, j^n)$; in formulas:

$$g^n = j^n g \quad \text{and} \quad n\, g^{n-1} = (\mathbf{D}\, j^n)\, g \text{ – briefly } \mathbf{D}\, j^n g.$$

Hence the Power Rule can be written in the form

$$\mathbf{D}\,(j^n g) = \mathbf{D}\, j^n g \cdot \mathbf{D}\, g;$$

in words: the derivative of the function obtained by substituting g into j^n equals the product of the function obtained by substituting g into the derivative of j^n times the derivative of g.

In this form, the Power Rule appears as a special case of the following all-important rule that is valid for any f (not only for j^n) :

The derivative of the function fg (obtained by substituting g into f) is the product of the function obtained by substituting g into the derivative of f times the derivative of g; in a formula, where $\mathbf{D}\, f g$ stands for $(\mathbf{D}\, f)\, g$,

SUBSTITUTION RULE. $\quad \mathbf{D}\,(fg) = \mathbf{D}\, f g \cdot \mathbf{D}\, g$.

In Fig. 35, there are two curves f and g, besides O and j. Above the points a and x on O, the points of fg have been determined graphically

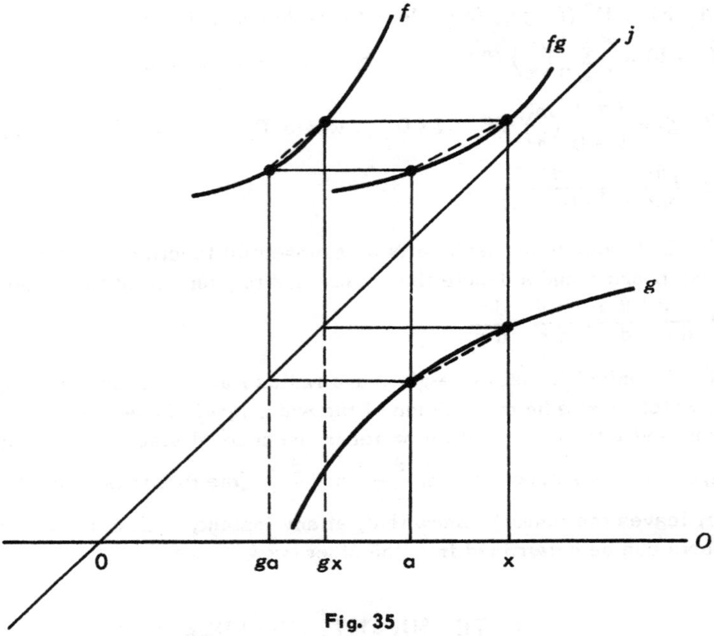

Fig. 35

by the construction explained in Fig. 27, and the rest of fg has been filled in.

The figure shows three (dotted) chords:

1. the chord of g between a and x; its slope is $\dfrac{g\,\mathrm{x} - g\,\mathrm{a}}{\mathrm{x} - \mathrm{a}}$;

2. the chord of f between $g\,\mathrm{a}$ and $g\,\mathrm{x}$; its slope is $\dfrac{fg\,\mathrm{x} - fg\,\mathrm{a}}{g\,\mathrm{x} - g\,\mathrm{a}}$;

3. the chord of fg between a and x; its slope is $\dfrac{fg\,\mathrm{x} - fg\,\mathrm{a}}{\mathrm{x} - \mathrm{a}}$.

Obviously, the third slope is the product of the first two. Now if x is sufficiently close to a, then

the first chord is close to *the tangent to g at* a whose slope is $\mathbf{D}\,g\,\mathrm{a}$;

the second chord is close to *the tangent to f at g* a whose slope is $\mathbf{D}f\,g\,\mathrm{a}$;

the third chord is close to *the tangent to fg at* a whose slope is $\mathbf{D}\,(fg)\,\mathrm{a}$.

This suggests that the slope of the curve fg at a is the product of the slope of the curve f at $g\,\mathrm{a}$, and the slope of the curve f at a or

(4) $$\mathbf{D}\,(fg)\,\mathrm{a} = \mathbf{D}f\,g\,\mathrm{a} \cdot \mathbf{D}\,g\,\mathrm{a}.$$

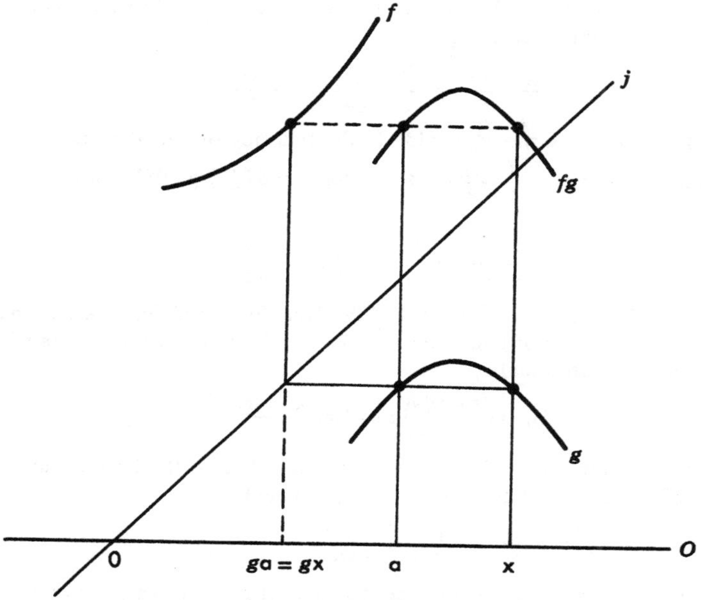

Fig. 36

(4) is valid if (a) g has a tangent at a; (b) f has a tangent at g a; that is to say, if

(5) a belongs to Dom [**D** g (into Dom **D** f)] .

A minor modification of the procedure is necessary if the chord of g between a and x is horizontal (Fig. 36). For, in this case, there is only one horizontal line through the two points on g, and the construction yields a single point on the curve f rather than a chord of f. But, in this case, it is also clear that the chord of fg between a and x is horizontal. The chords of both fg and g between a and x have the slope 0. If g has horizontal chords between a and points arbitrarily close to a, (which is not the case in Fig. 36), and if g has a tangent at a, then **D** g a = 0 and **D** (fg) a = 0. If f has a derivative at g a, equality (4) reads 0 = 0 and is thus again valid.

On account of the paramount importance of the Substitution Rule, **D** (fg) a will, under the assumption (5), also be determined by the Three-Step Rule.

Step I. The difference quotient of fg at a is $\frac{fg - fga}{j - a}$.

Step II. Divide and multiply the quotient by the same term, $g - ga$, and regroup factors thus obtaining

$$\frac{fg - fga}{j - a} = \frac{fg - fga}{g - ga} \cdot \frac{g - ga}{j - a} .$$

The first factor on the right side is the function obtained by substituting g into the function $\frac{f - fga}{j - ga}$, (see Exercise 37(j), p. 95). Hence

$$\frac{fg - fga}{j - a} = \left(\frac{f - fga}{j - ga}\right) g \cdot \frac{g - ga}{j - a} .$$

Step III. Compute the limit at a of the transformed difference quotient, which is the product of two functions. By the Law about the Limits of Products, its limit equals

$$\lim_{a} \left[\left(\frac{f - fga}{j - ga}\right) g\right] \cdot \lim_{a} \frac{g - ga}{j - a} .$$

The second limit equals $\mathbf{D} g\, a$; the first, according to the Law about Limits of Results of Substitution (p. 117), equals

$$\lim_{ga} \frac{f - fga}{j - ga} .$$

Since the quotient is the difference quotient of f at ga, its limit at $g\,a$ equals $\mathbf{D} f(g\,a)$. The latter number equals the value for a of the function obtained by substituting g into $\mathbf{D} f$, that is, $\mathbf{D} fg$ a. Thus altogether, in agreement with (4)

$$\lim_{a} \frac{fg - fga}{j - a} = \mathbf{D}\, fg\, a \cdot \mathbf{D}\, g\, a.$$

In applying the Law about the Limits of Results of Substitution, use has been made of the fact that $\lim_{a} g = g\,a$. The continuity of g at a is indeed a consequence of the assumption that g has a derivative at a. The same reasoning was applied in proving the Product Rule.

As a first application, replace f by *sin* and g by $2\,j$. Then $\mathbf{D}\, f = \cos$, and $\mathbf{D}\, g = 2$. Hence

$$\mathbf{D}\,[\sin(2\,j)] = \cos(2\,j) \cdot 2 \quad\text{or}\quad \mathbf{D}\,[\sin(2\,j)] = 2\cos(2\,j) .$$

This result confirms a consequence of the Product Rule (see Exercise 8). In the same way, for $f = \sin$ and $g = \mathrm{n}j$, where n is any integer, the Substitution Rule yields $\mathbf{D}[\sin(\mathrm{n}j)] = \mathrm{n}\cos(\mathrm{n}j)$. This result could be obtained from the Product Rule only in conjunction with rather complicated trigonometric laws. The power of the Substitution Rule is illustrated by the fact that, applied to $f = \sin$ and any linear function $g = \mathrm{a}j + b$, without any support from trigonometry, it yields

$$\mathbf{D}[\sin(\mathrm{a}j + b)] = \mathrm{a}\cos(\mathrm{a}j + b) .$$

If, in particular, $g = \frac{\pi}{2} - j$, then $\mathbf{D}\, g = -1$. Since $sin\,(\frac{\pi}{2} - j) = cos$ and $cos\,(\frac{\pi}{2} - j) = sin$, it follows that

$$\mathbf{D}\, cos = -sin\,.$$

More generally, if $f = sin$ and g is any function, then

$$\mathbf{D}(sin\, g) = cos\, g \cdot \mathbf{D} g \quad \text{and} \quad \mathbf{D}(cos\, g) = -sin\, g \cdot \mathbf{D} g\,.$$

Similarly, in view of $\mathbf{D}\, exp = exp$,

$$\mathbf{D}(exp\, g) = exp\, g \cdot \mathbf{D} g\,.$$

For instance, if $g = \mathrm{c} j$, then $\mathbf{D}[exp(\mathrm{c} j)] = \mathrm{c}\, exp(\mathrm{c} j)$. If $e^c = b$, and thus $c = log\, b$, then $exp(\mathrm{c} j)$ assumes, for any x, the value $e^{cx} = (e^c)^x = b^x$. Consequently, $exp(\mathrm{c} j) = exp_b$ and

$$(6) \qquad \mathbf{D}\, exp_b = (log\, b) \cdot exp_b\,.$$

In particular, it is now clear that the slope at 0 of the curve exp_b is $log\, b$. This number is greater or smaller than 1 according to whether b is $> e$ or $< e$.

Obviously, for any function f,

$$\mathbf{D}[f(\mathrm{a} j + b)] = \mathrm{a}\, \mathbf{D} f(\mathrm{a} j + b)\,.$$

Next consider three functions, f_1, f_2, f_3, and set, *ad hoc*, $g = f_2 \cdot f_3$. By the Substitution Rule, $\mathbf{D}(f_1 g) = \mathbf{D} f_1 g \cdot \mathbf{D} g$. Now $\mathbf{D} f_1 g = \mathbf{D} f_1 f_2 f_3$, which, in conformity with the basic convention of page 225, stands for $(\mathbf{D} f_1)(f_2 f_3)$, that is, the function obtained by substituting $f_2 f_3$ into $\mathbf{D} f_1$. By the Substitution Rule, $\mathbf{D} g = \mathbf{D} f_2 f_3 \cdot \mathbf{D} f_3$. Hence altogether

$$\mathbf{D}(f_1 f_2 f_3) = \mathbf{D} f_1 f_2 f_3 \cdot \mathbf{D} f_2 f_3 \cdot \mathbf{D} f_3\,.$$

For instance,

$$\mathbf{D}(exp\, exp\, g) = exp\, exp\, g \cdot exp\, g \cdot \mathbf{D} g \quad \text{and}$$
$$\mathbf{D}(exp\, exp\, sin) = exp\, exp\, sin \cdot exp\, sin \cdot cos\,.$$

In a similar way, by induction, one obtains the

CHAIN RULE.

$$\mathbf{D}(f_1 f_2 \ldots f_{n-1} f_n) = \mathbf{D} f_1 f_2 \ldots f_{n-1} f_n \cdot \mathbf{D} f_2 \ldots f_{n-1} f_n \cdot \ldots \cdot \mathbf{D} f_{n-1} f_n \cdot \mathbf{D} f_n\,.$$

Here, in agreement with the basic convention of page 225,

$$\mathbf{D} f_k f_{k+1} \ldots f_n \quad \text{stands for} \quad (\mathbf{D} f_k)(f_{k+1} \ldots f_n)\,,$$

that is, the function obtained by substituting $f_{k+1} \ldots f_n$ into $\mathbf{D} f_k$. Clearly, $f_{k+1} \ldots f_n$ is the result of substituting f_n into $f_{k+1} \ldots f_{n-1}$, not the product of the functions $f_{k+1}, \ldots, f_n$, which would be denoted by $f_{k+1} \cdot \ldots \cdot f_n$.

Suppose u, v, w are three consistent classes of quantities, and f and g two functions, such that relative to certain pairings of the domains

$w = f\,v$ and $v = g\,u$ and hence $w = f(g\,u) = (fg)\,u$.

According to the Basic Scheme for the Application of Derivatives to C.C.Q's, (see p. 215),

$$\frac{\mathbf{d}\,w}{\mathbf{d}\,u} = (\mathbf{D}\,(fg))\,u,\quad \frac{\mathbf{d}\,w}{\mathbf{d}\,v} = (\mathbf{D}\,f)\,v = (\mathbf{D}f)\,gu = (\mathbf{D}\,f\,g)\,u, \text{ and } \frac{\mathbf{d}\,v}{\mathbf{d}\,u} = (\mathbf{D}g)\,u.$$

By virtue of the Substitution Rule,

$$\mathbf{D}\,(fg)\,u \;=\; (\mathbf{D}\,fg)\,u \cdot (\mathbf{D}\,g)\,u.$$

Hence

$$\frac{\mathbf{d}\,w}{\mathbf{d}\,u} = \frac{\mathbf{d}\,w}{\mathbf{d}\,v} \cdot \frac{\mathbf{d}\,v}{\mathbf{d}\,u}\,.$$

This is the formulation of the Substitution Rule in Leibniz' notation. And every freshman who sees this formula says "Of course."

EXERCISES

17. Show that $exp\,(\mathrm{c}j) = exp^{\mathrm{c}}$. From the Substitution Rule, prove that $\mathbf{D}\,exp^{\mathrm{c}} = \mathrm{c}\,exp^{\mathrm{c}}$, for any number c. For any rational number c, confirm this result by the Power Rule (which, so far, has been proved for rational numbers only; see Exercise 11).

18. Find $\mathbf{D}(1/g)$ *(a)* by the Quotient Rule; *(b)* by the Substitution Rule ($1/g = j^{-1}\,g$). In particular, find the derivatives of $1/(1 + j^2)$ and $1/(1 + exp\,(j^{-1}))$.

19. From $\mathbf{D}\,tan = sec^2$, find $\mathbf{D}\,cot$ by the Substitution Rule. Then find $\mathbf{D}[\,tan\,(\mathrm{c}j)]$ and $\mathbf{D}[\,cot\,(\mathrm{c}j)]$ and confirm the last result by applying the Quotient Rule to $cot\,(\mathrm{c}j) = cos\,(\mathrm{c}j)/\,sin\,(\mathrm{c}j)$.

20. Express the derivatives of the six trigonometric functions (*sin, cos, tan, cot, sec, csc*) in terms of $trig = tan(\frac{1}{2}j)$.

21. Using the Substitution and Product Rules, find the derivatives of *(a)* $sinh\,(\mathrm{c}j)$ and $cosh\,(\mathrm{c}j)$; *(b)* $exp\,(\mathrm{c}j) \cdot sin\,(\mathrm{b}j)$; *(c)* $j^{\mathrm{n}} \cdot exp\,(\mathrm{c}j)$; *(d)* $exp\,(\mathrm{c}j) \cdot sinh\,(\mathrm{b}j)$; *(e)* $exp\,(-j^{-2})$; *(f)* $exp_{\mathrm{b}}\,(\mathrm{c}j + \mathrm{c})$; *(g)* $sin\,(j^{-1})$; *(h)* $j \cdot sin\,(j^{-1})$.

22. Find $\mathbf{D}^2\,(fg)$, the second derivative of the function fg.

23. Show that for any three numbers a, b, c,

(a) if $f = \mathrm{a}\,cos\,(\mathrm{c}j) + \mathrm{b}\,sin\,(\mathrm{c}j)$, then $\mathbf{D}^2 f = -\mathrm{c}^2 f$;

(*b*) if $f = a\,cosh(cj) + b\,sinh(cj)$, then $\mathbf{D}^2 f = c^2 f$;

(*c*) if $f = exp(\frac{a}{2}j) \cdot sin(cj)$, then $\mathbf{D}^2 f - a\,\mathbf{D} f + (c^2 + \frac{a^2}{4}) f = O$.

5. THE INVERSION RULE

The inverse $j /\!/ f$ of f satisfies the conditions (see p. 94):

(7) $f(j /\!/ f) = j$ (on Dom $j /\!/ f$) and $(j /\!/ f)f = j$ (on Ess f).

Applying the Substitution Rule to the functions on the left sides, one obtains

$$\mathbf{D}[f(j /\!/ f)] = \mathbf{D}f(j /\!/ f) \cdot \mathbf{D}(j /\!/ f) = \mathbf{D}f /\!/ f \cdot \mathbf{D}(j /\!/ f) \text{ and}$$
$$\mathbf{D}[(j /\!/ f)f] = \mathbf{D}(j /\!/ f)f \cdot \mathbf{D}f.$$

Hence (7) implies

$\mathbf{D}f /\!/ f \cdot \mathbf{D}(j /\!/ f) = 1$ (on Dom $j /\!/ f$) and $\mathbf{D}(j /\!/ f)f \cdot \mathbf{D}f = 1$ (on Ess f).

and the following

INVERSION RULES.

$$\mathbf{D}(j /\!/ f) = 1/(\mathbf{D}f /\!/ f) \quad \text{and} \quad \mathbf{D}(j /\!/ f)f = 1/\mathbf{D}f,$$

where the domains are restricted as before, and besides, by the requirement that the denominators assume values $\neq 0$.

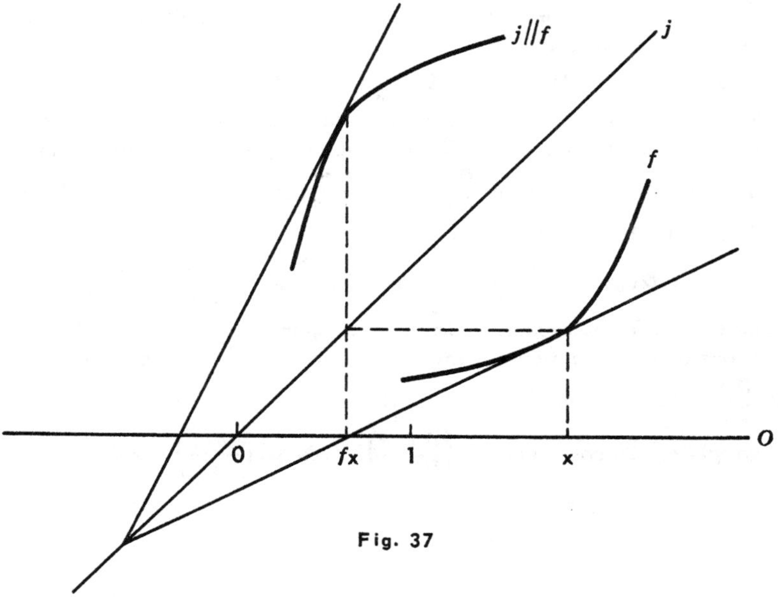

Fig. 37

Obviously, the second rule results from the first upon substitution of f on both sides; the first from the second, upon substitution of $j /\!/ f$.

Fig. 37 illustrates the geometrical meaning of the second Inversion Rule. Since f and $j /\!/ f$ are symmetric with regard to j, so are the tangents to f at x and to $j /\!/ f$ at $f\mathrm{x}$. This means that the slopes of these two lines are reciprocal. But the slope of the second is $\mathbf{D}(j /\!/ f) f\mathrm{x}$, that of the first is $\mathbf{D} f\mathrm{x}$.

Example 1. If $f = j^3$, then $j /\!/ f = j^{1/3}$ and $\mathbf{D} f = 3j^2$. According to the Inversion Rules

$$\mathbf{D} j^{1/3} = 1/(3j^2 /\!/ j^3) = 1/(3j^{2/3}) = \frac{1}{3} j^{-2/3}, \text{ and } \mathbf{D} j^{1/3} j^3 = \frac{1}{3} j^{-2}.$$

Example 2. If $f = j^{1/2}$ then $j /\!/ f = j^2$ (on $\{O < j\}$) and $\mathbf{D} f = \frac{1}{2} j^{-1/2}$. Thus

$\mathbf{D} j^2 = 1/(\frac{1}{2} j^{-1/2} j^2) = 1/(\frac{1}{2} j^{-1}) = 2j$ (on $\{O < j\}$). One finds $\mathbf{D} j^2 = 2j$ (on $\{j \leq O\}$) by applying the Inversion Rule to $f = -j^{1/2}$.

Example 3. $\mathbf{D}(j /\!/ tan) = 1/(sec^2 /\!/ tan)$ which equals $1/(1 + j^2)$ (see p. 95). Another way of writing this result is $\mathbf{D}\, arctan = 1/(1 + j^2)$.

Example 4. $\mathbf{D}(j /\!/ sin) = 1/(cos /\!/ sin) = 1/\sqrt{1 - j^2}$ or $\mathbf{D}\, arcsin = 1/\sqrt{1 - j^2}$.

Example 5. $\mathbf{D}(j /\!/ exp) = 1/(exp /\!/ exp)$ (on Dom $j /\!/ exp$) $= 1/j$ (on P) or $\mathbf{D}\, log = j^{-1}$ (on P).

To find the derivative of $g /\!/ f$, that is, of $g(j /\!/ f)$, apply the Substitution Rule to the latter function

$$\mathbf{D}[g(j /\!/ f)] = \mathbf{D} g(j /\!/ f) \cdot \mathbf{D}(j /\!/ f) = (\mathbf{D} g /\!/ f) \cdot \mathbf{D}(j /\!/ f).$$

The last expression, if the Inversion Rule is applied to $\mathbf{D}(j /\!/ f)$, is seen to be equal to $(\mathbf{D} g /\!/ f)/(\mathbf{D} f /\!/ f)$, which equals the result of substituting $j /\!/ f$ into the quotient $\frac{\mathbf{D} g}{\mathbf{D} f}$. Thus, altogether, one has the following

COMPLETE INVERSION RULE.

$$\mathbf{D}(g /\!/ f) = \left(\frac{\mathbf{D} g}{\mathbf{D} f}\right) /\!/ f \quad \text{and} \quad \mathbf{D}(g /\!/ f) f = \frac{\mathbf{D} g}{\mathbf{D} f}.$$

The second formula results from the first upon substitution of f. If $g = j$, one obtains the previous Inversion Rules. If $g = f$, then $g /\!/ f = j$ and $\frac{\mathbf{D} g}{\mathbf{D} f} = 1$. Clearly, $\mathbf{D} jf$, that is, $(\mathbf{D}\, j) f = 1$.

Example 6. $\mathbf{D}(cos /\!/ sin) = \left(\frac{\mathbf{D}\, cos}{\mathbf{D}\, sin}\right) /\!/ sin = \left(\frac{-sin}{cos}\right) /\!/ sin =$

$$\frac{-1}{cos /\!/ sin} = \frac{-j}{\sqrt{1 - j^2}}$$

EXERCISES

24. Find the derivatives of the following functions: *(a) arccos*; *(b) arccot*; *(c) arcsec*; *(d) arctrig*; *(e) arcsin + arccos*. Explain the result. Express the inverses of the six trigonometric functions in terms of *arctrig*. For instance, $arccos = arctrig \sqrt{\dfrac{1 - j^2}{1 + j^2}}$.

25. Find the inverses of *sinh*, *cosh* (on P), and *cosh* (on N). In what relations are these inverses to the functions defined by the conditions $f^2 - 2j \cdot f + 1 = O$ and $f^2 - 2j \cdot f - 1 = O$ of Exercise 13. Explain why the inverses of the two restrictions of *cosh* have the sum O.

26. Find the derivatives of the following functions: *(a)* $arcsin\ j^2$; *(b)* $arcsin\ j^{1/2}$; *(c)* $arctan\ j^n$; *(d)* $arcsin\ g$ and $arctan\ g$, for any function g; *(e)* $arctan\ (f/j)$.

27. Find the derivative of the (non-elementary) inverse of the function $j + exp_2$ (pp. 94 and 98).

28. $\mathbf{D} j^3 j^{1/3}$ and $\mathbf{D} j^{1/3}$ are reciprocal. Can the same be said about the second derivatives? In other words, are $\mathbf{D}^2 j^3 j^{1/3}$ and $\mathbf{D}^2 j^{1/3}$ reciprocal? Find $\mathbf{D}^2 (j /\!/ f)$ *(a)* by determining the derivative of $\mathbf{D}(j /\!/ f)$ *(b)* by observing that the product of $\mathbf{D}(j /\!/ f)$ and $\mathbf{D} f$ equals the constant function *1*, and hence has the derivative O.

6. LOGARITHMIC DERIVATION

The formula $\mathbf{D}\ log = j^{-1}$ (on P) of Example 5 on page 240 can be greatly extended by applying the Substitution Rule to *log abs*. Since

$$\mathbf{D}\ abs = \begin{cases} 1 \text{ (on P)} \\ -1 \text{ (on N)} \end{cases},$$

one finds

$$\mathbf{D}(log\ abs) = \mathbf{D}\ log\ abs \cdot \mathbf{D}\ abs = j^{-1}\ abs \cdot \mathbf{D}\ abs = \frac{\mathbf{D}\ abs}{abs} = \begin{cases} 1/j \text{ (on P)} \\ -1/-j \text{ (on N)} \end{cases}.$$

Thus

$$\mathbf{D}(log\ abs) = j^{-1} \text{ (on P + N)}. \tag{8}$$

This result is illustrated in Fig. 38. The trombone-shaped graph of *log abs* consists of *log* and the curve $log(-j)$ that is symmetric to *log* about the vertical line through 0. The hyperbola j^{-1} is dotted. The slope of *log abs* is $\frac{1}{2}$ at 2, and $-\frac{1}{2}$ at -2, as are the altitudes of j^{-1}. Only 0 is outside of Dom *log abs* and Dom j^{-1}.

The function *log abs*, which assumes for any $x \neq 0$ the value $log\,|\,x\,|$, is so important that it might warrant the introduction of a permanent symbol, perhaps *logabs*.

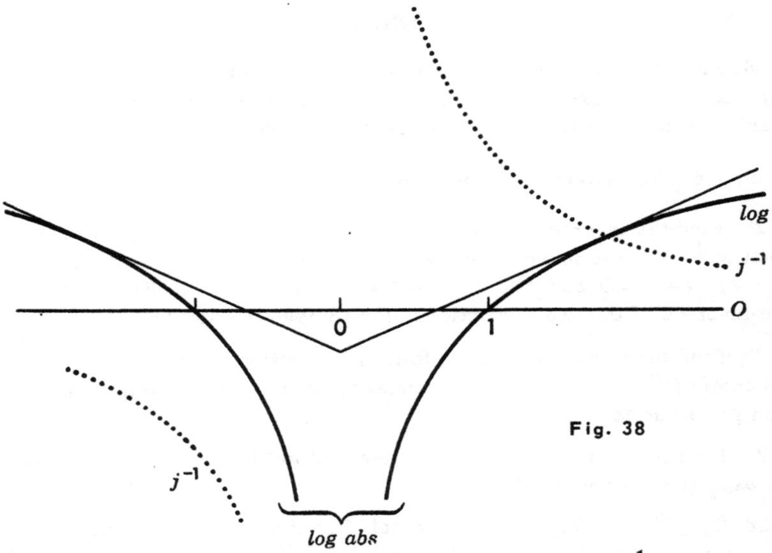

Fig. 38

Since $log_{10} = .4343\ldots log$ and, more generally, $log_b = \frac{1}{log\ b} \cdot log$, from (7) it follows by virtue of the Constant Factor Rule:

(9) $\mathbf{D}(log_{10}\ abs) = \frac{.4343\ldots}{j}$ and $\mathbf{D}\ log_b\ abs = \frac{1}{log\ b} \cdot j^{-1}$ (on P + N).

Only if b = e, is the nuisance denominator, *log* b, equal to 1. It is mainly this fact that has earned the function *log* (the logarithm to the base e) the name *natural logarithm*.

If a function g not assuming the value 0 is substituted into *log abs*, then from (8) it follows by the Substitution Rule

(10) $$\mathbf{D}(log\ abs\ g) = j^{-1} g \cdot \mathbf{D} g = \frac{\mathbf{D} g}{g}.$$

Of course, (10) is valid only for the restriction of any function g to the part of the domain where the values are $\neq 0$. In this sense, (10) is used in the following examples:

$$\mathbf{D}(log\ abs\ sin) = \frac{cos}{sin} = cot \text{ and } \mathbf{D}(log\ abs\ cos) = \frac{-sin}{cos} = -tan;$$

$$\mathbf{D}(log\ abs\ tan) = \frac{sec^2}{tan} = \frac{1}{sin \cdot cos} = 2\ csc(2j);$$

$$\mathbf{D}(log\ abs\ arctan) = \frac{1}{(1 + j^2) \cdot arctan}.$$

Since $x^c = (e^{log\ x})^c = e^{c\ log\ x}$ or $j^c = exp(c\ log)$, the Substitution Rule yields

$\mathbf{D} j^c = \mathbf{D}[exp(c\ log)] = \mathbf{D}\ exp(c\ log) \cdot \mathbf{D}(c\ log) = exp(c\ log) \cdot (c j^{-1}) = j^c \cdot (c j^{-1}) = c j^{c-1}$.

Thus the Power Rule is established for any exponent (not only, as on p. 231, for rational exponents).

The function $exp(j \cdot log)$ assuming for any $x > 0$ the value $e^{x \cdot log\, x} = x^x$ can be treated similarly. Denoting the function *ad hoc* by *f*, one finds

$$\mathbf{D} f = \mathbf{D}\, exp(j \cdot log) \cdot \mathbf{D}(j \cdot log) = f \cdot (log + j \cdot j^{-1}) = f \cdot (log + 1).$$

The reader may find the derivative of $g^h = exp(h \cdot log\, g)$.

More important than the derivation of the rather rare functions of the type g^h is the application of the preceding method to a quotient of products of functions,

$$f = \frac{g_1 \cdot g_2 \cdot \ldots \cdot g_m}{h_1 \cdot h_2 \cdot \ldots \cdot h_n}.$$

Of course, $\mathbf{D} f$ might be found by the Quotient and Product Rules. But it is much simpler to find $\mathbf{D} f$ after giving f the form

$$f = exp\, log \frac{g_1 \cdot g_2 \cdot \ldots \cdot g_m}{h_1 \cdot h_2 \cdot \ldots \cdot h_n} = exp(log\, g_1 + \ldots + log\, g_m - log\, h_1 - \ldots - log\, h_n).$$

For it is now clear that

$$\text{(11)} \qquad \mathbf{D} f = f \cdot \left[\frac{\mathbf{D} g_1}{g_1} + \ldots + \frac{\mathbf{D} g_m}{g_m} - \frac{\mathbf{D} h_1}{h_1} - \ldots - \frac{\mathbf{D} h_n}{h_n}\right].$$

For instance, to find the derivative of

$$f = \frac{(j^2 + 1)^{1/3} \cdot (j^3 + 1)^4 \cdot exp\, j^2}{(j^4 + 1)^{1/5} \cdot (j^2 + j + 1)^6},$$

set $f = exp\left[\frac{1}{3} log(j^2+1) + 4\, log(j^3+1) + j^2 - \frac{1}{5} log(j^4+1) - 6\, log(j^2+j+1)\right]$.

Then, clearly,

$$\mathbf{D} f = f \cdot \left[\frac{2}{3} \frac{j}{j^2+1} + 12 \frac{j^2}{j^3+1} + 2j - \frac{4}{5} \frac{j^3}{j^4+1} - 6 \frac{2j+1}{j^2+j+1}\right].$$

On the right side, f may be replaced by the quotient of products.

More generally, $f = exp(log\, f)$ implies

$$\mathbf{D} f = exp(log\, f) \cdot \mathbf{D}(log\, f) = f \cdot \mathbf{D}(log\, f).$$

It follows that $\mathbf{D}(log\, f) = \frac{\mathbf{D} f}{f}$, which is a special case of (10). But the formula

$$\text{(12)} \qquad \mathbf{D} f = f \cdot \mathbf{D}(log\, f)$$

is also of intrinsic interest. Its application to specific functions is usually phrased in the form of the following rule, known as

METHOD OF LOGARITHMIC DERIVATION. *To find* $\mathbf{D} f$, *determine* $\mathbf{D}(log\, f)$ *and multiply by* f (Johann Bernoulli, 1697).

Clearly, this method is useful only in cases where the determination of $\mathbf{D}(\mathit{log}\ f)$ is easier than that of $\mathbf{D}f$; for instance, if f is a quotient of products, where logarithmic derivation yields (11). The method obviously is not useful for functions such as *log abs log*. While (10) readily yields

$$\mathbf{D}(\mathit{log\ abs\ log}) = \frac{\mathbf{D}(\mathit{log\ abs})}{\mathit{log\ abs}} = \frac{1}{j \cdot \mathit{log\ abs}},$$

logarithmic derivation would complicate the problem:

$$\mathbf{D}(\mathit{log\ abs\ log}) = \mathit{log\ abs\ log} \cdot \mathbf{D}(\mathit{log\ log\ abs\ log}).$$

EXERCISES

29. Find the derivatives of *arcsinh* g and *arctanh* g, where *arctanh* $= j \,//\, \mathit{tanh} = j \,//\, \dfrac{\mathit{sinh}}{\mathit{cosh}}$.

30. Find the derivatives of
(a) $\mathit{log}_{10}\,\mathit{log}_{10}$; (b) $\mathit{log}(\mathit{exp} + 1)$; (c) $j \cdot \mathit{exp}(-j^2) \,/\, \sqrt{j^2 + 1}$;
(d) *exp sin*; (e) $\mathit{exp\ arctan}\ j^2$; (f) $\mathit{exp}(\mathit{arctan} \cdot j^2)$.

31. Prove $\mathbf{D}f = \dfrac{\mathbf{D}(\mathit{exp}\ f)}{\mathit{exp}\ f}$ and formulate a corresponding "Method of Exponential Derivation." For what functions is this method useful? Examine $f = \mathit{log\ log}(j^2 + 1)$.

32. Apply logarithmic derivation to the function j. Does the method yield $\mathbf{D}j = 1$ without restriction? Then apply the method to $f \cdot g$ and f/g and formulate the restrictions under which the Product and Quotient Rules can in this way be obtained.

33. Find the derivative of the function f assuming for any x between 0 and π the value $(\mathit{sin}\ x)^{\mathit{cos}\ x}$.

34. Some books ask the reader to find $\mathbf{D}(\mathit{log\ log\ sin})$. What is the domain of *log log sin*? Sketch the graph of *log abs log abs sin* and find the domain of and derivative of the latter function.

7. ELEMENTARY FUNCTIONS

In view of the simple interrelation of the exponential functions to various bases and the logarithmic functions to various bases, any of them, in particular, the natural exponential and logarithmic functions, may be used in the definition of elementary functions.

Elementary then are the functions that can be obtained from constant functions, *exp* and *log*, the trigonometric and the arc functions by repeated additions, multiplications, and substitutions. Since the constant functions, *exp*, *log*, the trigonometric and the arc functions have elementary derivatives, the Sum, Product, and Substitution Rules imply that *the derivative of any elementary function is elementary.*

But there are also non-elementary functions having elementary derivatives. Examples are the functions

$$\int_0 \sqrt{1-.64\,sin^2}\,,\quad \int_0 \sqrt{(j^2-1)(j^2-4)}\,,\quad \int_0 [exp(-j^2)]\,,\quad \int_1 (1/log)$$

whose graphs can be approximately constructed and whose values can be approximately computed by the methods of Chapters II and III. But these functions can be proved to be non-elementary, whereas their derivatives

$$\sqrt{1-.64\,sin^2}\,,\quad \sqrt{(j^2-1)\cdot(j^2-4)}\,,\quad exp(-j^2)\,,\quad 1/log$$

obviously are elementary.

CHAPTER IX

THE CALCULUS OF ANTIDERIVATIVES

1. GENERAL REMARKS. STANDARD FORMULAS

In the present chapter, function variables f, g, ... stand for functions *having antiderivatives*. In particular, every continuous function f has antiderivatives. For, if a belongs to Dom f, then according to (16), p. 150,

$$\int_a f = \mathbf{D}^{-1}_{(a,\,0)} f \quad \text{and} \quad b + \int_a f = \mathbf{D}^{-1}_{(a,\,b)} f.$$

It will not, in general, be assumed that function variables stand for elementary functions. In fact, some of the laws proved in this chapter will be applied to non-elementary functions.

The statement (p. 245) that the antiderivatives of $\sqrt{1 - .64\ sin^2}$ are non-elementary has two parts: that there are functions having the derivative $\sqrt{1 - .64\ sin^2}$, but that those functions are non-elementary. The proof of the negative part, which is quite complicated, will not here be given. That of the positive part is based on the First Reciprocity Law, according to which

$$\int_0 \sqrt{1 - .64\ sin^2} = \mathbf{D}^{-1}_{(0,\,0)} \sqrt{1 - .64\ sin^2}.$$

An approximate graph of this (non-elementary) function (an approximate 0-area curve) can be sketched by the methods of Chapter II. The situation concerning the other two non-elementary functions mentioned on p. 245 is similar.

All three types of results of the antiderivation of a function f will be studied:

the *definite antiderivative* $\mathbf{D}^{-1}_{(a,\,b)} f$ assuming the value b at a;

the *antiderivative class* $\{\mathbf{D}^{-1} f\}$ consisting of all definite antiderivatives of f;

the *indefinite antiderivative* $\mathbf{D}^{-1} f$, a function variable with the scope $\{\mathbf{D}^{-1} f\}$.

The handling of antiderivative classes will be facilitated by the introduction of some general operations on classes of functions: the connection of a function with a class of functions, and certain connections of two classes of functions.

If $\mathcal{F}$ is any class of functions and g is a function, then

$g + \mathcal{F}$ $(= \mathcal{F} + g)$, $g \cdot \mathcal{F}$ $(= \mathcal{F} \cdot g)$, $3\mathcal{F}$, a $\mathcal{F}$, $\mathcal{F}g$, $g\mathcal{F}$, and $\mathbf{D}\mathcal{F}$

will denote the classes of all functions

$$g + f,\quad g \cdot f,\quad 3f,\quad a\,f,\quad fg,\quad gf,\quad \text{and } \mathbf{D}\,f$$

for any f belonging to $\mathcal{F}$, respectively. For instance,

$cos \cdot \{\mathbf{D}^{-1} cos\}$ is the class of all functions $cos \cdot (sin + c) = cos \cdot sin + cos \cdot c = cos \cdot sin + c\ cos$ for any constant function c;

$3\{\mathbf{D}^{-1} cos\}$ is the class of all functions $3(sin + c)$ or $3\ sin + c$;

$\{\mathbf{D}^{-1} cos\}\ exp$ is the class of all functions $(sin + c)\ exp = sin\ exp + c$, for any constant function c.

The class $\mathbf{D}\{\mathbf{D}^{-1} g\}$ consists of only one function, namely, g.

More generally, if $\mathcal{F}_1$ and $\mathcal{F}_2$ are two classes of functions, there are reasons for studying the class of all functions each of which is the sum of a function belonging to $\mathcal{F}_1$ and a function belonging to $\mathcal{F}_2$. This class might be called the *function sum* of $\mathcal{F}_1$ and $\mathcal{F}_2$. It will be denoted by $\mathcal{F}_1 \oplus \mathcal{F}_2$. For instance,

(1) $$\{\mathbf{D}^{-1} g\} = \{\mathbf{D}^{-1} g\} \oplus \mathcal{C},$$

(2) $$\mathcal{C} \oplus \mathcal{C} = \mathcal{C}.$$

$\mathcal{F}_1 \oplus \mathcal{F}_2$ must not be confused with the class $\mathcal{F}_1 + \mathcal{F}_2$ of all functions that belong to at least one of the classes $\mathcal{F}_1$ and $\mathcal{F}_2$. Clearly, one might define also a function product $\mathcal{F}_1 \odot \mathcal{F}_2$, to be distinguished from the class $\mathcal{F}_1 \cdot \mathcal{F}_2$ of all functions belonging to both $\mathcal{F}_1$ and $\mathcal{F}_2$.

If A and B are two classes of numbers, one might denote by $A \odot B$ the class of all numbers that are the product of a number belonging to A and a number belonging to B. If C is the class consisting of the two numbers 1 and -1, then, in analogy to (2) and (1),

$$C = C \odot C \quad \text{and} \quad \{\pm\sqrt{9}\} = \{\pm\sqrt{9}\} \odot C.$$

Note that $\{\pm\sqrt{9}\} \odot \{\pm\sqrt{9}\} = \{\pm\sqrt{81}\}$, whereas $j^2(\pm\sqrt{9}) = 9$.

The basic formula of the Calculus of Antiderivatives (p. 136) is

(3) f belongs to $\{\mathbf{D}^{-1} g\}$ if and only if $\mathbf{D}\,f = g$.

In view of the importance of this law, its two halves will be quoted separately. Instead of f and g, h and k will be used as function variables:

(3′) If h belongs to $\{\mathbf{D}^{-1} k\}$, then $\mathbf{D}\,h = k$.

(3″) If $\mathbf{D}\,h = k$, then h belongs to $\{\mathbf{D}^{-1} k\}$.

(3′) will also be used in the form

(3*) $\mathbf{D}(\mathbf{D}^{-1} k) = k$ for any function $\mathbf{D}^{-1} k$ belonging to $\{\mathbf{D}^{-1} k\}$.

While (3) makes it possible to test whether a given function f is an antiderivative of a given function g, it does not contain any clue as to how, if only g is given, an antiderivative f may be found.

The simplest way to find an antiderivative of g is by remembering that g has been encountered as the derivative of a certain function. For instance, if one remembers that j^{-1} is the derivative of *log abs*, then he can conclude that

$$log\ abs \text{ is in } \{ \mathbf{D}^{-1} j^{-1} \} \text{ or } \{ \mathbf{D}^{-1} j^{-1} \} = log\ abs + \mathcal{C}$$
$$\text{or } \mathbf{D}^{-1} j^{-1} + \mathcal{C} = log\ abs + \mathcal{C}$$

Similarly, j^n is in $\{ \mathbf{D}^{-1} (n\, j^{n-1}) \}$. If one is more interested in the antiderivatives of j^n, he readily finds

$$\{ \mathbf{D}^{-1} j^n \} = \begin{cases} \dfrac{1}{1+n}\ j^{n+1} + \mathcal{C} \text{ if } n \neq -1 \\ log\ abs + \mathcal{C} \text{ if } n = -1 \end{cases}.$$

Since $j^{-1} = 1/j$ is a rational function, whereas *log abs* can be shown not to be rational, it appears that not every rational function has rational antiderivatives (whereas it always has a rational derivative). Thus, at its very outset, the calculus of antiderivatives differs from the calculus of derivatives in an ominous way.

On the other hand, the difference between the antiderivatives of the -1st power and those of the other powers can be bridged by the following consideration. Clearly,

$$\mathbf{D}^{-1}_{(1,0)}\ j^{-1} = log\ abs \text{ and } \mathbf{D}^{-1}_{(1,0)}\ j^{-1+k} = \frac{1}{k}(j^k - 1).$$

The value that the latter function assumes, say, for 2 is $\frac{1}{k}(2^k - 1)$. What becomes of this value as k approaches 0 (in which case j^{-1+k} approaches j^{-1})? According to (6) page 237, $\lim_{k \to 0} \dfrac{2^k - 1}{k} = log\ 2$, which is the value of *log abs* for 2.

The following formulas from the Calculus of Derivatives	yield the following **STANDARD ANTIDERIVATIVES**
$\mathbf{D}\ exp = exp$	$\{ \mathbf{D}^{-1}\ exp \} = exp + \mathcal{C}$
$\mathbf{D}\ sin = cos$	$\{ \mathbf{D}^{-1}\ cos \} = sin + \mathcal{C}$
$\mathbf{D}\ (-cos) = sin$	$\{ \mathbf{D}^{-1}\ sin \} = -cos + \mathcal{C}$
$\mathbf{D}\ tan = sec^2$	$\{ \mathbf{D}^{-1}\ sec^2 \} = tan + \mathcal{C}$
$\mathbf{D}\ arcsin = \dfrac{1}{\sqrt{1 - j^2}}$	$\left\{ \mathbf{D}^{-1} \left(\dfrac{1}{\sqrt{1-j^2}} \right) \right\} = arcsin + \mathcal{C}$
$\mathbf{D}\ arctan = \dfrac{1}{1 + j^2}$	$\left\{ \mathbf{D}^{-1} \left(\dfrac{1}{1+j^2} \right) \right\} = arctan + \mathcal{C}$
$\mathbf{D}\ (log\ abs\ sin) = cot$	$\{ \mathbf{D}^{-1}\ cot \} = log\ abs\ sin + \mathcal{C}$

EXERCISES

1. Which of the following functions is in the antiderivative class below?

$exp\,(3\,j)$	$exp\,(2\,j)$	$exp\,(2\,j)$
$3\{\mathbf{D}^{-1}[exp\,(3\,j)]\}$	$\left\{\mathbf{D}^{-1}\left[\frac{1}{2}\,exp\,(2\,j)\right]\right\}$	$\frac{1}{2}\{\mathbf{D}^{-1}[exp\,(2\,j)]\}$

exp_2	log	$log\ abs$	sec^2
$\{\mathbf{D}^{-1}\,exp_2\}$	$\{\mathbf{D}^{-1}\,j^{-1}\}$	$\{\mathbf{D}^{-1}\,j^{-1}\}$	$\{\mathbf{D}^{-1}\,tan\}$

2. By (3), test which of the following are correct:

(a) $\{\mathbf{D}^{-1}(3\,sin^2 \cdot cos)\} = sin^3 + \mathcal{C}$;

(b) $\{\mathbf{D}^{-1}(3\,f^2 \cdot \mathbf{D}\,f)\} = f^3 + \mathcal{C}$ for any f;

(c) $\{\mathbf{D}^{-1}[(j^2+1)^2 \cdot 2\,j]\} = \frac{1}{3}(j^2+1)^3 + \mathcal{C}$;

(d) $\{\mathbf{D}^{-1}[(j^2+1)^2 \cdot \mathbf{D}\,(j^2+1)^2]\} = \frac{1}{3}(j^2+1)^3 + \mathcal{C}$;

(e) $\{\mathbf{D}^{-1}[(j^4+1)^3 \cdot \mathbf{D}\,(j^4+1)^3]\} = \frac{1}{2}(j^4+1)^6 + \mathcal{C}$.

3. Find the classes of all antiderivatives of *tan, csc, arccos, arccot.*

2. THE ADDITION AND CONSTANT FACTOR RULES

ADDITION RULE

The rule will be formulated in three different forms:

I. *For Indefinite Antiderivatives.*

$$\mathbf{D}^{-1}(f+g) + \mathcal{C} = \mathbf{D}^{-1}f + \mathbf{D}^{-1}g + \mathcal{C} \text{ for any } f \text{ and } g$$

II. *For Antiderivative Classes.*

$$\{\mathbf{D}^{-1}(f+g)\} = \{\mathbf{D}^{-1}f\} \oplus \{\mathbf{D}^{-1}\,g\}$$

III. *For Particular Antiderivatives.*

$$\mathbf{D}^{-1}_{(a,0)}(f+g) = \mathbf{D}^{-1}_{(a,0)}f + \mathbf{D}^{-1}_{(a,0)}g \quad \text{and}$$

$$\mathbf{D}^{-1}_{(a,b+c)}(f+g) = \mathbf{D}^{-1}_{(a,b)}f + \mathbf{D}^{-1}_{(a,c)}g$$

Applying (3*) to f and to g and adding the results yields

$$f + g = \mathbf{D}\,\mathbf{D}^{-1}f + \mathbf{D}\,\mathbf{D}^{-1}g.$$

The function on the right side, according to the Sum Rule for Derivatives, equals $\mathbf{D}(\mathbf{D}^{-1}f + \mathbf{D}^{-1}g)$. From

$$f + g = \mathbf{D}(\mathbf{D}^{-1}f + \mathbf{D}^{-1}g)$$

it follows by (3″) that $\mathbf{D}^{-1} f + \mathbf{D}^{-1} g$ belongs to $\{\mathbf{D}^{-1}(f+g)\}$. Thus the sum of any antiderivative of f and any antiderivative of g is an antiderivative of $f + g$. Since any two antiderivatives of $f + g$ differ by a constant function, also the converse is true. This completes the proof of I and II.

To prove III, one has to verify that the values for a of the functions on both sides are equal.

In a similar way, the reader should prove the

CONSTANT FACTOR RULE

I. $\mathbf{D}^{-1}(\mathrm{k}\, g) + \mathcal{C} = \mathrm{k}\, \mathbf{D}^{-1} g + \mathcal{C}$ or $\mathbf{D}^{-1} g + \mathcal{C} = \frac{1}{\mathrm{k}} \mathbf{D}^{-1}(\mathrm{k}\, g) + \mathcal{C}$,

for any k and g.

II. $\{\mathbf{D}^{-1}(\mathrm{k}\, g)\} = \mathrm{k}\{\mathbf{D}^{-1} g\}$ or $\{\mathbf{D}^{-1} g\} = \frac{1}{\mathrm{k}}\{\mathbf{D}^{-1}(\mathrm{k}\, g)\}$.

III. $\mathbf{D}^{-1}_{(a,\,b)}(\mathrm{k}\, g) = \mathrm{k}\, \mathbf{D}^{-1}_{(a,\,b/k)} g$ or $\mathbf{D}^{-1}_{(a,\,b)} g = \frac{1}{\mathrm{k}} \mathbf{D}^{-1}_{(a,\,kb)}(\mathrm{k}\, g)$,

for any a, b, k, g.

(In the second alternatives, $\mathrm{k} \neq 0$.)

EXERCISES

4. Find all antiderivatives of the following functions:

(a) $j^{-2} + 2\, j^{-3/2} - j^{-1} + 3\, j^{-1/2} + 5 - j^{1/2} - 3\, j$;

(b) $\exp$; *(c)* $\exp_2$; *(d)* $\log \operatorname{abs} \cos$; *(e)* $\sin^2 + \cos^2$; *(f)* O;

(g) abs; *(h)* sgn. Then determine $\mathbf{D}^{-1}_{(a,\,b)} \exp$; $\mathbf{D}^{-1}_{(0,\,1)} \operatorname{sgn}$; $\mathbf{D}^{-1}_{(0,\,0)} \operatorname{abs}$.

5. Show that all antiderivatives of any polynomial are polynomials. Give examples of rational functions that are not polynomials and whose antiderivatives are *(a)* rational, *(b)* irrational functions.

6. Show that *(a)* $\{\mathbf{D}^{-1}(j \cdot g)\} \neq j \cdot \{\mathbf{D}^{-1} g\}$ for any g, even for O. *(⋆b)* if $g \,||\, O$, then $\{\mathbf{D}^{-1}(f \cdot g)\} = f \cdot \{\mathbf{D}^{-1} g\}$ if and only if f is constant.

7. Assuming that $\mathbf{D}^{-1} g + \mathcal{C} = f + \mathcal{C}$ and that h is a linear function (that is, of the form a $j + b$ for two numbers a and b), what can one say about $\mathbf{D}^{-1}(hg)$?

8. Show that $f + g$ may have antiderivatives without either f or g having antiderivatives. What can be said about g if $\mathbf{D}^{-1}(f+g)$ and $\mathbf{D}^{-1} f$ exist?

9. For what n are the following limits finite: *(a)* $\lim_{\infty} \mathbf{D}^{-1}_{(1,\,1)} j^n$; *(b)* $\lim_{-\infty} \mathbf{D}^{-1}_{(-1,\,-1)} j^n$?

Using the relation between integrals and antiderivatives, determine

(c) $\int_a^b j^n$; (d) $\int_{-\infty}^{-a} j^{-2}$; (e) $\int_{-\infty}^{a} exp\ (a > 0)$. Here, $\int_{-\infty}^{a} exp$ designates, of course, the limit for $-\infty$ of the function $\int^{a} exp$, which for b assumes the value $\int_b^a exp$.

★10. Show that the analogue of the Addition Rule II for square roots is $\{\pm\sqrt{a}\}\odot\{\pm\sqrt{b}\} = \{\pm\sqrt{a \cdot b}\}$ for any two positive numbers a and b. What are the analogues of I and III and of the Constant Factor Law? Formulate and answer questions analogous to Exercise 8.

3. THE TRANSFORMATION RULE

$\mathbf{D}^{-1}(f \cdot g)$, $\mathbf{D}^{-1}(f/g)$, and $\mathbf{D}^{-1}(fg)$, in an unfortunate contrast to $\mathbf{D}(f \cdot g)$, $\mathbf{D}(f/g)$, and $\mathbf{D}(fg)$, cannot in general be expressed by applying addition, multiplication, and substitution to f, g, $\mathbf{D}^{-1} f$, $\mathbf{D}^{-1} g$, $\mathbf{D} f$, $\mathbf{D} g$. For instance, $\mathbf{D}^{-1}\left(\frac{1}{log}\right)$ and $\mathbf{D}^{-1} exp(-j^2)$, being non-elementary functions, cannot thus be expressed.

However, if h is any function having a derivative, then the antiderivatives of $gh \cdot \mathbf{D} h$ can be obtained simply by substituting h into $\mathbf{D}^{-1}g$. This important fact will be referred to as the

TRANSFORMATION RULE.

I. $\mathbf{D}^{-1}(gh \cdot \mathbf{D} h) + \mathcal{C} = (\mathbf{D}^{-1}g)\, h + \mathcal{C}$.

II. $\{\mathbf{D}^{-1}(gh \cdot \mathbf{D} h)\} = \{\mathbf{D}^{-1} g\}\, h$.

III. $\mathbf{D}^{-1}_{(a,\, b)}(gh \cdot \mathbf{D} h) = (\mathbf{D}^{-1}_{(h\,a,\, b)}\, g)\, h$, for any a, b, g, h.

To prove I it is sufficient to show that the derivative of any function on the right side is equal to $gh \cdot \mathbf{D} h$, which is the derivative of any function on the left side. The Substitution Rule for Derivatives implies indeed that

$$\mathbf{D}[(\mathbf{D}^{-1} g)\, h] = (\mathbf{D}\,\mathbf{D}^{-1} g)\, h \cdot \mathbf{D} h = gh \cdot \mathbf{D} h.$$

III follows from the fact that the value for a of the function on the right side equals the value of $\mathbf{D}^{-1}_{(h\,a,\, b)}\, g$ for h a, that is, b; and that b is also the value for a of the function on the left side.

If h is a linear function a $j + b$, so that $\mathbf{D} h = a$, then Rule I implies

$$\mathbf{D}^{-1}[g(a\, j + b) \cdot a] + \mathcal{C} = \mathbf{D}^{-1} g(a\, j + b) + \mathcal{C}.$$

Hence, the Constant Factor Rule yields the

LINEAR TRANSFORMATION RULE.

$$\mathbf{D}^{-1}[g(\mathrm{a}\,j+b)]+\mathcal{C}=\frac{1}{\mathrm{a}}(\mathbf{D}^{-1}g)(\mathrm{a}\,j+b)+\mathcal{C} \text{ for any a, b, } g.$$

Thus, if one knows $\mathbf{D}^{-1}g$, he can determine $\mathbf{D}^{-1}[g(\mathrm{a}\,j+b)]$. For instance, using the Standard Antiderivatives of *sin* and *exp*, one finds

$$\mathbf{D}^{-1}[\sin(\mathrm{a}\,j+b)]+\mathcal{C}=\frac{1}{\mathrm{a}}\cos(\mathrm{a}\,j+b)+\mathcal{C},$$

$$\mathbf{D}^{-1}[\exp(\mathrm{a}\,j+\mathrm{a}')]+\mathcal{C}=\frac{1}{\mathrm{a}}\exp(\mathrm{a}\,j+a')+\mathcal{C}.$$

In particular, for $\mathrm{a}'=0$ and $\mathrm{a}=\log \mathrm{b}$,

$$\mathbf{D}^{-1}\exp_{\mathrm{b}}+\mathcal{C}=\frac{1}{\log \mathrm{b}}\exp_{\mathrm{b}}+\mathcal{C}.$$

Of special interest is the case $\mathrm{a}=-1$ and $\mathrm{b}=0$ (i.e., the transformation by $-j$) of functions such as *log* or $j^{1/2}$ whose domain is P or P_0. For instance, $j^{1/2}(-j)=(-j)^{1/2}$ has the domain N_0; and

$$\{\mathbf{D}^{-1}(-j)^{1/2}\}=-1\cdot\frac{2}{3}\,j^{3/2}(-j)+\mathcal{C}=-\frac{2}{3}(-j)^{3/2}+\mathcal{C}.$$

If a non-linear function h, even as simple as j^2 or $-j^2$, is substituted into g, then the knowledge of $\mathbf{D}^{-1}g$ may be of no avail in finding $\mathbf{D}^{-1}(gh)$, as demonstrated by $\mathbf{D}^{-1}[\exp(-j^2)]$.

If $g=j^{\mathrm{n}}$, then, in view of the Standard Antiderivative of j^{n}, the Transformation Rule I yields (if h is replaced by f),

$$\text{POWER RULE}\quad \{\mathbf{D}^{-1}(f^{\mathrm{n}}\cdot\mathbf{D}f)\}=\begin{cases}\dfrac{1}{\mathrm{n}+1}f^{\mathrm{n}+1}+\mathcal{C} & \text{if } \mathrm{n}\neq-1\\ \log \operatorname{abs} f+\mathcal{C} & \text{if } \mathrm{n}=-1\end{cases}$$

For instance, $\{\mathbf{D}^{-1}(\sin^{\mathrm{n}}\cdot\cos)\}=\dfrac{1}{\mathrm{n}+1}\sin^{\mathrm{n}+1}+\mathcal{C}$. (The case $\mathrm{n}=-1$ is covered by the Standard Antiderivative of *cot*.) Another example is

$$\mathbf{D}^{-1}\frac{2\,j}{1+j^2}+\mathcal{C}=\mathbf{D}^{-1}[(1+j^2)^{-1}\cdot\mathbf{D}(1+j^2)]+\mathcal{C}=\log(1+j^2)+\mathcal{C}$$

(For functions such as $1+j^2$ with a positive range, one may replace *log abs* by *log*.)

The last example demonstrates that, to find an antiderivative by the Power Rule, one must not only know the rule, he must realize the applicability of the rule where it is applicable. If one does not happen to notice that the quotient $2\,j/(1+j^2)$ is the product of the -1st power of $1+j^2$ and the derivative of $1+j^2$, then, in spite of a complete knowledge of the Power Rule, he will not find the antiderivative.

The function $j/(1+j^2)$ actually is not a product of a power of any function times the derivative of that function; only after multiplication by 2 it is. If one notices this fact, then (and only then) by applying the Constant Factor Rule to $\frac{1}{2}\,[(1+j^2)^{-1}\cdot 2\,j\,]$ he can reduce the problem of finding $\mathbf{D}^{-1}[j\,/\,(1+j^2)]$ to the preceding problem.

Many functions have to be more extensively transformed to make the Power Rule applicable; for instance,

$$\sqrt[5]{\frac{1+j^{4/5}}{j}}$$

If this function is simplified to $(j^{-1}+j^{-1/5})^{1/5}$, the Power Rule seems to be altogether inapplicable. Only in the form

$$\frac{5}{4}\left[(1+j^{4/5})^{1/5}\cdot\frac{4}{5}\,j^{-1/5}\right]$$

is the function (except for the factor 5/4, which can be handled by the Constant Factor Rule) the product of a power of a function times its derivative. Hence its antiderivative class is

$$\frac{5}{4}\,\frac{(1+j^{4/5})^{6/5}}{6/5}+\mathcal{C}=\frac{25}{24}\,(1+j^{4/5})^{6/5}+\mathcal{C}.$$

In the case of $\mathbf{D}^{-1}\,tan$, the necessary transformation is

$$tan=\frac{sin}{cos}=-\frac{-sin}{cos}=-\frac{1}{cos}\cdot\mathbf{D}\,cos,$$

wherefore $\{\mathbf{D}^{-1}\,tan\}=-\,log\;abs\;cos+\mathcal{C}$.

It is even more complicated to make the Power Rule applicable to the important function $sec=1/cos$. One has to multiply by $\frac{sin}{cos}+\frac{1}{cos}$ and to divide by $\frac{1}{cos}+\frac{sin}{cos}$. The result is

$$\frac{1}{cos}=\frac{\frac{sin}{cos^2}+\frac{1}{cos^2}}{\frac{1}{cos}+\frac{sin}{cos}},\ \text{which equals}\ \frac{\mathbf{D}\left(\frac{1}{cos}+\frac{sin}{cos}\right)}{\frac{1}{cos}+\frac{sin}{cos}}=\frac{\mathbf{D}\,(sec+tan)}{sec+tan}.$$

Hence

$$\text{(4)}\qquad \{\mathbf{D}^{-1}\,sec\}=log\;abs\;(sec+tan)+\mathcal{C}.$$

To remember the necessary transformation of *sec* is as hard as (if not harder than) to remember the antiderivative itself. This example illustrates the unfortunate amount of material that must be memorized for the actual determination of antiderivatives.

If f is elementary, then so are the powers of f and $log\,f$. Hence the functions $\mathbf{D}^{-1}\,(f^n\cdot\mathbf{D}\,f)$ are elementary. But the Power Rule may also be applied, e.g., to the square of the non-elementary function $D^{-1}_{(2,0)} \frac{1}{log}$. Since the derivative of the latter function is $1/log$, the result is

$$D^{-1}\left[\left(D^{-1}_{(2,0)} \frac{1}{log}\right)^2 \cdot \frac{1}{log}\right] + \mathcal{C} = \frac{1}{3}\left(D^{-1}_{(2,0)} \frac{1}{log}\right)^3 + \mathcal{C}.$$

$D^{-1}_{(2,0)}(1/log) = \int_2 \frac{1}{log}$ is called the *integral logarithm* and denoted by Li. Thus

$$D^{-1} \frac{Li^2}{log} + \mathcal{C} = \frac{1}{3} Li^3 + \mathcal{C}.$$

The function Li is important in number theory since its value for large x is close to the number of prime numbers $\leq$ x. For instance,

$Li(500{,}000) = 41{,}606$ and $Li(10^7) = 664{,}918$,

while there are 41,539 and 664,580 prime numbers $\leq 500{,}000$ and $\leq 10^7$, respectively.

The Power Rule may even be applied to unknown functions. Suppose one wishes to determine all functions f satisfying the condition

(5) $$f^2 \cdot D f = 2j.$$

(Such a problem is called a *differential equation.*) If f is a solution, that is, a function satisfying (5), then clearly the antiderivative classes of the functions on either side of (5) are equal; that is to say,

(6) $$D^{-1}(f^2 \cdot D f) + \mathcal{C} = D^{-1}(2j) + \mathcal{C}.$$

By the Power Rule, even though f is as yet unknown, the antiderivatives on the left side can be expressed in terms of f. They are $\frac{1}{3} f^3 + \mathcal{C}$ (thus apart from an additive constant, one third of the cube of whatever f is). But since, according to (6), those antiderivatives are equal to $j^2 + \mathcal{C}$, it follows that, for any function f satisfying (5),

(7) $$f = (3j^2 + c)^{1/3} \text{ for some constant function c.}$$

Conversely, if f satisfies (7), then

$$f^2 = (3j^2 + c)^{2/3} \text{ and } D f = \frac{1}{3}(3j^2 + c)^{-2/3} \cdot 6j$$

and consequently f satisfies (5).

Thus, by means of the Power Rule, the class of all functions satisfying (5) has been determined or, as one says, the differential equation (5) has been solved. The class of all functions (7) is the class of all solutions.

In view of the importance of these remarks for the understanding of differential equations, the main idea may be restated in a somewhat more general form.

If f is an unknown function, then so are, in general, the antiderivatives of $f^n \cdot \mathbf{D} f$. The Power Rule relates those antiderivatives to f: apart from added constants they are powers of f or the logarithm of f. (Incidentally, this very statement demonstrates that, in general, those antiderivatives are unknown functions. For if one knew, say, that $f^3 = j^2$, then one would know that $f = j^{2/3}$.) Now consider the case that f satisfies a condition of the following type

$$f^n \cdot \mathbf{D} f = g, \tag{8}$$

where n is a given number, g a given function (such as 2 and $2j$ in (5)). By virtue of the very condition (8) the antiderivatives of $f^n \cdot \mathbf{D} f$ are known: they are equal to the antiderivatives of g. Hence, by the Power Rule, apart from constant terms, a power or the logarithm of f equals an antiderivative of g. Hence, f itself can be determined: it is a root or exponential of that antiderivative of g.

On the other hand, the antiderivatives of f^n itself (without the factor $\mathbf{D} f$) are, if f is unknown, in general not only unknown but not even in a known relation to the unknown function f. For instance, the antiderivatives of $1/log$ and the $\frac{1}{2}$th power of $j^3 + j + 1$ can be proved to be non-elementary functions and therefore certainly are not expressible by addition, multiplication, or substitution in terms of log and/or other elementary functions.

Next replace in the Transformation Rule g by exp and apply the Standard Antiderivative of exp. The result is the

EXPONENTIAL RULE. $\mathbf{D}^{-1}(exp\ h \cdot \mathbf{D}\ h) + \mathcal{C} = exp\ h + \mathcal{C}$.

For instance,

$$\mathbf{D}^{-1}(j \cdot exp(-j^2)) + \mathcal{C} = -\frac{1}{2}\mathbf{D}^{-1}(exp(-j^2) \cdot (-2j)) + \mathcal{C}$$
$$= -\frac{1}{2}\, exp(-j^2) + \mathcal{C}.$$

The reader may apply the Transformation Rule to trigonometric functions of h. Combining the rule with the two Standard Antiderivatives before the last on page 248 one obtains

$$\mathbf{D}^{-1}\left(\frac{\mathbf{D}\,h}{\sqrt{1-h^2}}\right) + \mathcal{C} = arcsin\ h + \mathcal{C} \text{ and } \mathbf{D}^{-1}\left(\frac{\mathbf{D}\,h}{1+h^2}\right) + \mathcal{C} = arctan\ h + \mathcal{C};$$

$$\mathbf{D}^{-1}\frac{a}{\sqrt{1-a^2j^2}} + \mathcal{C} = arcsin(a\,j) + \mathcal{C} \text{ and } \mathbf{D}^{-1}\frac{a}{1+a^2j^2} = arctan\,(a\ j) + \mathcal{C};$$

$$\mathbf{D}^{-1}\frac{1}{\sqrt{b^2-j^2}} + \mathcal{C} = arcsin\left(\frac{1}{b}\,j\right) + \mathcal{C} \text{ and } \mathbf{D}^{-1}\frac{1}{b^2+j^2} = \frac{1}{b}\ arctan\left(\frac{1}{b}\,j\right) + \mathcal{C}.$$

The second line results from the first by setting $h = a\ j$; the third from the second, by setting $a = 1/b$. The reader should carry out this substitution and thereby verify that the factor $1/b$ appears only in connection with $arctan\,(\frac{1}{b}\,j)$ and not with $arcsin\,(\frac{1}{b}\,j)$. Also a comparison of the following two results is instructive:

$\mathbf{D}^{-1}\dfrac{\mathbf{D}h}{1+h^2} + \mathcal{C} = \arctan h + \mathcal{C}$ and $\mathbf{D}^{-1}\dfrac{2h\cdot \mathbf{D}h}{1+h^2} + \mathcal{C} = \log(1+h^2) + \mathcal{C}.$

In the latter case, the numerator is the derivative of the entire denominator; hence the *logarithmic* antiderivative by virtue of

$\mathbf{D}^{-1}\dfrac{\mathbf{D}g}{g} + \mathcal{C} = \log \text{ abs } g + \mathcal{C}$. In the former case, the numerator is the derivative of a function whose square increased by *1* forms the denominator; hence the *arctangent*.

EXERCISES

11. Find the antiderivatives of the following functions:

(a) $\dfrac{\sin j^{1/2}}{j^{1/2}}$; (b) $j^{-2/3} \cdot \exp j^{1/3}$; (c) $\log/j$; (d) $1/(j \cdot \log)$;

(e) $j^{-2} \cdot \sin j^{-1}$; (f) $j \cdot \tan j^2$.

12. From the Linear Transformation Rule, express for given numbers m, s, and a:

(a) $\mathbf{D}^{-1}_{(0,\,a)}[\exp(-\frac{1}{2}j^2)]$ in terms of $\mathbf{D}^{-1}_{(0,\,0)}[\exp(-j^2)]$, and vice versa;

(b) $\mathbf{D}^{-1}_{(0,\,a)}[\exp(-\frac{1}{2s^2}(j-m)^2)]$ in terms of $\mathbf{D}^{-1}_{(0,\,0)}[\exp(-j^2)]$ and vice versa. (Problems of this kind are important in statistics.)

13. For any numbers m and n, find the antiderivatives of $\sin(\text{m } j)\cdot\sin(\text{n } j)$, $\sin(\text{m } j)\cdot\cos(\text{n } j)$, $\cos(\text{m } j)\cdot\sin(\text{n } j)$. Hint. Replace the products by sums of trigonometric functions. In the cases of $[\sin(\text{m } j)]^2$ and $[\cos(\text{m } j)]^2$, introduce $\cos(2\text{m } j)$. Instead of $[\sin(\text{m } j)]^2$, one often writes $\sin^2(\text{m } j)$. Using the results, find

$$\int_0^{\pi}[\sin(\text{m } j)\cdot\sin(\text{n } j)], \quad \int_0^{2\pi}[\sin(\text{m } j)\cdot\sin(\text{n } j)]$$ and the corresponding integrals for the other two products. What are the values of the integrals if m and n are integers? These integrals are of basic importance in Fourier's wave analysis.

14. Assuming that $\{\mathbf{D}^{-1}(j^{-1} \text{ on P})\} = \log + \mathcal{C}$, find the antiderivatives of j^{-1} transformed by $-j$. Does the result contradict $\{\mathbf{D}^{-1}j^{-1}\} = \log \text{abs} + \mathcal{C}$?

15. Find the antiderivatives of the following functions:
(a) $\sinh(\text{a } j)$ and $\cosh(\text{a } j)$; (b) $1/(\text{a } j + b)$; (c) $\dfrac{1}{j^2 - a^2}$. (Hint. Express the last function as the sum of "partial fractions" of the form b.)

16. Compare the antiderivatives of the following functions:

(a) $\frac{j}{a^2 - j^2}$ and $\frac{1}{a^2 - j^2}$; (b) $\frac{h \cdot \mathbf{D}\, h}{1 - h^2}$ and $\frac{\mathbf{D}\, h}{1 - h^2}$.

17. The antiderivatives of $1/(j^2 + c^2)$, $1/(j^2 - c^2)$, and $1/j^2$ differ considerably. Find the antiderivatives of $1/(j^2 + a\,j + b)$ by applying the Transformation Rule to the above three functions according to whether $b - \frac{a^2}{4}$ is positive, negative, or 0.

18. Using the result of Exercise 17 and the Power Rule, find the antiderivatives of $\frac{c\,j + d}{j^2 + a\,j + b}$. Hint. Decompose the numerator into a constant multiple of $2j + a$ (that is, the derivative of the denominator) and a constant.

19. Solve the three differential equations:

$$f^2 \cdot \mathbf{D}\, f = exp; \quad f^4 \cdot \mathbf{D} f = j^4; \quad (f + 1) \cdot \mathbf{D}\, f = 2\,j.$$

20. What are the graphs of the solutions of the three differential equations:

$$f \cdot \mathbf{D} f = -j; \quad f \cdot \mathbf{D}\, f = j; \quad f \cdot \mathbf{D}\, f = 1?$$

Is the difference between any two solutions of the same differential equation a constant function?

21. Determine c so that $\int_0^\infty exp\,(c\,j) = 1$; then relate a and c so that $\int_0^\infty a\,exp\,(c\,j) = 1$. Of course, $\int_0^\infty f = \lim_\infty \int_0 f$.

22. Suppose the instantaneous velocity of a particle moving along a straight line is connected with the time by

(a) $v = 5 + 32\,t$; (b) $v = 3\,sin(2\,t)$; (c) $v = 4\,exp\,(-t)$.

How is the position of the particle related to the time assuming that it occupies the point 3 at the instant 0?

4. ANTIDERIVATION BY SUBSTITUTION

On both sides of the Transformation Rule in Form I, substitute $j /\!/ h$, the inverse of h. On the right side, $(\mathbf{D}^{-1} g)h$ becomes $\mathbf{D}^{-1} g$; on the left side, the result is $[\mathbf{D}^{-1}(gh \cdot \mathbf{D} h)] /\!/ h$. Reading the formula from right to left, one obtains the following important result, referred to as

ANTIDERIVATION OF g BY THE SUBSTITUTION OF h.

(9) $$\mathbf{D}^{-1} g + \mathcal{C} = [\mathbf{D}^{-1}(gh \cdot \mathbf{D} h)] /\!/ h + \mathcal{C}.$$

As a first example, find the antiderivatives of $(1+j^2)^{-1/2}$ by the substitution of *tan*.

$$\mathbf{D}^{-1}(1+j^2)^{-1/2}+\mathcal{C}=[\mathbf{D}^{-1}((1+tan^2)^{-1/2}\cdot \sec^2)]//tan+\mathcal{C}$$
$$=[\mathbf{D}^{-1}\sec]//tan+\mathcal{C}.$$

Hence, according to (4), the antiderivatives are

$$=[log\ abs\ (\sec+tan)]//tan+\mathcal{C}=log\ abs\ (\sqrt{1+j^2}+j)+\mathcal{C}.$$

This result can be verified by showing that the derivative of the last function indeed equals $(1+j^2)^{-1/2}$.

As (9) indicates and the preceding example illustrates, the antiderivation of g by the substitution of h is performed in four steps:

Step I. Substitution of h into g; in the example, $gh=(1+tan^2)^{-1/2}=\sec^{-1}$;

Step II. Multiplication of the resulting function by $\mathbf{D}h$; in the example, $\sec^{-1}$ is to be multiplied by $\sec^2$;

Step III. Antiderivation of the product; in the example, antiderivation of sec, which yields *log abs* (sec + *tan*);

Step IV. Substitution of the inverse of h into the antiderivative; in the example, substitution of *arctan* into *log abs* (sec + *tan*) + $\mathcal{C}$.

In (9), the right side looks more complicated than the left side. In some cases it actually is. For instance, in the antiderivation of $g=(1+j^2)^{-1/2}$ by the substitution of $h=exp$ or $h=sin$, the antiderivative in Step III is more complicated than that of g itself. For $h=tan$, Step III consists in finding $\mathbf{D}^{-1}\sec$. Since this problem had been solved in (4), the antiderivation of g by the substitution of *tan* could be completed. Had $\mathbf{D}^{-1}\sec$ not been known, even the substitution of *tan* would have been of no avail.

But in other cases, $\mathbf{D}^{-1}(gh\cdot\mathbf{D}h)$ is much simpler than $\mathbf{D}^{-1}g$. A striking example is $g=(b^2-j^2)^{-1/2}$ and $h=\mathrm{b}\ sin$. For, in this case, $gh\cdot\mathbf{D}h$ $=\frac{1}{\mathrm{b}\cos}\cdot\mathrm{b}\cos=1$. Certainly, the antiderivation of 1 is simpler than that of $(b^2-j^2)^{-1/2}$. Since $\{\mathbf{D}^{-1}1\}=j+\mathcal{C}$, all that has to be done to obtain $\mathbf{D}^{-1}g$ is to substitute into j the inverse of b *sin*, that is, $arcsin\ (\frac{1}{\mathrm{b}}j)$; and one finds

$$\mathbf{D}^{-1}(b^2-j^2)^{-1/2}+\mathcal{C}=arcsin\ (\tfrac{1}{\mathrm{b}}j)+\mathcal{C},$$

a result that could be also obtained from the fifth Standard Antiderivative in conjunction with the Linear Transformation Rule.

The example suggests that $h=\mathrm{b}\ sin$ might be substituted with benefit whenever g contains $\sqrt{b^2-j^2}$. In some cases that substitution is indeed helpful, in others it is not.

For instance, if $g = (1 - j^2)^{n/2}$, then

$$g\ sin \cdot \mathbf{D}\ sin = cos^n \cdot cos = cos^{n+1}.$$

If $\{\mathbf{D}^{-1} cos^{n+1}\}$ is known, then substitution of *arcsin* into that antiderivative solves the original problem. $\{\mathbf{D}^{-1} cos^{n+1}\}$ will be determined on page 264 for any n. So far, it has been determined for n = −3 and 1:

$$\{\mathbf{D}^{-1} cos^{-2}\} = \{\mathbf{D}^{-1} sec^2\} = tan + \mathcal{C} \text{ and } \{\mathbf{D}^{-1} cos^2\} = \frac{1}{2}(j + sin \cdot cos) + \mathcal{C}$$

(from Exercise 13). Substituting *arcsin* into these antiderivatives one obtains

$$\{\mathbf{D}^{-1}(1-j^2)^{-3/2}\} = j \cdot (1-j^2)^{-1/2} + \mathcal{C} \text{ and}$$

$$\{\mathbf{D}^{-1}(1-j^2)^{1/2}\} = \frac{1}{2}[arcsin + j \cdot (1-j^2)^{1/2}] + \mathcal{C}.$$

If, on the other hand, one attempts the antiderivation of $1/\sqrt{(1-j^2)\cdot(1-.64\,j^2)}$ by the substitution of *sin*, he arrives at

$$[\mathbf{D}^{-1}(1/\sqrt{1-.64\,sin^2})]\,//\,sin + \mathcal{C},$$

and thus replaces one non-elementary antiderivative by another one. Of course, this is all that the substitution of an elementary function into a non-elementary antiderivative can possibly achieve.

While (9) is valid for any g and any h, the practical value of (9) in determining $\mathbf{D}^{-1} g$ for a given function g depends entirely upon the discovery of an *efficient* function h, that is, one for which $\mathbf{D}^{-1}(gh \cdot \mathbf{D}h)$ is known or can be determined. If there is such an h, its discovery is a creative act of mathematical invention. One may have to try several functions h until he finds one that is efficient. If several attempts fail, one may give up, discouraged by the idea that an efficient h may not exist (for instance, because the antiderivatives of g may be non-elementary).

With this warning in mind, one may use the following rules of thumb:

If g contains	try $h =$	The expression in the first column becomes	Instead of $\mathbf{D}^{-1} g$, find $\mathbf{D}^{-1}$ of	and substitute into it $j \,//\, h$:
$b^2 - j^2$	$b\ sin$	$b^2 cos^2$	$g(b\ sin) \cdot b\ cos$	$arcsin(\frac{1}{b}j)$
$b^2 + j^2$	$b\ tan$	$b^2 sec^2$	$g(b\ tan) \cdot b\ sec^2$	$arctan(\frac{1}{b}j)$
$j^2 - b^2$	$b\ sec$	$b^2 tan^2$	$g(b\ sec) \cdot b\ sec \cdot tan$	$arcsec(\frac{1}{b}j)$
$j^{1/n}$	j^n	j	$gj^n \cdot n\,j^{n-1}$	$j^{1/n}$
$(a\,j + b)^{1/n}$	$\frac{1}{a}(j^n - b)$	j	$g\left[\frac{1}{a}(j^n - b)\right] \cdot \frac{n}{a} j^{n-1}$	$(a\,j + b)^{1/n}$

As an application of the last line, find the antiderivatives of

$$g = \frac{1}{(2j+3)^{1/3} + (2j+3)^{1/2}} = \frac{1}{(2j+3)^{2/6} + (2j+3)^{3/6}} .$$

Substitute $\frac{1}{2}(j^6 - 3)$ which, substituted into $(2j+3)^{1/6}$, yields j. Then, instead of $\{\mathbf{D}^{-1} g\}$, one has to find the antiderivatives of

$$\frac{1}{j^2 + j^3} \cdot 3j^5 = \frac{3j^3}{1+j} = 3\left[j^2 - j + 1 - \frac{1}{1+j}\right].$$

All one has to do is to substitute $(2j+3)^{1/6}$ into

$$j^3 - \frac{3}{2}j^2 + 3j - 3 \log abs\,(1+j) + \mathcal{C}.$$

In $\{\mathbf{D}^{-1}(1/cosh)\}$, the substitution of log (not mentioned in the table) yields $\frac{2j}{j^2+1} \cdot \mathbf{D}\,log = \frac{2}{j^2+1}$. Into the resulting antiderivative, $2\,arctan + C$, one has to substitute $j /\!/ log = exp$.

The following application to integration is important. Since $\int_a^b g$ is the rise of any $\mathbf{D}^{-1} g$ from a to b, one has

$$\int_a^b g = [\mathbf{D}^{-1} g]_a^b = [(\mathbf{D}^{-1}(gh \cdot \mathbf{D}h)) /\!/ h]_a^b .$$

Clearly, the last rise equals the rise of $\mathbf{D}^{-1}(gh \cdot \mathbf{D}h)$ from $(j /\!/ h)$a to $(j /\!/ h)$b; and

$$\Big[\mathbf{D}^{-1}(gh \cdot \mathbf{D}h)\Big]_{(j/\!/h)a}^{(j/\!/h)b}$$

in turn, equals

$$\int_{(j/\!/h)\,a}^{(j/\!/h)\,b} (gh \circ \mathbf{D}h) .$$

Hence the following

INTEGRATION OF g BY THE SUBSTITUTION OF h.

(10) $$\int_a^b g = \int_{(j/\!/h)\,a}^{(j/\!/h)\,b} (gh \cdot \mathbf{D}h) .$$

Thus, $\int_a^b g$ is found by the substitution of h in four steps:

Step I. Substitution of h.

Step II. Multiplication by $\mathbf{D}h$.

Step III. Antiderivation of the product.

Step IV. Computation of the rise of this antiderivative from $(j // h)\,\mathrm{a}$ to $(j // h)\,\mathrm{b}$.

The reader will note the parallelism of (10) with the application following the basic scheme of p. 206

of the integral $\int_{h\,\mathrm{a}}^{h\,\mathrm{b}} g$ to the cumulation $\int_{\mathrm{a}}^{\mathrm{b}} gh \, \mathrm{d}\, h$.

EXERCISES

23. Find $\mathbf{D}^{-1}(\sqrt{a^2 - j^2})$ and compute $\int_{\mathrm{c}}^{\mathrm{c}'} \sqrt{r^2 - j^2}$, the area under a portion of a circle of radius r. Adding to $\int_{\mathrm{b}}^{1} \sqrt{1 - j^2}$ the triangular area $\int_0^{\mathrm{b}} \frac{\sqrt{1 - \mathrm{b}^2}}{\mathrm{b}}\, j$, one obtains the area of the sector of the circle of radius 1 shaded in Fig. 39 (a). Verify that if a denotes twice this area, then $\mathrm{b} = \cos \mathrm{a}$, and $\sqrt{1 - \mathrm{b}^2} = \sin \mathrm{a}$.

24. Find $\mathbf{D}^{-1}(\sqrt{j^2 - a^2})$ and compute $\int_1^{\mathrm{b}} \sqrt{j^2 - 1}$ the area under a portion of the hyperbola $\sqrt{j^2 - 1}$. Subtracting this area from the triangular area $\int_0^{\mathrm{b}} \frac{\sqrt{\mathrm{b}^2 - 1}}{\mathrm{b}}\, j$ one obtains the area of the hyperbolic sector shaded in Fig. 39 (b). Verify that if a denotes twice this area, then

$$\mathrm{b} = \cosh \mathrm{a} \quad \text{and} \quad \sqrt{\mathrm{b}^2 - 1} = \sinh \mathrm{a}\,.$$

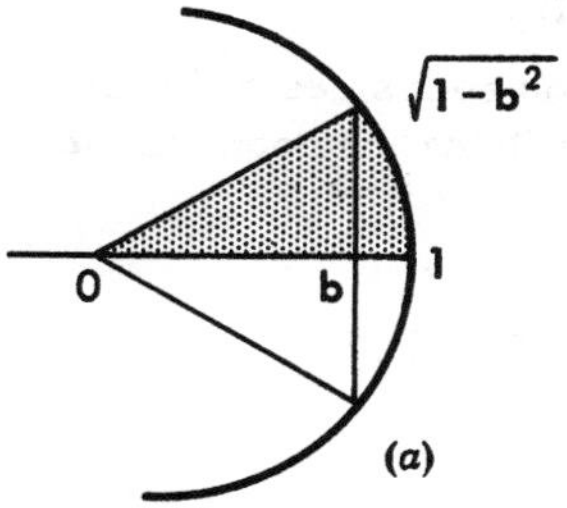

Fig. 39 (a)

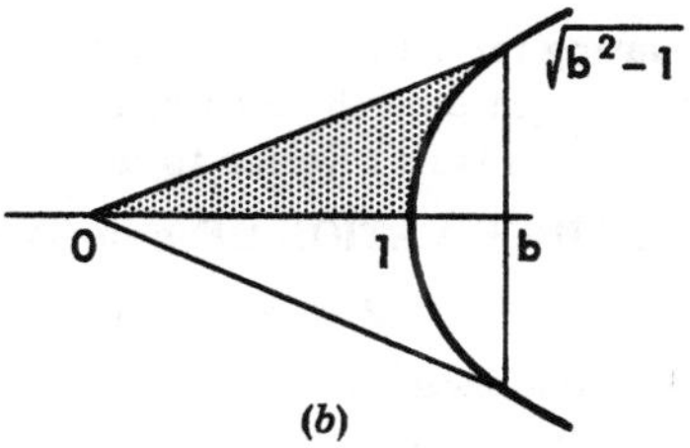

Fig. 39 (b)

Comparison of the results of Exercises 24 and 23 proves that *cosh* and *sinh* are related to the hyperbola $\sqrt{j^2 - 1}$ as are *cos* and *sin* to the circle $\sqrt{1 - j^2}$. This accounts for the terms hyperbolic cosine and sine.

25. The fact that the antiderivatives of $\sqrt{j^2 - a^2}$ and $\sqrt{j^2 + a^2}$ contain logarithms of simpler functions suggests that they might be obtained by means of the antiderivation of a −1st power. What are the necessary substitutions? (Compare the transformation of *sec* on p. 253).

26. Find $\mathbf{D}^{-1} \dfrac{cj + d}{\sqrt{j^2 + aj + b}}$ (Compare with Exercise 18).

27. Find the antiderivatives of the following functions:
(a) $1/\sqrt{j^2 - 1}$; (b) $1/(j \cdot \sqrt{a^2 - j^2})$; (c) $\sqrt{1 - j^2}/j^2$; (d) $j^2/\sqrt{1 - j^2}$. In view of $\sqrt{1 - j^2} = (1 - j^2)/\sqrt{1 - j^2}$, combine result (d) with previous results.

28. Find the antiderivatives of $1/(j^2 \cdot \sqrt{21 j^2 + 4j - 1})$ by substituting j^{-1}.

29. Transform the (non-elementary) function $\mathbf{D}^{-1}_{(0,0)}[exp(-j^2)]$ by the substitution of $j^{1/2}$ into another non-elementary function. How can the latter be transformed into the former? How can $\mathbf{D}^{-1}(1/\sqrt{1 - .64\, sin^2})$ be transformed into the $-\frac{1}{2}$-power of a polynomial of degree 4?

30. Formula (9) corresponds to form I of the Transformation Rule. Formulate the analogues of forms II and III.

5. ANTIDERIVATION BY PARTS

The Product Rule for Derivatives implies

$$g \cdot \mathbf{D}h = \mathbf{D}(g \cdot h) - h \cdot \mathbf{D}g .$$

Equating the antiderivatives of the functions on both sides and applying on the right side the Sum and Constant Factor Rules, one obtains the following result referred to as

ANTIDERIVATION OF $g \cdot h$ BY PARTS.

(11) $$\{\mathbf{D}^{-1}(g \cdot \mathbf{D}h)\} = g \cdot h - \mathbf{D}^{-1}(h \cdot \mathbf{D}g) + \mathscr{C}.$$

For instance,

(12) $$\{\mathbf{D}^{-1}(j \cdot cos)\} = j \cdot sin - \{\mathbf{D}^{-1}(sin \cdot 1)\} = j \cdot sin + cos + \mathscr{C};$$

$$\{\mathbf{D}^{-1}(j \cdot exp)\} = j \cdot exp - \{\mathbf{D}^{-1}(exp \cdot 1)\} = j \cdot exp - exp + \mathscr{C}.$$

In these examples, the knowledge of the antiderivatives on the right side makes it possible to compute those on the left side. If (11) is applied to the same factors in the opposite order, the antiderivatives on the right side are more complicated than those on the left:

$$\{ \mathbf{D}^{-1}(cos \circ j)\} = cos \cdot \tfrac{1}{2} j^2 - \mathbf{D}^{-1}(\tfrac{1}{2} j^2 \cdot (-sin)) + \mathcal{C}$$
$$= \tfrac{1}{2} cos \cdot j^2 + \tfrac{1}{2} \mathbf{D}^{-1}(j^2 \cdot sin) + \mathcal{C}.$$

In this case, the application of (11), though correct, is of no avail for the determination of $\{ \mathbf{D}^{-1}(cos \cdot j)\}$.

One readily verifies the equivalence of (11) with the following statements:

$$\{ \mathbf{D}^{-1}(g \cdot \mathbf{D} h)\} = g \cdot h - \{ \mathbf{D}^{-1}(h \cdot \mathbf{D} g)\}$$

(13) $$\mathbf{D}^{-1}_{(a,c)}(g \cdot \mathbf{D} h) = g \cdot h - g\,a \cdot h\,a - \mathbf{D}^{-1}_{(a,-c)}(h \cdot \mathbf{D} g)$$

(13) may yield the antiderivatives of a given function f. First one has to factor f. Obviously, there are infinitely many factorizations:

$$f = f^{1/2} \cdot f^{1/2} = f^{1/3} \cdot f^{2/3} = \frac{f}{j} \cdot j = f^2 \cdot \frac{1}{f} = f \cdot 1 = \ldots$$

To each factorization $f = f_1 \cdot f_2$ one can apply (11) by setting $f_2 = \mathbf{D} h$. This makes h an antiderivative of f_2.

ANTIDERIVATION OF f BY PARTS.

(14) $$\mathbf{D}^{-1} f + \mathcal{C} = f_1 \cdot \mathbf{D}^{-1} f_2 - \mathbf{D}^{-1}(\mathbf{D} f_1 \cdot \mathbf{D}^{-1} f_2) + \mathcal{C}.$$

Here, $\mathbf{D}^{-1} f_2$ is *any* antiderivative of f_2; *however, it must be the same in both places on the right side.* Consider, for instance,

$$f = 3j^3, \quad f_1 = j \quad \text{and} \quad f_2 = 3j^2. \quad \text{According to (11):}$$
$$\mathbf{D}^{-1}(3j^3) + \mathcal{C} = j \cdot \mathbf{D}^{-1}(3j^2) - \mathbf{D}^{-1}(1 \cdot \mathbf{D}^{-1}(3j^2)) + \mathcal{C}.$$

If here $\mathbf{D}^{-1}(3j^2)$ is replaced by $j^3 + a$ the first time, and by $j^3 + b$ the second time, then

$$\mathbf{D}^{-1}(3j^3) + \mathcal{C} = j \cdot (j^3 + a) - \mathbf{D}^{-1}(j^3 + b) + \mathcal{C} = \frac{3}{4} j^4 + (a - b) \cdot j + \mathcal{C},$$

which is valid only if $a = b$.

Formula (14) explains the name of the method. To find $\mathbf{D}^{-1} f$ according to (14), first one determines the antiderivative of, as it were, a part of f, namely, of f_2, and then the antiderivative of the product $\mathbf{D}^{-1} f_2 \cdot \mathbf{D} f_1$.

While (14) is valid for *any* factorization of f, the formula is helpful in finding $\mathbf{D}^{-1} f$ only if the factorization satisfies the following conditions:

1. $\mathbf{D}^{-1} f_2$ is known,
2. $\mathbf{D}^{-1}(\mathbf{D} f_1 \cdot \mathbf{D}^{-1} f_2)$ is known or can be determined.

In some cases, rather unexpected factorizations meet these requirements. For instance, if f is *log* or an arcfunction, $f = f \cdot 1$ is helpful.

$\{\mathbf{D}^{-1}\, log\} = \{\mathbf{D}^{-1}\,(log \cdot 1)\} = log \cdot j - \mathbf{D}^{-1}(\frac{1}{j} \cdot j) + \mathcal{C} = log \cdot j - j + \mathcal{C},$

$\{\mathbf{D}^{-1}\, arcsin\} = \{\mathbf{D}^{-1}\,(arcsin \cdot 1)\} = arcsin \cdot j - \mathbf{D}^{-1}\left(\frac{1}{\sqrt{1-j^2}} \cdot j\right) =$

$arcsin \cdot j + \sqrt{1-j^2} + \mathcal{C}.$

In many cases, $D^{-1}(\mathbf{D} f_1 \cdot \mathbf{D}^{-1} f_2)$ is *simpler* than $\mathbf{D}^{-1} f$. For instance,

$$\{\mathbf{D}^{-1}(j^2 \cdot sin)\} = -j^2 \cdot cos + 2\,\mathbf{D}^{-1}(j \cdot cos) + \mathcal{C}.$$

The antiderivative on the right side was computed in (12). Thus, altogether, $\mathbf{D}^{-1}(j^2 \cdot sin)$ can be found by two successive antiderivations by parts. The reader should verify that, for any positive integer n, the antiderivatives of $j^n \cdot sin$ and $j^n \cdot cos$ can be found by n successive applications of this method. One obtains the following

RECURSION FORMULAS

(15) $\{\mathbf{D}^{-1}[j^n \cdot cos(aj)]\} = \frac{1}{a^2} j^{n-1}[aj \cdot cos(aj) + n\, sin(aj)] -$

$\frac{n(n-1)}{a^2}\mathbf{D}^{-1}[j^{n-2} \cdot cos(aj)] + \mathcal{C}.$

$\{\mathbf{D}^{-1}[j^n \cdot sin(aj)]\} = \frac{1}{a^2} j^{n-1}[-aj \cdot cos(aj) + n\, sin(aj)] -$

$\frac{n(n-1)}{a^2}\mathbf{D}^{-1}[j^{n-2} \cdot sin(aj)] + \mathcal{C}.$

$\{\mathbf{D}^{-1}(sin^n)\} = -\frac{1}{n}\, sin^{n-1} \cdot cos + \frac{n-1}{n}\mathbf{D}^{-1}(sin^{n-2}) + \mathcal{C}.$

(16) $\{\mathbf{D}^{-1}(cos^n)\} = \frac{1}{n}\, cos^{n-1} \cdot sin + \frac{n-1}{n}\mathbf{D}^{-1}(cos^{n-2}) + \mathcal{C}.$

(16) can be used to find the antiderivatives of $(j^2 + a^2)^{-m}$ by the substitution of the function (a *tan*), whose inverse is $arctan\,(\frac{1}{a} j)$.

$\{\mathbf{D}^{-1}(j^2 + a^2)^{-m}\} = \mathbf{D}^{-1}(a^{-2m}\, sec^{-2m} \cdot a\, sec^2) /\!/ (a\, tan) + \mathcal{C}$

$= a^{1-2m}\,\mathbf{D}^{-1}(cos^{2m-2}) /\!/ (a\, tan) + \mathcal{C}.$

The last antiderivative is found by applying (16) for n = 2m – 2. The first term on the right side of (16) becomes $\frac{1}{2(m-1)} cos^{2m-3} \cdot sin$. It may be replaced by $\frac{1}{4(m-1)} cos^{2(m-2)} \circ sin(2j)$. Then one uses

$cos^2 /\!/ (a\, tan) = cos^2\, arctan(\frac{1}{a} j) = \frac{a^2}{j^2 + a^2}$ and

$[sin(2j)] /\!/ (a\, tan) = sin\, arctan(\frac{1}{a} j) = \frac{2aj}{j^2 + a^2}.$

Thus altogether one obtains the important recursion formula

(17) $$\mathbf{D}^{-1}(j^2+a^2)^{-m}+\mathcal{C}$$
$$=\frac{1}{2(m-1)a^2}\,j\cdot(j^2+a^2)^{1-m}+\frac{2m-3}{2(m-1)a^2}\,\mathbf{D}^{-1}(j^2+a^2)^{-(m-1)}+\mathcal{C}.$$

Since $\mathbf{D}^{-1}(j^2+a^2)^{-1}=\frac{1}{a}\,arctan(\frac{1}{a}j)$, one can for any integer m, compute $\mathbf{D}^{-1}(j^2+a^2)^{-m}$ by repeated application of the above formula.

To every recursion formula one can apply the Transformation Rule. For instance, the last formula implies

(18) $$\left\{\mathbf{D}^{-1}\left(\frac{\mathbf{D}h}{(h^2+a^2)^m}\right)\right\}$$
$$=\frac{1}{2(m-1)a^2}\,\frac{h}{(h^2+a^2)^{m-1}}+\frac{2m-3}{2(m-1)a^2}\,\mathbf{D}^{-1}\left(\frac{\mathbf{D}h}{(h^2+a^2)^{m-1}}\right)+\mathcal{C}.$$

For some f, the antiderivative on the right side of (14) equals $k\,\mathbf{D}^{-1}f$ for some number $k \neq 1$. In such cases, subtracting $k\,\mathbf{D}^{-1}f$ on both sides, one finds $(1-k)\mathbf{D}^{-1}f$. This situation may materialize after iterated antiderivation by parts.

For instance, applying the method to $exp(a\,j)\cdot sin(b\,j)$ one obtains

$$\{\mathbf{D}^{-1}(exp(a\,j)\cdot sin(b\,j))\}$$
$$=\frac{-1}{b}\,exp(a\,j)\cdot cos(b\,j)+\frac{a}{b}\,\mathbf{D}^{-1}(exp(a\,j)\cdot cos(b\,j)+\mathcal{C}.$$

The antiderivative on the right side is just as complicated as that on the left side. Applying antiderivation by parts to it, one obtains

$$\{\mathbf{D}^{-1}(exp(a\,j)\cdot cos(b\,j))\}$$
$$=\frac{1}{b}\,exp(a\,j)\cdot sin(b\,j)-\frac{a}{b}\,\mathbf{D}^{-1}(exp(a\,j)\cdot sin(b\,j))+\mathcal{C}.$$

Substituting the expression for $\mathbf{D}^{-1}(exp(a\,j)\cdot cos(b\,j))$ from the latter equality into the former, one obtains on the right side a multiple of the antiderivative that he wishes to find. Subtracting this multiple on both sides, one readily finds

(19) $$\{\mathbf{D}^{-1}(exp(a\,j)\cdot sin(b\,j))\}$$
$$=\frac{1}{a^2+b^2}\,exp(a\,j)\cdot(a\,sin(b\,j)-b\,cos(b\,j))+\mathcal{C}.$$

EXERCISES

31. Find $D^{-1}\, sin^2$ by parts. Hint: $cos^2 = 1 - sin^2$.

32. Find the antiderivatives of the following functions:
(a) $arctan$; *(b)* $arcsec$; *(c)* $arcsinh = log(j + \sqrt{j^2+1})$; *(d)* $j \cdot arcsin$; *(e)* $arcsin(a j)$; *(f)* $arctan(a j)$; *(g)* $arctan\, j^{1/2}$; *(h)* $j^2/exp = j^2 \cdot exp(-j)$.

33. Formula (15), in the case a = 1, reads

$$\{D^{-1}(j^n \cdot cos)\} = j^{n-1} \cdot (j \cdot sin + n\, cos) - n(n-1)\, D^{-1}(j^{n-2} \cdot cos) + \mathcal{C}.$$

From the latter formula, obtain (15) using the Linear Transformation Rule.

34. As a counterpart to (19), find $D^{-1}(exp(a j) \cdot cos(b j))$. The result can be obtained either by retracing the entire proof of (19) or by using (19) in a phase of the proof of (19).

35. Derive the recursion formula

$$D^{-1} \frac{j^n}{exp} + \mathcal{C} = -\frac{j^n}{exp} + n\, D^{-1} \frac{j^{n-1}}{exp} + \mathcal{C}.$$

From it, using the Linear Transformation Rule as in Exercise 33, obtain a recursion formula for

$$\left\{ D^{-1} \frac{j^n}{exp(a j)} \right\} = \{ D^{-1}(exp(-a j) \cdot j^n)\}.$$

36. Derive recursion formulas for the antiderivatives of *(a)* tan^n; *(b)* $j^n \cdot log$.

37. Apply the recursion formula for $D^{-1}\, sin^n$ to $D^{-1}(a^2 - j^2)^{-n/2}$.

38. By substitution and the development of the proper trigonometric recursion formulas, set up recursion formulas for $D^{-1}(j^2 + a^2)^{n/2}$, one for positive and one for negative integers n.

39. For any function *f*, apply antiderivation by parts to the following factorizations: *(a)* $f = 1 \cdot f$; *(b)* $f = f \cdot 1$; *(c)* $f = j \cdot \frac{f}{j}$. For what functions *f* is *(c)* promising?

6. THE ANTIDERIVATIVES OF RATIONAL FUNCTIONS

Throughout this book, *number* means *finite real number*. Accordingly, in this section it should be clearly understood that the coefficients of all polynomials under consideration are real numbers.

The quadratic polynomials, $a_0 + a_1 j + a_2 j^2$ (where $a_2 \neq 0$), fall into three classes according to whether $a_1^2 - 4 a_0 a_2$ is > 0, $= 0$, or < 0. Those of the first type are products of two different linear polynomials; those of the second type are the square of a linear polynomial; those of the third type cannot be factored into two linear polynomials.

For instance, $1 + j^2$ is of the third type. At this point the restriction to real numbers plays an essential role. If imaginary and complex numbers were admitted, then also the polynomials of the third type would be products of two different linear polynomials; for instance, $1 + j^2 = (i + j) \cdot (-i + j)$.

The following statements are important results obtained in the theory of complex functions and in algebra. Any polynomial p of a degree $n > 0$ is the product of k (≥ 0) linear polynomials and m (≥ 0) quadratic polynomials of the third type, where $k + 2m = n$. If in p the coefficient of j^n is 1, then there is only one system of k linear polynomials of the form $-c + j$, and m quadratic polynomials of the form $a + b\,j + j^2$ (where $b^2 < 4a$) such that p is the product of these polynomials. Among the k linear, as well as among the m quadratic, factors of p there may, of course, be equal ones. It can be assumed that

(20) $p =$

$$(-c_1 + j)^{k_1} \cdot \ldots \cdot (-c_r + j)^{k_r} \cdot (a_1 + b_1\, j + j^2)^{m_1} \cdot \ldots \cdot (a_s + b_s\, j + j^2)^{m_s},$$

where $\qquad k_1 + \ldots + k_r + 2\,(m_1 + \ldots + m_s) = n,$

and where any two of the $r + s$ factors are different.

Any rational function r (the quotient of two polynomials) can always be represented (and in only one way) as the sum of a polynomial s and the quotient q/p of two polynomials such that the degree of the denominator p exceeds that of the numerator q. Thus $r = s + q/p$. If here p is the polynomial (20), then q/p admits the following decomposition into so-called *partial fractions;* that is to say, n numbers $u_{11}, \ldots u_{rk_r}, v_{11}, \ldots w_{sm_s}$ can be determined so that

$$\frac{q}{p} = \frac{u_{11}}{-c_1 + j} + \ldots + \frac{u_{1k_1}}{(-c_1 + j)^{k_1}} + \ldots + \frac{u_{r1}}{-c_r + j} + \ldots + \frac{u_{rk_r}}{(-c_r + j)^{k_r}}$$

$$+ \frac{v_{11} + w_{11}\, j}{a_1 + b_1\, j + j^2} + \ldots + \frac{v_{1m_1} + w_{1m_1}\, j}{(a_1 + b_1\, j + j^2)^{m_1}} + \ldots + \frac{v_{s1} + w_{s1}\, j}{a_s + b_s\, j + j^2} + \ldots$$

$$+ \frac{v_{sm_s} + w_{sm_s}\, j}{(a_s + b_s\, j + j^2)^{m_s}} .$$

The above mentioned results yield the following

Theorem. *The antiderivatives of any rational function r are elementary; more precisely, they are sums of rational functions, logarithms of linear or quadratic polynomials (factors of the denominator of r), and arctangents of linear polynomials (derivatives of quadratic factors of the denominator* *of r). The antiderivatives of r can actually be determined provided that the partial fractions of r have been computed.*

Indeed $\{\mathbf{D}^{-1} r\} = \mathbf{D}^{-1} s + \mathbf{D}^{-1}(q/p) + \mathcal{C}$. Here, any $\mathbf{D}^{-1} s$ is a polynomial, and $\mathbf{D}^{-1}(q/p)$ is the sum of the antiderivatives of the partial fractions.

1. Any antiderivative of a fraction of the form $\dfrac{u}{-c + j}$ is of the form $u \log (j - c)$.

2. Any antiderivative of a fraction of the form $\dfrac{u}{(-c + j)^t}$ for $t > 1$ is a rational function, more specifically, of the form $\dfrac{u'}{(-c + j)^{t-1}}$.

Any fraction of the form $\dfrac{v + w j}{(a + b j + j^2)^t}$ is equal to the sum

$$\frac{\frac{1}{2} w (2j + b)}{(a + b j + j^2)^t} + \frac{v - \frac{1}{2} wb}{(a + b j + j^2)^t}.$$

3. Any antiderivative of a fraction of the form $\dfrac{\frac{1}{2} w (2 j + b)}{(a + b j + j^2)^t}$ fot $t > 1$ is of the form $\dfrac{w'}{(a + b j + j^2)^{t-1}}$, thus rational.

4. Any antiderivative of a fraction of the form $\dfrac{\frac{1}{2} w (2 j + b)}{a + b j + j^2}$ is of the form $w' \log (a + b j + j^2)$.

5. Any fraction of the form $\dfrac{u}{(a + b j + j^2)^t}$ is equal to one of the form $\dfrac{u}{(h^2 + d^2)^t}$, where $h = j + b/2$ and $d = a - b^2/4$, which is > 0. Since $\mathbf{D}\, h = 1$, the antiderivatives of the last fraction can be determined by virtue of (18). They are sums of rational functions and *arctan h,* where h is $\frac{1}{2} \mathbf{D} (a + b j + j^2)$.

This completes the proof of the theorem.

EXERCISE

★40. Assuming that *log* and *arctan* are irrational functions, formulate conditions that are necessary and sufficient for a function to have rational antiderivatives.

7. RÉSUMÉ OF THE CALCULUS OF ANTIDERIVATIVES

Except for the theory of rational functions, the calculus of antiderivatives lacks the systematic character of the calculus of derivatives. There is a process of derivation; there is no process of antiderivation. Ultimately, the antiderivatives of a given function f are found by guessing which functions have the derivative f.

This guess is easy in cases where f is the derivative of a well-known function; and the results are embodied in the Standard Formulas. As aids in more complicated cases, three methods have been discussed:

1. Reduction of $\mathbf{D}^{-1} f$ to a standard or otherwise known antiderivative by the Transformation Rule, often in conjunction with the Constant Factor Rule;

2. Antiderivation by Substitution;

3. Antiderivation by Parts.

In (*2*) and (*3*), discovering (or guessing) *efficient* substitutions and factorizations is a matter of mathematical invention. Another method is

4. The use of tables.

Antiderivative Tables (usually called Integral Tables) include the antiderivatives of numerous functions as well as recursion formulas. The reduction, for a given function f, of $\mathbf{D}^{-1} f$ to an antiderivative listed in a table may well require the application of the Sum and Constant Factor Rules; it often requires the application of the Transformation Rule. For instance, one may be looking for the antiderivative treated in (15), while the table at his disposal contains only the formula mentioned in Exercise 33. Exercises 33 and 35 illustrate the procedure of reducing more complicated to simpler antiderivatives.

Failure to find $\{\mathbf{D}^{-1} f\}$ may have one of two reasons: lack of skill or non-existence of elementary antiderivatives of f. It would be desirable to know a simple criterion for an elementary function to have an elementary antiderivative. There are criteria, but they are not simple. The reader who worked on Exercise 40 knows that even the conditions for a rational function to have rational antiderivatives are not altogether simple.

Some waste of time and energy in looking for elementary antiderivatives where they do not exist can be forestalled by keeping in mind:

1. The following important types of functions with non-elementary antiderivatives:

(a) $\exp j^n$, $\sin j^n$, $\cos j^n$ for $n > 1$;

(b) $1/\log$, $\exp/j$, $\sin/j$;

(c) fractional powers of most polynomials of degrees > 2; in particular, square roots of most cubic, quartic, and higher polynomials, as well as the reciprocals of those square roots.

2. That $\mathbf{D}^{-1} f$ is certainly non-elementary if antiderivation by an elementary substitution or by parts reduces $\mathbf{D}^{-1} f$ to an antiderivative known to be non-elementary.

An approximate qualitative idea of non-elementary antiderivatives can be obtained by the graphical methods of Chapter II, which naturally do not discriminate between elementary and non-elementary cases. Another approximate method will be discussed in Chapter X.

8. APPLICATIONS TO INTEGRALS

(This section may be skipped at a first reading.)

Application of (11) to integrals yields the following formula for

$$\int_a^b (g \cdot \mathbf{D}\, h) = [g \cdot h]_a^b - \int_a^b (h \cdot \mathbf{D}\, g). \tag{21}$$

The reader will note the parallelism of (21) with the fact studied in Exercise 41, p. 211.

Example 1. From the recursion formula in Exercise 35 it follows that

$$\int_a^b \frac{j^n}{exp} = -\left(\frac{b^n}{e^b} - \frac{a^n}{e^a}\right) + n \int_a^b \frac{j^{n-1}}{exp};$$

in particular, for a = 0,

$$\int_0^b \frac{j^n}{exp} = -\frac{b^n}{e^b} + n \int_0^b \frac{j^{n-1}}{exp} \quad \text{or} \quad \int_0 \frac{j^n}{exp} = -\frac{j^n}{exp} + n \int_0 \frac{j^{n-1}}{exp}.$$

Equating the limits at ∞ of the functions on both sides of the last formula, in view of $\lim\limits_{\infty} (j^n/exp) = 0$, one obtains

$$\int_0^\infty \frac{j^n}{exp} = n \int_0^\infty \frac{j^{n-1}}{exp}, \text{ in particular, } \int_0^\infty \frac{j}{exp} = \int_0^\infty \frac{1}{exp} = 1$$

(see Exercise 21). Hence, by induction,

$$\int_0^\infty \frac{j^n}{exp} = n\,(n-1) \cdot \ldots \cdot 2 \cdot 1 = n!. \tag{22}$$

Euler's Gamma function, Γ, is defined as the function assuming for any number $c > 0$ the value $\Gamma(c) = \int_0^\infty \frac{j^{c-1}}{exp}$. According to (22), $\Gamma(n+1) = n!$

Example 2. The second recursion formula that had to be established in Exercise 35 reads

$$\mathbf{D}^{-1}\,\frac{j^n}{exp\,(a\,j)} + \mathcal{C} = -\frac{1}{a}\,\frac{j^n}{exp\,(a\,j)} + \frac{n}{a}\,\mathbf{D}^{-1}\,\frac{j^{n-1}}{exp\,(a\,j)} + \mathcal{C}.$$

If $a > 0$, then $\lim_{\infty} (j^n/exp\,(a\,j)) = 0$, and, by the method used in Example 1, the last formula yields the following generalization of (22):

$$(23)\qquad \int_0^\infty \frac{j^n}{exp\,(a\,j)} = \frac{n!}{a^{n+1}} \text{ for any } a > 0.$$

Example 3. From (19), for $b = 1$, it follows that

$$\left\{\mathbf{D}^{-1}\,\frac{sin}{exp\,(a\,j)}\right\} = \frac{-1}{a^2+1}\;\frac{a\,sin \pm cos}{exp\,(a\,j)} + \mathcal{C}.$$

If $a > 0$, then $\lim_{\infty} \dfrac{sin}{exp\,(a\,j)} = 0$ and $\lim_{\infty} \dfrac{cos}{exp\,(a\,j)} = 0$. Hence, from the last formula,

$$(24)\qquad \int_0^\infty \frac{sin}{exp\,(a\,j)} = \frac{1}{a^2+1}\,; \text{ and } \int_0^\infty \frac{cos}{exp\,(a\,j)} = \frac{a}{a^2+1} \text{ for any } a > 0.$$

(The last formula is a consequence of the recursion formula to be established in Exercise 34.)

If f is a function whose domain includes all positive numbers, then $\int_0^\infty f$ is the area under the graph of f above all points to the right of 0. Division by $exp\,(a\,j)$ or, which amounts to the same, multiplication by $exp\,(-a\,j)$ dampens the graph of f exponentially – the more rapidly, the larger a is. (In this section, a is assumed to be > 0).

Clearly, the area $\int_0^\infty j$ is not finite. But the area under j dampened by the division by exp, according to (22) is 1. If j is dampened by the division by $exp\,(.1\,j)$, according to (23), the area is 10. No matter how slightly j is exponentially dampened, according to (23), the area under the dampened curve is finite.

According to (24) the area under an exponentially dampened sine or cosine curve is finite.

The area $\int_0^\infty \dfrac{f}{exp\,(a\,j)}$ under the curve f dampened by the division by $exp\,(a\,j)$ is called the value for a of the Laplace Transform of f. It will here be denoted by $\mathbf{L}\,f\,a$. Thus, according to (23) and (24),

(25) $\mathbf{L}\ j^n\ a = \dfrac{n!}{a^{n+1}}$, $\mathbf{L}\ sin\ a = \dfrac{1}{a^2+1}$, $\mathbf{L}\ cos\ a = \dfrac{a}{a^2+1}$, for any $a > 0$.

Just as the formulas

$\mathbf{D}\ j^n\ a = n\ j^{n-1}\ a$, $\mathbf{D}\ sin\ a = cos\ a$, $\mathbf{D}\ cos\ a = -sin\ a$ for any a are written

(26) $\mathbf{D}\ j^n = n\ j^{n-1}$, $\mathbf{D}\ sin = cos$, $\mathbf{D}\ cos = -sin$,

and the functions $\mathbf{D}\ j^n$, $\mathbf{D}\ sin$, and $\mathbf{D}\ cos$ are called the derivatives of *j*, *sin*, and *cos*, respectively, (25) can be written

$$\mathbf{L}\ j^n = \frac{n!}{j^{n+1}}, \quad \mathbf{L}\ sin = \frac{1}{j^2+1}, \quad \mathbf{L}\ cos = \frac{j}{j^2+1}.$$

$\mathbf{L}\ j^n$, $\mathbf{L}\ sin$, and $\mathbf{L}\ cos$ are called the *Laplace Transforms* of j^n, *sin*, and *cos*, respectively.

$\mathbf{L}\ f$ expresses the reaction of the area under the curve *f* to exponential dampening of various degrees. To the beginner, this appears to be a rather far-fetched concept. The great importance of Laplace transforms lies in the fact that $\mathbf{L}\ f$ characterizes *f* somewhat like $\mathbf{D}\ f$ characterizes *f* – and even better. If, about a function *f* it is known that $\mathbf{D}\ f = cos$, then *f* is confined to a relatively small class of functions, namely, to $sin + \mathcal{C}$. Still, there are infinitely many functions in that class. If, however, about a continuous function *f* it is known that

$\mathbf{L}\ f = \dfrac{1}{j^2+1}$, then it can be proved that *f* cannot be anything but *sin*.

Traditionally, the formulas (26) (with all letters italicized) are written in the form $\mathbf{L}\ (t^n) = \dfrac{n!}{s^n+1}$, $\mathbf{L}\ (sin\ t) = \dfrac{1}{s^2+1}$, $\mathbf{L}\ (cos\ t) = \dfrac{s}{s^2+1}$. It is rather fortunate that the formulas (26) are not traditionally written in the form

$$\mathbf{D}\ (t^n) = n\ s^{n-1}, \quad \mathbf{D}\ (sin\ t) = cos\ s, \quad \mathbf{D}\ (cos\ t) = -sin\ s.$$

EXERCISES

41. From (19) and the formula established in Exercise 34, prove

$\mathbf{L}\ sin\ (b\ j) = \dfrac{b}{j^2+b^2}$ and $\mathbf{L}\ cos\ (b\ j) = \dfrac{1}{j^2+b^2}$.

42. Prove $\mathbf{L}\ exp\ (-b\ j) = \dfrac{1}{j+b}$, $\mathbf{L}\ sinh = \dfrac{1}{j^2-1}$, $\mathbf{L}\ cosh = \dfrac{j}{j^2-1}$.

43. Prove that

$\mathbf{L}\ (c\ f) = c\ \mathbf{L}\ f$, $\mathbf{L}\ (f+g) = \mathbf{L}\ f + \mathbf{L}\ g$, and $\mathbf{L}\ (\mathbf{D}\ f) = j \cdot \mathbf{L}\ f - f\ 0$.

44. If f and g have derivatives between $-\pi$ and π, and g assumes the value 0 at these two places, prove that

$$\int_{-\pi}^{\pi} (\mathbf{D}\, f \cdot g) = - \int_{-\pi}^{\pi} (f \cdot \mathbf{D}\, g).$$

If f is the step function assuming the value 0 for all numbers ≤ 0, and the value a (> 0) for all positive numbers, then neither $\mathbf{D}\, f\ 0$ nor the integral on the left side is defined. Prove that the integral on the right side equals a g 0. This fact is of basic importance for a theory of derivatives recently developed by L. Schwartz.

CHAPTER X

THE MEAN VALUE THEOREM AND ITS CONSEQUENCES

1. THE MEAN VALUE THEOREM

The approximate equalities (20) and (21) on page 159 were based on the mere assumption that f has a tangent *at the point* a or, in other words, that Dom **D** f includes a. In what follows it will be assumed that f has a tangent *at every point between* a *and* b or that Dom **D** f includes the interval $\{ a \leq j \leq b \}$.

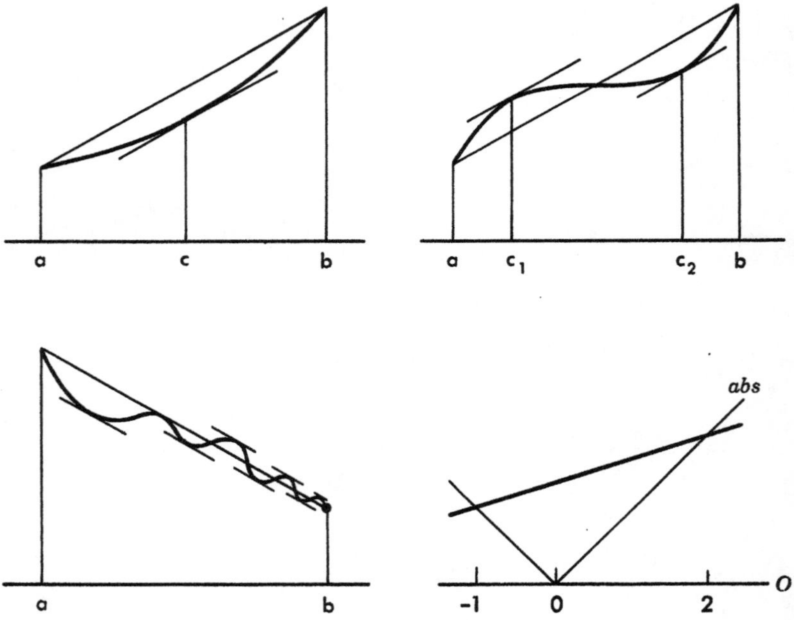

Fig. 40

The first three curves in Fig 40 possess one or more tangents that are parallel to their chord between a and b. The curve *abs* has no tangent parallel to its chord between −1 and 2, even though there is only one single point, namely 0, at which *abs* fails to have a tangent.

Parallelism of lines means equality of their slopes. The slope of the chord is $\dfrac{f\,\mathrm{b} - f\,\mathrm{a}}{\mathrm{b} - \mathrm{a}}$, that of the tangent at x is **D** f x. Thus the following statement claims that the observation concerning the first three curves in Fig. 40 is generally true.

MEAN VALUE THEOREM. *If* Dom **D** *f includes* $\{ a \leq j \leq b \}$, *then*

(1) $\frac{f\,b - f\,a}{b - a} = \mathbf{D}\,f\,c$ *for at least one "mean value"* c *such that* $a < c < b$.

If, for instance, $f = j^3$, $a = 1$, $b = 3$, then the M. V. Theorem claims that $\frac{27 - 1}{3 - 1} = 3\ c^2$ for some c between 1 and 3. Indeed, the equality holds for $c = \sqrt{13/3} = 2.08\ldots$. (It also holds for $-2.08\ldots$, which is not between 1 and 3.)

For many functions it is very difficult to determine the mean value or values accurately. Fortunately, however, this problem hardly ever arises. The importance of the M.V. Theorem will be found to lie in inferences that can be drawn *without a precise knowledge of the mean values.*

A special case of the M. V. Theorem was discovered by Rolle in 1690.

Theorem of Rolle. *If* Dom **D** *g includes* $\{ a \leq j \leq b \}$, *and* $g\ a = g\ b = 0$, *then* **D** $g\ c = 0$ *for some* c *between* a *and* b.

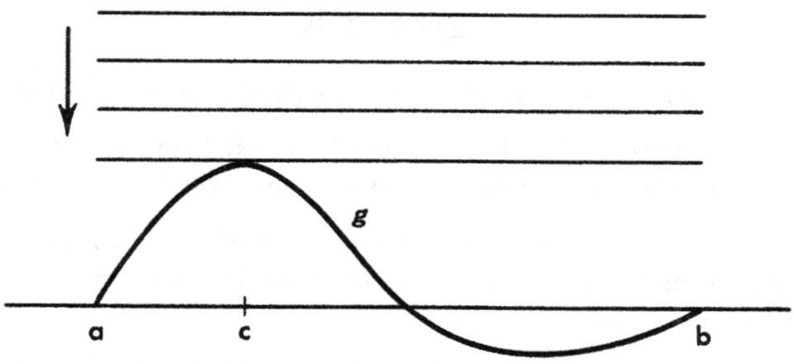

Fig. 41

This theorem can be made plausible in the following way. Unless Ran $g = \{ 0 \}$ (in which case $g = O$ and **D** $g = O$ and *each* number between a and b is a mean value), Ran g contains a number $\neq 0$, say, > 0. Consider a horizontal line above the curve g. Let that line slide downward, parallel to itself (Fig. 41). Upon reaching a certain level, the line will, for the first time, meet a point on g. Clearly, this point is the highest point (or one of several equally high peaks) on g; that is to say, the projection of that point on O is a place of a maximum of g. Hence, if c denotes that projection, **D** $g\ c = 0$ according to Theorem III on page 164.

If Dom **D** g = Dom g, then g is continuous (p. 134). If g possesses a maximum at c, and c is an interior element of Dom g, then **D** $g\ c = 0$ (see p. 164). Hence, to prove Rolle's Theorem, it is sufficient to show:

If (1) Dom $g = \{a \leq j \leq b\}$, (2) g is continuous, (3) g a = g b = 0, (4) Ran g contains a number > 0,

then g possesses a maximum at an interior element.

This, in turn, readily follows from the Theorem: If g satisfies (1) and (2), then g possesses a maximum. A proof of this profound theorem (which presupposes lengthy discussions of the concepts of number and of continuity) will not here be given.

Even though Rolle's Theorem is a special case of the M.V. Theorem, the latter can be deduced from the former. For suppose that Dom **D** f includes $\{a \leq j \leq b\}$, then, if g is defined by

$$g = f - fa - \frac{f\,\mathrm{b} - f\,\mathrm{a}}{\mathrm{b} - \mathrm{a}}(j - a),$$

also Dom **D** g includes $\{a \leq j \leq b\}$. Moreover, g a = g b = 0. Hence, by Rolle's Theorem, **D** g c = 0 for some c between a and b. But **D** g equals **D** f minus the constant function whose value is $\frac{f\,\mathrm{b} - f\,\mathrm{a}}{\mathrm{b} - \mathrm{a}}$. Hence

$$\mathbf{D}\, f\, \mathrm{c} - \frac{f\,\mathrm{b} - f\,\mathrm{a}}{\mathrm{b} - \mathrm{a}} = 0,$$

as the M.V. Theorem claims.

EXERCISES

1. Determine the mean value (or mean values) for *(a)* j^3 between $-\frac{1}{2}$ and 1; *(b)* j^{-2} between 2 and 4; *(c) exp* between 1 and 2; *(d)* $j^{1/2}$ between a and b. But what about $j^3 + exp$ or $sin + log$ between 1 and 2?

2. Is it *necessary* for a function to have a derivative everywhere between a and b (or even to be continuous in the interval) in order that some tangent be parallel to the chord?

3. Make Rolle's Theorem plausible for a function assuming no value > 0. Why is it certain that some mean value is > a and < b (not only ≥ a and ≤ b)? Can it happen that **D** g $\mathrm{c}_1 = 0$ for a point c_1 *(a)* other than that where the sliding line first reaches the curve g? *(b)* where g does not even possess a relative minimum? *(c)* for a and/or b? *(d)* for all points of an open interval?

4. At −1 and 1, the following two functions assume the value 0. Yet they have nowhere between −1 and 1 the derivative 0: *(a)* $j^{-2} - 1$; *(b)* the function f on p. 163 for which f x = x − [x]. Why are these examples not at variance with Rolle's Theorem? What is the relation of a sliding horizontal line to the graphs of these two functions? In the second example, there is a level at which, for the last time, the downward sliding line fails to meet g. The subtle difference between this possibility and that of a level on which the sliding line, for the first time, does meet g

is what makes geometric and intuitive considerations insufficient for a proof of Rolle's Theorem. This is why only deeper studies of the concepts of number and continuity yield the theorem that a continuous function with the domain $\{a \leq j \leq b\}$ possesses a maximum.

2. INDETERMINATE EQUALITIES AND DETERMINATE INEQUALITIES

Multiplying (1) by b − a and adding f a on both sides one obtains an equality that is precise (and not approximate as that on p. 159), in which, however, the number c is not determined beyond the stipulation that c belongs to a certain interval. Regardless of which is larger, b or a, one can claim:

INDETERMINATE EQUALITY. *If* Dom **D** f *includes the closed interval between* a *and* b, *then*

(2) $f\,\mathrm{b} = f\,\mathrm{a} + (\mathrm{b} - \mathrm{a}) \cdot \mathbf{D} f\,\mathrm{c}$ *for some* c *in the open interval between* a *and* b.

(This paragraph should be skipped at a first reading.) Suppose that a < b. If Dom **D** f includes $\{a \leq j \leq b\}$, then obviously Dom **D** f also includes $\{a \leq j \leq b'\}$ for any number b′ between a and b. Hence, according to the M.V. Theorem, there is a mean value between a and b′ at which the tangent is parallel to the chord between a and b′. Consequently, there is a function m with the following properties:

(a) Dom m includes $\{a \leq j \leq b\}$; (in particular, $m\,\mathrm{b} = \mathrm{c}$);

(b) Ran m is included in $\{a < j < b\}$; more precisely, $\mathrm{a} < m\,\mathrm{b}' < \mathrm{b}'$ for any b′ that is > a and ≤ b;

(c) $f\,\mathrm{b}' = f\,\mathrm{a} + (\mathrm{b}' - \mathrm{a}) \cdot \mathbf{D} fm\,\mathrm{b}'$ for $\mathrm{a} < \mathrm{b}' \leq \mathrm{b}$.

Thus, if Dom **D** f includes $\{a \leq j \leq b\}$, then there is a function m with the properties (a) and (b) such that, between a and b,

(3) $$f = fa + (j - a) \cdot \mathbf{D} f m.$$

There may exist several such functions, m, and all of them may be very complicated.

The enormous practical importance of (2) lies in the fact that the indeterminate equality can readily be transformed into inequalities. The number f b, being equal to $f\,\mathrm{a} + (\mathrm{b} - \mathrm{a}) \cdot \mathbf{D} f\,\mathrm{c}$ for some number c between a and b, obviously satisfies the

DETERMINATE INEQUALITIES.

(4) $$f\,\mathrm{a} + \underset{[\mathrm{a},\,\mathrm{b}]}{\mathrm{Min}}\,[(\mathrm{b} - \mathrm{a}) \cdot \mathbf{D} f] \leq f\,\mathrm{b} \leq f\,\mathrm{a} + \underset{[\mathrm{a},\,\mathrm{b}]}{\mathrm{Max}}\,[(\mathrm{b} - \mathrm{a}) \cdot \mathbf{D} f],$$

that is,

(4⁺) $f\,a + (b - a) \cdot \underset{[a,\,b]}{\text{Min}}\ \mathbf{D} f \leq f\,b \leq f\,a + (b - a) \cdot \underset{[a,\,b]}{\text{Max}}\ \mathbf{D} f$ if $a < b$,

(4⁻) $f\,a + (b - a) \cdot \underset{[a,\,b]}{\text{Max}}\ \mathbf{D} f \leq f\,b \leq f\,a + (b - a) \cdot \underset{[a,\,b]}{\text{Min}}\ \mathbf{D} f$ if $b < a$.

Here, $\underset{[a,\,b]}{\text{Min}}$ and $\underset{[a,\,b]}{\text{Max}}$ have the meaning defined on page 143. Clearly, the preceding inequalities yield an approximate value f b in terms of f a and $\mathbf{D} f$ a. The determination is the more accurate, the shorter the interval to which f b is confined, that is to say, the smaller the difference between the upper and the lower bounds.

First consider the functions j^3, *exp*, $j^{-1/2}$. Their derivatives ($3\,j^2$, *exp*, $-\frac{1}{2}\,j^{-3/2}$) are increasing functions and therefore, in each interval, assume their minimum at the initial point, their maximum at the terminal point. Hence if a < b we find

$$a^3 + 3a^2(b - a) < b^3 < a^3 + 3b^2(b - a),$$

$$e^a + e^a(b - a) < e^b < e^a + e^b(b - a),$$

$$\frac{1}{\sqrt{a}} - \frac{\frac{1}{2}}{\sqrt{a^3}}(b - a) < \frac{1}{\sqrt{b}} < \frac{1}{\sqrt{a}} - \frac{\frac{1}{2}}{\sqrt{b^3}}(b - a).$$

Here, if the values at a are known, it is easy to find lower bounds for the values at b, and for b^3 also the upper bound. For instance, if a = 2 and b = 2.1, then

$$8 + 3 \cdot 4 \cdot .1 < 2.1^3 < 8 + 3 \cdot 4.41 \cdot .1 \text{ or } 9.1 < 2.1^3 < 9.323.$$

The upper bound for e^b seems to be useless since it involves the very number e^b. However $e^b < e^a + e^b(b - a)$ implies $e^b(1 - b + a) < e^a$ and thus leads also to a workable upper bound. All in all,

$$e^a(1 + b - a) < e^b < e^a/(1 - b + a).$$

For $1/\sqrt{4.3}$, the lower bound is $1/\sqrt{4} - .3/(2 \cdot \sqrt{4^3}) = .48125$. But the upper bound is harder to find than $1/\sqrt{4.3}$ itself. Hence, *it will be replaced by a larger number that can be determined easily.* For instance, the upper bound $1/\sqrt{4} - .3/(2 \cdot \sqrt{4.3^3})$ is increased if the number 4.3 in the denominator of the negative term is replaced by 4.41. Then $\sqrt{4.41^3} = 2.1^3 = 9.282$. In view of the denominator .3, the last number may well be replaced by the still larger number 9.3. Then the upper bound becomes –.5–.1/6.2 and

$$.48125 < 1/\sqrt{4.3} < .4839 \text{ (the true value is .48225...).}$$

The last example suggests the following

AMENDED INEQUALITIES. *If* Dom **D** *f includes the interval* $\{a \leq j \leq b\}$, *then* $f\,a + (b - a) \cdot \underset{(a,\,b)}{\text{L.B.}}\ \mathbf{D} f \leq f\,b \leq f\,a + (b - a) \cdot \underset{(a,\,b)}{\text{U.B.}}\ \mathbf{D} f$ *if* $a < b$ *and a similar inequality if* $b < a$.

Here, U.B. **D** *f* (upper bound) stands for any number that exceeds **D** *f* x for any x between a and b. The lower bound is defined analogously.

Suppose 10.01^3 is to be found by applying the preceding results to $f = j^3$, a = 10, b = 10.01. The inequalities (4^+) yield

$$1003 \leq 10.01^3 \leq 1000 + 3 \cdot 10.01^2 \cdot .01 = 1003.006003.$$

Should one feel that computing 10.01^2 involves more work than the determination of an upper bound for 10.01^3 warrants, then, by sacrificing accuracy, he can reduce the amount of work. Since

$10.01 < 10.1 < 11$, one might say

$10.01^3 < 1000 + 3 \cdot 10.1^2 \cdot .01 = 1003.0603$ or

$10.01^3 < 1000 + 3 \cdot 11^2 \cdot .01 = 1003.63$.

Thus with hardly any work one obtains the upper bounds 1003.0603 and 1003.63.

Next, the reader should treat the functions $j^{1/2}$, $j^{1/3}$, and *log*, whose derivatives are decreasing and therefore assume their minima at the terminal, their maxima at the initial point of any interval.

The results of this section remedy the two shortcomings of the approximate equalities on page 159. For it is now possible to answer the following two questions:

1. How much, at the very most, can *f* b differ from $f\,a + \mathbf{D} f\,a \cdot (b - a)$? Answer: By $\left[\underset{[a,\,b]}{\text{Max}}\ \mathbf{D} f - \underset{[a,\,b]}{\text{Min}}\ \mathbf{D} f\right] \cdot (b - a)$.

2. How far, at the very most, may b be from a without endangering the preceding result? Answer: Any distance, provided that $\{a \leq j \leq b\}$ is included in Dom **D** *f*.

EXERCISES

6. Enclose between bounds *(a)* $\sqrt{4.3}$; *(b)* $\sqrt{3.9}$; *(c)* $\sqrt[3]{8.1}$; *(d)* $\sqrt[3]{7.9}$; *(e)* *log* 1.1; *(f)* *log* 1.024; *(g)* *log* .5.

7. If b < a, find a lower bound for $\sqrt{b}$ in terms of a. Hint: Use the fact that an upper bound for $\sqrt{b}$ in terms of a is known.

8. For $\sqrt{4.4}$, find bounds on the basis of the information that $\sqrt{1} = 1$; then bounds on the basis of $\sqrt{4} = 2$; finally, bounds on the basis of $\sqrt{4.41} = 2.1$. The answers can be summarized in:

$$1.81 < 2.095 < 2.0975 < \sqrt{4.4} < 2.09762 < 2.2 < 2.7.$$

Note how the length of the interval to which one can confine $\sqrt{4.4}$ depends upon the information available at the outset.

9. Show that, if $\mathbf{D}\, f$ is either increasing or decreasing between a and b, then one of the two bounds in (4) is the number $f\,a + \mathbf{D}\,f\,a \cdot (b - a)$ which occurred in the approximate equality (20) on page 159. Give examples of f, a, b for which

(a) $\underset{[a,\,b]}{\text{Min}}\ \mathbf{D}\,f = \mathbf{D}\,f\,a$ and $\underset{[a,\,b]}{\text{Max}}\ \mathbf{D}\,f \neq \mathbf{D}\,f\,b$;

(b) both $\underset{[a,\,b]}{\text{Min}}\ \mathbf{D}\,f$ and $\underset{[a,\,b]}{\text{Max}}\ \mathbf{D}\,f$ are $\neq\ \mathbf{D}\,f\,a$ and $\neq\ \mathbf{D}\,f\,b$.

3. TAYLOR'S EXPANSION

Let $g, g_1, g_2, \ldots, g_{n-1}, g_n$ be n + 1 functions whose derivatives have a common domain, and let b be a number. What is the derivative of the function

$$g + (b - j) \cdot g_1 + (b - j)^2 \cdot g_2 + \ldots + (b - j)^n \cdot g_n ,$$

which may be denoted by f? Writing the two terms of the derivative of each product in a column, one obtains

$$\mathbf{D}f =$$

$$\begin{aligned}\mathbf{D}g + (b-j)\cdot\mathbf{D}g_1 &+ (b-j)^2\cdot\mathbf{D}g_2 + \ldots + (b-j)^{n-1}\cdot\mathbf{D}g_{n-1} + (b-j)^n\cdot\mathbf{D}g_n\\ - g_1 &- 2(b-j)\cdot g_2 - \ldots - (n-1)(b-j)^{n-2}\cdot g_{n-1} - n(b-j)^{n-1}\cdot g_n\end{aligned}$$

In the special case where

$$g_1 = \mathbf{D}\,g,\ g_2 = \frac{1}{2}\,\mathbf{D}\,g_1,\ g_3 = \frac{1}{3}\,\mathbf{D}\,g_2,\ \ldots,\ g_n = \frac{1}{n}\,\mathbf{D}\,g_{n-1},$$

each term, except the upper term in the last column, is cancelled by the lower term in the following column, and the sum reduces to

$$\mathbf{D}\,f = (b - j)^n \cdot \mathbf{D}\,g_n .$$

Clearly, this special case is present if

$$g_1 = \frac{1}{1!}\,\mathbf{D}\,g,\ g_2 = \frac{1}{2!}\,\mathbf{D}^2\,g,\ g_3 = \frac{1}{3!}\,\mathbf{D}^3\,g,\ \ldots,\ g_n = \frac{1}{n!}\,\mathbf{D}^n\,g,$$

and hence $$\mathbf{D}\,g_n = \frac{1}{n!}\,\mathbf{D}^{n+1}\,g.$$

The preceding result may be formulated as the following Lemma: If

(5) $f = g + (b - j) \cdot \mathbf{D}\, g + \frac{1}{2!}(b-j)^2 \cdot \mathbf{D}^2\, g + \dots + \frac{1}{n!}(b-j)^n \cdot \mathbf{D}^n\, g,$

then Dom $\mathbf{D}\, f$ = Dom $\mathbf{D}^{n+1}\, g$ and

(6) $$\mathbf{D}\, f = \frac{1}{n!}(b-j)^n \cdot \mathbf{D}^{n+1}\, g.$$

If a is a number such that the closed interval between a and b belongs to Dom $\mathbf{D}\, f$, that is, to Dom $\mathbf{D}^{n+1}\, g$, then, by the Mean Value Theorem,

(7) $f\,b = f\,a + (b - a) \cdot \mathbf{D}\, f\, c$ for some c between a and b.

Now express f b, f a, and $\mathbf{D}\, f$ c in terms of g. Using (5) and (6) one finds

$f\,b = g\,b;$

$f\,a = g\,a + (b-a) \cdot \mathbf{D}\, g\, a + \frac{1}{2!}(b-a)^2 \cdot \mathbf{D}^2\, g\, a + \dots$

$\qquad + \frac{1}{n!}(b-a)^n \cdot \mathbf{D}^n\, g\, a;$

$\mathbf{D}\, f\, c = \frac{1}{n!}(b-c)^n \cdot \mathbf{D}^{n+1} g\, c$, for some c between a and b.

Since c is between a and b, it follows that (b – c)/(b – a) is a fraction between 0 and 1, say t. Consequently,

(8) $\mathbf{D}\, f\, c = \frac{1}{n!} t^n (b-a)^n \cdot \mathbf{D}^{n+1}\, g\, c = \frac{1}{n!} u\, (b-a)^n \cdot \mathbf{D}^{n+1} g\, c$ where

$$0 < u < 1.$$

Thus (7) becomes

TAYLOR'S FORMULA.

$g\,b = g\,a + \frac{1}{1!}(b-a) \cdot \mathbf{D}\, g\, a + \frac{1}{2!}(b-a)^2 \cdot \mathbf{D}^2\, g\, a + \dots$

$\qquad + \frac{1}{n!}(b-a)^n \cdot \mathbf{D}^n\, g\, a + r_{n+1},$

where r_{n+1}, called the $(n+1)^{th}$ *remainder*, is the number

(9) $r_{n+1} = \frac{1}{n!} u\, (b-a)^{n+1} \cdot \mathbf{D}^{n+1}\, g\, c$, for some c between a and b and some u between 0 and 1.

This celebrated formula is valid provided Dom $\mathbf{D}^{n+1}\, g$ includes the interval between a and b. It expresses the value that g assumes at b in terms of the values of g, $\mathbf{D}\, g$, $\mathbf{D}^2\, g$, ... , $\mathbf{D}^n\, g$ at a, plus a remainder.

As an example, replace g by *log*. By induction, one readily finds $\mathbf{D}^n\, log = (-1)^{n-1}(n-1)!\; j^{-n}$ (on P) for n = 1, 2, ... , where 0! = 1. Thus if a > 0, then

$$\mathbf{D}^n\, log\, a = (-1)^{n-1}(n-1)!\,/a^n.$$

Hence, for any $b > 0$, Taylor's Formula yields

$$\log b = \log a + \frac{1}{a}(b-a) - \frac{1}{2a^2}(b-a)^2 + \frac{1}{3a^3}(b-a)^3 - + \ldots$$

$$+ \frac{(-1)^{n-1}}{na^n}(b-a)^n + r_{n+1},$$

(if one notes that $(k-1)!/k! = 1/k$), where

$$r_{n+1} = \frac{1}{n!}\, u(b-a)^{n+1}\, \frac{(-1)^n\, n!}{c^{n+1}} = (-1)^n\, u\left(\frac{b-a}{c}\right)^{n+1} \quad (0 < u < 1,$$

c between a and b).

If, here, b is replaced by 1, one obtains

$$\log b = (b-1) - \frac{1}{2}(b-1)^2 + \frac{1}{3}(b-1)^3 - + \ldots \frac{(-1)^{n-1}}{n}(b-1)^n + r_{n+1},$$

where $r_{n+1} = (-1)^n\, u\left(\frac{b-1}{c}\right)^{n+1}$ $(0 < u < 1$; c between 1 and b).

If b is replaced by 1.1 and .8, one finds

$$\log 1.1 = .1 - \frac{1}{2}(.01) + \frac{1}{3}(.001) - + \ldots + \frac{(-1)^n}{n}(.1)^n + (-1)^n\, u\left(\frac{.1}{c}\right)^{n+1},$$

$$\log .8 = -.2 - \frac{1}{2}(.04) - \frac{1}{3}(.008) - \ldots - \frac{(-1)^n}{n}(-.2)^n + (-1)^n\, u\left(\frac{-.2}{c}\right)^{n+1}.$$

The first c is between 1 and 1.1; the second c is between .8 and 1. The second remainder equals $-u\,(.2/c)^{n+1}$.

The reader should replace the preceding indeterminate equalities and Taylor's Formula by *determinate inequalities*. (He will have to formulate different inequalities according to whether n is even or odd.)

The case $a = 0$ of Taylor's Formula (in which, besides, g will be replaced by f) is known as

MACLAURIN'S FORMULA. If Dom $\mathbf{D}^{n+1} f$ includes the interval between 0 and b, then

$$f\,b = f\,0 + \frac{1}{1!}\,\mathbf{D} f\,0 \cdot b + \frac{1}{2!}\,\mathbf{D}^2 f\,0 \cdot b^2 + \ldots + \frac{1}{n!}\,\mathbf{D}^n f\,0 \cdot b^n + r_{n+1},$$

where $r_{n+1} = \frac{1}{n!}\, u\, \mathbf{D}^{n+1} f\, c \cdot b^{n+1}$ $(0 < u < 1$; c between 0 and b).

Clearly, for any x between 0 and b, Dom $\mathbf{D}^{n+1} f$ includes the interval between 0 and x. Hence

$$f\,x \doteq f\,0 + \frac{1}{1!}\,\mathbf{D} f\,0 \cdot x + \frac{1}{2!}\,\mathbf{D} f\,0 \cdot x^2 + \ldots + \frac{1}{n!}\,\mathbf{D}^n f\,0 \cdot x^n + r_{n+1}\,x.$$

Here, r_{n+1} is a function that can be shown to be of the form

$$r_{n+1} = \frac{1}{n!}\, t^n \cdot j^n \cdot \mathbf{D}^{n+1} f(\,(1-t)\cdot j\,)\,,$$

where t is a function whose range is the interval between 0 and 1. (The reader may now go over the paragraph on page 277 following the Indeterminate Equality.)

If $f = exp$, then $\mathbf{D}^k f = exp$ and $\mathbf{D}^k f\, 0 = 1$ for every k. Hence

$$e^x = 1 + x + x^2/2! + x^3/3! + \ldots + x^n/n! + r_{n+1}\,.$$

For x = 1, on the left side one obtains e. If n = 8, then

$$e = 1 + 1 + 1/2! + 1/3! + \ldots + 1/8! + r_9\,.$$

For $f = sin$ and cos, the reader should verify

$$sin\, x = x - x^3/3! + x^5/5! - + \ldots + (-1)^{k+1} x^{2k-1}/(2k-1)! + r_{2k+1}\,.$$

$$cos\, x = 1 - x^2/2! + x^4/4! - + \ldots + (-1)^k x^{2k}/(2k)! + r_{2k+2}\,.$$

The reader should find the first coefficients of the Maclaurin series of *tan* and *sec*. (To set up general recursion formulas for them is a very difficult problem.)

EXERCISES

10. Find the first four terms of the Maclaurin Expansions of *tan, sec, arctan, arcsin.*

11. Find the Maclaurin Expansions of *(a)* $j^3 - 3j + 2$; *(b)* any polynomial; *(c)* $(j - 1)^c$ for any number c. For what numbers x can $(x - 1)^c$ be computed from the expansion?

12. Have the functions *log*, j^{-n}, $j^{1/n}$ for $n > 1$ Maclaurin Expansions? What can be said about $j^{3/2}$, $j^{5/2}$, $j^{p/q}$ for integers p and q, and more generally, about j^c?

13. For every positive integer n, find a function f such that all values of $\mathbf{D} f$, $\mathbf{D}^2 f$, $\mathbf{D}^3 f$, ... , $\mathbf{D}^n f$ on $\{-1 \leq j \leq 1\}$ lie between 0 and 1 and for which $\mathbf{D}^{n+1} f\, 0$ does not exist. What about the Maclaurin Expansions of such a function f?

14. Find and compare the Maclaurin Expansions of $1/(j - a)$ and $log\,(j - a)$. Compute the (0, 0)–antiderivative of the Maclaurin Expansions of *cos, sin,* and $1/(j^2 + 1)$, and compare the last antiderivative with the expansion of *arctan.* (In this exercise ignore the remainder.)

15. Obtain the Maclaurin Expansion of $exp\,(-j^2)$ *(a)* directly, *(b)* by substituting $-j^2$ into the expansion of *exp.* What is the easiest way of finding the expansions of $sin\, j^2$ and $cos\, j^2$? Could $sin\, j^{1/2}$ be

handled in the same way? Find the Maclaurin Expansions of *exp* $(-j)$, *cosh*, and *sinh*.

16. Find the Taylor Expansions at c of *(a)* $j^2 + p\,j + q$; *(b)* j^n; *(c)* any polynomial $a_0 + a_1 j + a_2 j^2 + \dots + a_n j^n$.

17. Compare the Taylor Expansion of g at a with the Maclaurin expansion of the function $g\,(j + a)$. In particular, find the latter for *log* $(j + 1)$ and compute, by means of it, *log* 1.1 and *log* .8. Then apply the formula to the computations of *log* 2 and *log* 3. Why is the latter result less significant than the former?

4. REMARKS CONCERNING TAYLOR'S FORMULA

Remark 1. Taylor's Formula, discovered in 1715, is one of the great marvels of mathematics. Ignoring for a moment the remainder, one can, by means of that formula, express the value that a function assumes at one place in terms of the values of the function and its derivatives at another place. This is something like a mathematical action at a distance. For instance, in Maclaurin's case, one expresses the value f x in terms of the values of f and its derivatives at 0. This procedure corresponds to determining the value of a phenomenon at some future or past instant on the basis of thorough information concerning the phenomenon at present; it corresponds, in other words, to prediction.

Remark 2. Of course, the remainder must not be ignored. Suppose all that is known about a function f are the values of f, $\mathbf{D}\,f$ and $\mathbf{D}^2\,f$ at 0, say

$$f\,0 = 4,\ \ \mathbf{D}\,f\,0 = -2,\ \mathbf{D}^2 f\,0 = 3.$$

Then nothing at all can be said about the value of f at .1; in other words, $f\,(.1)$ may be any given number c. Indeed, the function

$$4 - 2\,j + 1.5\,j^2 + \mathrm{k}\,j^3,$$

which will *ad hoc* be denoted by f, and its first two derivatives assume at 0 the values 4, −2, and 3, while the value of f at .1 is 3.815 + .001k, which (for properly chosen k) equals the given number c. (For instance, $f\,(.1) = 1003.815$ if $\mathrm{k} = 10^6$ and $f\,(.1) = -996.195$ if $\mathrm{k} = -10^6$.)

Suppose, however, it is further known that

$$4 \le \mathbf{D}^3 f\,\mathrm{x} \le 6 \ \text{ for any x such that } 0 < \mathrm{x} < .1,$$

then from Maclaurin's Formula

$$f(.1) = f\,0 + .1\,\mathbf{D}\,f\,0 + \frac{.01}{2}\,\mathbf{D}^2 f\,0 + \frac{.001}{6}\,\mathrm{u}\ \ \mathbf{D}^3 f\,\mathrm{c}\ (0 < \mathrm{u} < 1 \text{ and } 0 < \mathrm{c} < .1)$$

one can infer the inequality

$3.815 + .001 \cdot 4/3! < f(.1) < 3.815 + .001 \cdot 6/3!$, that is,

$$3.8156\dot{6} < f(.1) < 3.816$$

More generally, knowing

1. the values of f, $\mathbf{D} f$, $\mathbf{D}^2 f$, ..., $\mathbf{D}^n f$ at one point a,

2. bounds for the values which $\mathbf{D}^{n+1} f$ assumes between a and b,

one can, from Taylor's Formula, obtain bounds for the number f b.

Remark 3. Beginners are tempted to ignore the remainder mainly because, except for it, the expansion is of great mathematical beauty and they get fascinated with its pattern. Inserting factors 0! and b^0 (which equal 1) and writing $\mathbf{D}^1 f$ and $\mathbf{D}^0 f$ instead of $\mathbf{D} f$ and f, respectively, one can make Maclaurin's expansion aesthetically perfect:

$$\frac{1}{0!}\mathbf{D}^0 f\,0 \cdot b^0 + \frac{1}{1!}\mathbf{D}^1 f\,0 \cdot b^1 + \frac{1}{2!}\mathbf{D}^2 f\,0 \cdot b^2 + \dots + \frac{1}{n!}\mathbf{D}^n f\,0 \cdot b^n .$$

Unfortunately, as Remark 2 shows, the beautiful expansion

$$f\,0 + \frac{.1}{1!}\mathbf{D} f\,0 + \frac{.1}{2!}\mathbf{D}^2 f\,0$$

is of no use in computing $f(.1)$ unless one is sure that $\frac{.001}{6}\mathbf{D}^3 f c$ is close to 0. Should, anywhere between 0 and .1, the function $\mathbf{D}^3 f$ assume the value $6 \cdot 10^6$, then $f(.1)$ might be any number between 4 and about 1000.

Or consider the formula

$$\text{(10)} \qquad e = 1 + 1/1! + 1/2! + 1/3! + \dots + 1/8! + r_9 .$$

It is beautiful, but so are

$$\text{(*)} \qquad \begin{aligned} e &= 1/1^2 + 1/2^2 + 1/3^3 + \dots + 1/8^2 + s_9 \quad \text{and} \\ e &= 0/1! + 0/2! + 0/3! + \dots + 0/8! + t_9 . \end{aligned}$$

Yet the formulas (*) are without any interest since $s_9 \sim e/2$ and $t_9 = e$.

The usefulness of (10) is therefore entirely contingent on the smallness of r_9. How small is r_9? Since $1^9 = 1$ and $\mathbf{D}^9$ *exp* = *exp*, it follows that

$$r_9 = \frac{u}{9!}\, e^c \text{ where } 0 < u < 1 \text{ and } 0 < c < 1.$$

Now Max_0^1 *exp* = $e < 3$. Hence $r_9 < 3/9!$, which is $< .00001$, wherefore at least the first four decimals of $1 + 1/1! + \dots + 1/8!$ equal those of e.

Remark 4. $\mathbf{D}^n$ *exp* exists for any n. Hence Maclaurin's Expansion is applicable to *exp* for any n. For $x = 1$,

$$e = 1/0! + 1/1! + 1/2! + \dots + 1/n! + r_{n+1}$$

and one can show $r_{n+1} < 3/(n+1)!$, just as $r_9 < 3/9!$. Since by choosing n sufficiently large, one can make r_{n+1} as close to 0 as he pleases, e is the sum of the series

$$1/0! + 1/1! + 1/2! + \dots + 1/n! + \dots$$

(whose terms are the values of the function $1/\Gamma$ for the positive integers (see p. 270). Hence in the notation of page 128, one can write

$$e = \sum_{1}^{\infty} 1/\Gamma .$$

Remark 5. Also $\mathbf{D}^n\, sin$ exists for any n and equals *sin* or *cos* or $-sin$ or $-cos$. Hence any value of any $\mathbf{D}^n\, sin$ lies between -1 and 1, and one can easily show that Maclaurin's remainder r_{n+1} is as close to 0 as he pleases if n is sufficiently large. For angles between 0 and 1 radian $(= 57° \; 17.758')$, $-1/n! < r_{n+1} < 1/n!$. That is

$$|r_5| < .01, \; |r_7| < .0002, \; |r_9| < .000003 .$$

Thus by choosing n = 8 one obtains

$sin\, x \sim x - x^3/3! + x^5/5! - x^7/7!$ to 5 decimals for $0 \leq x \leq 1$.

If inaccuracies not exceeding .0002 are permitted, then

$$sin\, x \sim x - x^3/3! + x^5/5! .$$

For instance, $sin\, 1 \sim 1 - 1/6 + 1/120 = .84166$ (according to tables, $sin\, 1 = .8415$.)

Remark 6. In the interval between $-\pi/2$ and $\pi/2$, each of the curves

$$j,\; j - \frac{1}{6}\, j^3,\; j - \frac{1}{6}\, j^3 + \frac{1}{120}\, j^5,\; j - \frac{1}{6}\, j^3 + \frac{1}{120}\, j^5 - \frac{1}{5040}\, j^7$$

is closer to the sine curve than is the preceding curve. The fourth curve is practically indistinguishable from the curve *sin*. Outside of the interval mentioned, each of these polynomial curves differs tremendously from the sine curve (which oscillates between the horizontal lines *1* and *−1*), for it crosses the entire plane. If more and more terms are included in Maclaurin's expansion, then to the right and left of 0, the resulting polynomial curves include more and more waves resembling those of a sine curve.

In a similar way, the Maclaurin Expansion of any function *f* (where the remainder is small) approximates *f* by polynomials.

Remark 7. Can for every *f* the value that *f* assumes at b be expressed, with a prescribed degree of accuracy, in terms of the values of *f* and its derivatives at a? The answer is negative for two reasons. First of all, the non-existence of $\mathbf{D}^{n+1} f$ at a single point between a and b makes the case n of Maclaurin's Formula inapplicable to f, just as the non-existence of $\mathbf{D}\ abs\ 0$ makes the Mean Value Theorem inapplicable to *abs* between -1 and 2. But secondly, even if the entire interval between a and b belongs to Dom $\mathbf{D}^{n+1} f$, the remainder need not be small. An interesting example is the extension of $exp(-j^{-2})$, *ad hoc* denoted by g, for which $g\ 0 = 0$ (see p. 112). It is easy to show that g has everywhere derivatives of any order and that, at 0, all of them assume the value 0. Hence in Maclaurin's formula

$$g\,x = g\,0 + \frac{1}{1!}\mathbf{D}\,g\,0 \cdot x + \frac{1}{2!}\mathbf{D}^2 g\,0 \cdot x^2 + \ldots + \frac{1}{n!}\mathbf{D}^n g\,0 \cdot x^n + r_{n+1},$$

all terms on the right side, except r_{n+1}, equal 0. But this implies that $r_{n+1} = g\,x$, for every x and every n. Far from becoming insignificant with increasing n, the remainder equals the value of the function, just as $t_9 = e$ in (*).

Remark 8. Another function having everywhere derivatives of any order which, at 0, all assume the value 0, is the constant function O. It, too, has a Maclaurin Expansion for every n but, for it, $r_{n+1} = 0$.

It follows that the constant function O and the function g in Remark 7 have derivatives of any order everywhere. At 0, all derivatives of both functions assume the value 0. Yet, everywhere except at 0, the values of the two functions as well as those of all their derivatives differ from one another. This shows that a function is not determined by the values of all its derivatives at one place.

If, when a resting particle starts moving, not only velocity and acceleration but also all fluxions of higher order should be free of abrupt changes, then the position of the particle would have to be connected with the time by a function of the type of $exp(-j^{-2})$.

EXERCISES

18. Find $\sqrt{e}$ to 5 decimals. How many terms of the Maclaurin Expansion have to be considered to determine e^2 to 3 decimals?

19. Suppose that the following information is available about a function f:

$$f\,1 = 3,\quad \mathbf{D} f\,1 = -2,\quad \mathbf{D}^2 f\,1 = 3,\quad \mathbf{D}^3 f\,1 = 0,\quad \mathbf{D}^4 f\,1 = -1,$$

and $$3 \le \mathbf{D}^5 f\,x \le 5 \quad \text{for} \quad .5 \le x \le 1.5.$$

What inference can be drawn concerning the values of f at

(a) 1.4; *(b)* 1.2; *(c)* 1.01; *(d)* .9; *(e)* .6; *(f)* .4?

20. For what angles is

(a) $\sin x \sim x - x^3/3! + x^5/5!$ to 5 decimals?

(b) $\sin x \sim x - x^3/3!$ to 5 decimals?

(c) $\sin x \sim x - x^3/6$ to 3 decimals?

(d) $\sin x \sim x$ to 3 decimals?

(e) $\sin x \sim x$ to 2 decimals?

(f) Substitute $\pi/2$ into $x - x^3/3! + x^5/5! - x^7/7!$.

21. Sketch the curves $1 + 1\,j + \frac{1}{2!}\,j^2 + \frac{1}{3!}\,j^3 + \frac{1}{4!}\,j^4 + \frac{1}{5!}\,j^5$ and $1 - j + \frac{1}{2!}\,j^2 - \frac{1}{3!}\,j^3 + \frac{1}{4!}\,j^4 - \frac{1}{5!}\,j^5$.

22. Find to 5 decimals $\sqrt{2}$, $\sqrt[3]{2}$, $\sqrt{5}$, $\sqrt{10}$, $\sqrt[3]{10}$.

23. By adding the Maclaurin Expansions of $\log\,(1 + j)$ and $\log\,(1 - j)$, find the Expansion of $\log \dfrac{1 + j}{1 - j}$. For 1/3 and 2/3, one obtains $\log 2$ and $\log 5$, and by adding them, $\log 10$.

24. On page 23, $\log_{10} 2$ was approximately computed by equating $2^{10} = 1024$ and $10^3 = 1000$. Clearly, $\log_{10} 2 = .3 + \frac{1}{10} \log_{10} 1.024$. By computing $\frac{.4343}{10} \log 1.024$, show that the corrective term is .00103 and $\log_{10} 2 = .30103$.

25. Find functions different from those in Remark 8 all of whose derivatives assume the value 0 at 0. Find two unequal functions all of whose derivatives assume at 3 the value 1.

5. MAXIMA AND MINIMA

Let a be an interior element of Dom g, and m an integer ≥ 0, and assume that $\mathbf{D}^{m+1} g\,a$ exists. Then a is an interior element of Dom $\mathbf{D}^m g$. (Otherwise $\mathbf{D}^{m+1} g$ could not be defined.)

If in Taylor's Formula one sets $b = a + h$ and $n = m - 1$, he obtains the alternate formulation

$$(11)\quad g\,(a + h) = g\,a + \frac{\mathbf{D} g\,a}{1!}\,h + \frac{\mathbf{D}^2 g\,a}{2!}\,h^2 + \dots + \frac{\mathbf{D}^{m-1} g\,a}{(m-1)!}\,h^{m-1}$$

$$+ \frac{u}{m!}\,h^m\,\mathbf{D}^m g\,c \;(0 < u < 1,\ c \text{ between a and } a + h).$$

Sufficient for g to have a relative minimum at a (a relative maximum at a) is

(12) $\mathbf{D}\, g\, a = \mathbf{D}^2\, g\, a = \ldots = \mathbf{D}^m\, g\, a = 0$ and $\mathbf{D}^{m+1}\, g\, a > 0$ ($\mathbf{D}^{m+1}\, g\, a < 0$),

provided m + 1 is even. In other words, sufficient for g to have a relative extremum at an interior element a of Dom g is *that the derivative of g of lowest order assuming a value* $\neq 0$ *at* a *be of even order.* According to whether this value is *positive* or *negative,* the extremum is a *minimum* or a *maximum.*

Proof: In view of $\mathbf{D}^{m+1}\, g\, a > 0$, it follows from $\mathbf{D}^m\, g\, a = 0$ that

(13) $$\mathbf{D}^m\, g\, c > 0 \text{ for } c \sim a + 0 \text{ and } \mathbf{D}^m\, g\, c < 0 \text{ for } c \sim a - 0.$$

Since m is odd, $h > 0$ implies $h^m > 0$. If $h \sim +0$, then $a < c < a + h$ and, by (13), $\mathbf{D}^m\, g\, c > 0$. Similarly, if $h < 0$, then $h^m < 0$ and $\mathbf{D}^m\, g\, c < 0$. In both cases, thus, for any small h ($\neq 0$),

$$h^m\, \mathbf{D}^m\, g\, c > 0 \text{ and hence } \frac{u}{m!}\, h^m\, \mathbf{D}^m\, g\, c > 0.$$

Now from (11) and (12) it follows that the preceding number equals $g\,(a + h) - g\, a$. Thus, if m + 1 is even, then, under the assumption (12), g has a relative minimum at a, (and, if $\mathbf{D}^{m+1}\, g\, a < 0$, a relative maximum).

In the same way one proves that (12) is sufficient for g *not having a relative extremum* at a provided m + 1 is odd. Indeed, under these assumptions,

$g\,(a + h) > g\, a$, one one side, and $g\,(a + h) < g\, a$, on the other side.

It follows that, if a derivative of g at a is $\neq 0$, for g to have a relative extremum at a it is necessary that the derivative of g of lowest order whose value at a is $\neq 0$ be of an even order.

For instance,

$\mathbf{D}\, g\, a = 0, \mathbf{D}^2\, g\, a > 0$: Minimum at a,
$\mathbf{D}\, g\, a = 0, \mathbf{D}^2\, g\, a < 0$: Maximum at a,
$\mathbf{D}\, g\, a = 0, \mathbf{D}^2\, g\, a = 0$; $\mathbf{D}^3\, g\, a \neq 0$: no extremum at a,
$\mathbf{D}\, g\, a = \mathbf{D}^2\, g\, a = \mathbf{D}^3\, g\, a = 0$, $\mathbf{D}^4\, g\, a > 0$: Minimum at a,
$\mathbf{D}\, g\, a = \mathbf{D}^2\, g\, a = \mathbf{D}^3\, g\, a = 0$, $\mathbf{D}^4\, g\, a < 0$: Maximum at a,
$\mathbf{D}\, g\, a = \mathbf{D}^2\, g\, a = \mathbf{D}^3\, g\, a = \mathbf{D}^4\, g\, a = 0$, $\mathbf{D}^5\, g\, a \neq 0$: no extremum at a

Examples of the six cases just mentioned are a = 0 and $g = j^2, -j^2, j^3, j^4, -j^4, j^5$, respectively.

The function g in Remark 7, page 287, has a relative minimum at 0 although $\mathbf{D}^n\, g\, 0 = 0$ for any n. The function *abs* has a minimum at 0 although $\mathbf{D}$ *abs* 0 does not exist.

EXERCISE

26. Find the relative extrema of

(a) $(j-1)^4 \cdot (j-2)^3$; (b) $(j^2-1)^3$; (c) $j^2 \cdot \sin$.

27. In what relations are the necessary and the sufficient conditions for a relative extremum at a non-interior element mentioned on page 165 to the preceding conditions. Extend the latter to the initial and the terminal elements of Dom g.

6. THE APPROXIMATE COMPUTATION OF AREAS BY EXPANSIONS

One of the great triumphs of the Calculus of Derivatives is its contribution to the approximate computation of integrals $\int_a^b f$ even in cases where an antiderivative of f is not available. Simply replace f by a Taylor or Maclaurin polynomial p and compute $\int_a^b p$ by determining the rise of an antiderivative of p from a to b. If p and f differ by not more than d (that is to say, if p x and f x differ by not more than d for any x between a and b), then, according to the Second Fundamental Inequality (p. 143),

$$\int_a^b f \text{ and } \int_a^b p \text{ differ by not more than } d \cdot (b - a).$$

As an example, consider the important function $\exp(-j^2/2)$. By substituting $-j^2/2$ into

$$\exp \sim 1 + j + j^2/2! + j^3/3! + j^4/4!\,,$$

one obtains

$$\exp(-j^2/2) \sim 1 - j^2/2 + j^4/8 - j^6/48 + j^8/384.$$

Thus $\int_0^a [\exp(-j^2/2)]$ is approximately equal to the integral from 0 to a of the polynomial on the right side. The latter integral equals the rise of any antiderivative of the polynomial from 0 to a.

Hence $$\int_0^a [\exp(-j^2/2)] \sim [j - j^3/6 + j^5/40 - j^7/336 + j^9/3456]_0^a\,.$$

For a = 1, we obtain

$$\int_0^1 [\exp(-j^2/2)] \sim 1 - .16667 + .02500 - .00297 + .00029 = .85565,$$

which is very near the true value.

EXERCISES

27. What inaccuracy is committed by replacing *(a)* $\int_0^a \exp$ by $\left[\mathbf{D}^{-1}\left(1 + j + \frac{1}{2!}\, j^2 + \frac{1}{3!}\, j^3 + \frac{1}{4!}\, j^4\right)\right]_0^a$, in particular, if a = 1;

(b) $\int_0^1 [\exp(-j^{\cdot 2})]$ by .85565?

28. Find approximately $\int_0^{\pi/4} \sqrt{1 - .64\, \sin^2}$ and $\int_0^{\pi/2} [\sin/j]$.

29. Derive Leibniz' famous expression for $\pi/4$ by evaluating

$$\int_0^1 \frac{1}{1 + j^2} \sim \int_0^1 [1 - j^2 + j^4 - \dots + (-j^2)^n]$$

and using $\mathbf{D}^{-1}_{(0,0)}\, \dfrac{1}{1 + j^2} = \arctan$.

CHAPTER XI

TWO-PLACE FUNCTIONS

1. SIMPLE SURFACES. VOLUMES AND TANGENTIAL PLANES

In Fig. 42, there is a horizontal plane with a Cartesian frame of reference. Relative to this frame, each point in the plane can be described by an ordered pair of numbers. The positive halves of the axes (marked 1st and 2nd) are shown. The said plane will be referred to as the *basic* plane.

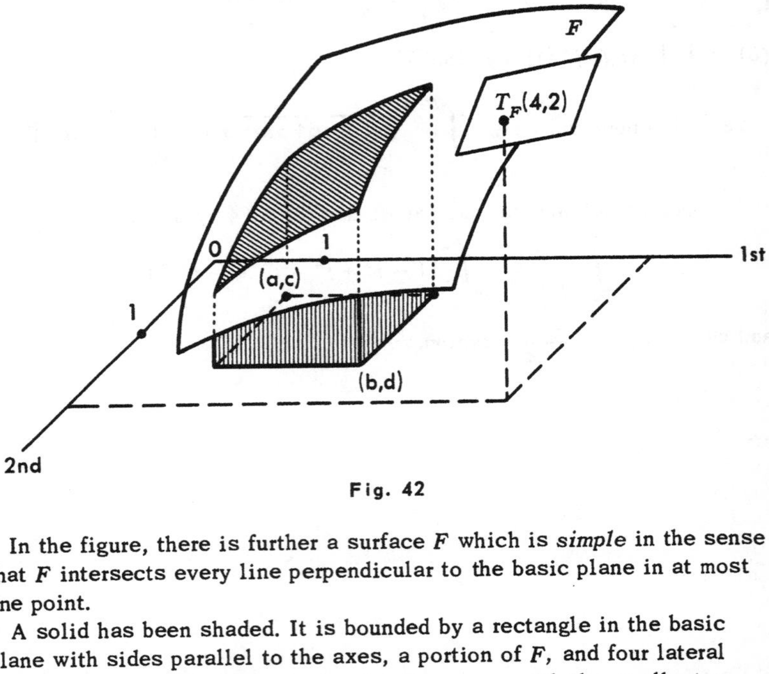

Fig. 42

In the figure, there is further a surface F which is *simple* in the sense that F intersects every line perpendicular to the basic plane in at most one point.

A solid has been shaded. It is bounded by a rectangle in the basic plane with sides parallel to the axes, a portion of F, and four lateral vertical faces. In the basic rectangle, the corner with the smallest coordinates is the point (a, c); the diagonally opposite corner, with the largest coordinates, is (b, d). These two corners fully determine the rectangle. The volume of the shaded solid will be denoted by $\int_{(a,\,c)}^{(b,\,d)} F$.

Above the point (4, 2), the tangential plane to F has been drawn. It will be denoted by $T_F\ (4, 2)$.

How can the volume under F and the tangential plane to F be determined? If H is a horizontal plane of the altitude h above the basic plane, then $\int_{(a,c)}^{(b,d)} H$ is the volume of a right prism with a rectangular base, and thus equal to the product of the altitude times the area of the base,

$$\int_{(a,c)}^{(b,d)} H = h \cdot (b - a) \cdot (d - c) .$$

If S^* is a step surface made up of several horizontal rectangles, then $\int_{(a,c)}^{(b,d)} S^*$ is the sum of the volumes of several such prisms.

A simple surface such as F in Fig. 42 may be replaced by a step surface the volume under which is close to that under F. In particular, one may, for any integer n, divide the basic rectangle into n^2 mutually congruent rectangles and choose above each of them the horizontal rectangle whose center lies on F. What is thus obtained may be called *the* n^{th} *regular step surface adapted to* F and be denoted by $R_n^{(F)}$. The volume under F may be approximately *computed* since

$$\int_{(a,c)}^{(b,d)} F \underset{\text{arb}}{\sim} \int_{(a,c)}^{(b,d)} R_n^{(F)} \quad \text{if n sufficiently large,}$$

or that volume may be *defined* as the limit of the sequence of the volumes under

$$R_1^{(F)}, R_2^{(F)}, \dots, R_n^{(F)}, \dots .$$

If S is a polyhedral surface, (Reread the discussion on p. 33.) each plane containing one of the faces of S is, except for the edges, tangential to S, just as a line containing one of the sides of a polygon p is, except for the vertices, tangential to p.

But the approximate determination of a tangential plane to a curved surface at a point is more difficult than that of a tangent to a curve at, say, P. For a curve, one has simply to join P to any sufficiently close neighbor point by a line. But on a hemisphere, if P is joined to two neighbor points P′ and P″ (no matter how close to P) by a plane, the latter (unless P′ and P″ are properly chosen) is not close to the tangential plane at P. If P′ and P″ are chosen on a great circle through P, then the plane through P, P′, P″ passes through the center of the hemisphere and is perpendicular to the tangential plane at P.

The *exact* volume under F and (notwithstanding the indicated difficulties) the *exact* position of a tangential plane to F can, in many cases, be found in terms of the two basic concepts of calculus, although the latter primarily deal with curves. The applicability of integrals and derivatives to volumes and tangential planes is one of the great triumphs of calculus.

EXERCISE

1. Let P be the surface whose altitude above any point (x, y) is $P(x, y) = x \cdot y$, and determine $\int_{(0,0)}^{(1,1)} R_m^{(P)}$ for m = 1 and m = 3. Set up an expression for $m = 3^n$. Do the same for S, where $S(x, y) = x + y$.

★2. By means of triangles inscribed in a cylinder that are almost perpendicular to its tangential planes, one can inscribe in the cylinder a polyhedron whose area exceeds that of the cylinder by as much as he pleases.

Choose two positive integers, m and n, and consider two circles of radius 1 and calibrated in radians, one of blue, the other of yellow, wire. Place the blue circle on top of the yellow, the 0 marks coinciding. Into the blue circle, inscribe a blue polygon from 0 to $2\pi/n$ to $4\pi/n$... to $2(n-1)\pi/n$ and back to 0; in the yellow circle, a yellow polygon from π/n to $3\pi/n$... to $(2n-1)\pi/n$ and back to π/n. Then draw a red rubber band from blue 0 to yellow π/n to blue $2\pi/n$... to yellow $(2n-1)\pi/n$ and back to blue 0. Determine the combined area, A_n, of the 2n blue-red-red and yellow-red-red triangles. Then, stretching the rubber and thereby increasing the areas of the 2n triangles, lift the blue circle 1/m units above the yellow circle. The 2n triangles constitute a polyhedron inscribed into the lateral surface of a cylinder of radius 1 and height 1/m. The area, B_n, of that polyhedron exceeds A_n. By stacking m such polyhedra one on top of the other, one obtains an accordion-like polyhedron P inscribed in a cylinder of height 1. Determine m and n in such a way that the area of P is more than 1000 times the area of the cylinder and that each edge of the polyhedron (i.e., each side of any of the 2n triangles) is less than .01. (H. A. Schwarz, 1870.)

3. How should, on a cylinder or on a sphere, two neighbor points of a point P be chosen to make the plane determined by the three points reasonably close to the tangential plane at P? Note, however, that no precaution that is valid on a cylinder or a sphere will be of any avail when applied to the vertex P of a quarter of a cone, since that surface has no tangential plane at P.

2. TWO-PLACE FUNCTIONS

a. The Basic Concepts. In space, if a plane with a Cartesian frame of reference has been chosen, the quantity of value 5 (or *z*) whose object is the ordered pair (4, 3) (or (x, y)) may be represented by the point of altitude 5 above the point (4, 3) (or the point of altitude *z* above the point (x, y)). Conversely, any point in space represents a quantity whose object is an ordered pair of numbers.

Any simple surface represents a consistent class of such quantities and any consistent class of such quantities is represented by a simple surface. Such a class will be called a *2-place function.* Thus a 2-place function is a class of pairs ((x, y), z) in each of which the 1st member is an ordered pair of numbers, the 2nd member is a number, and such that no two pairs belonging to the class have equal first, and unequal second, members.

The number *z* is called the *value* of the function for (x, y). If the 2-place function (i.e., the aforementioned class of pairs) is denoted by *F,* then the value *z* is denoted by *F* (x, y); the class of all values (i.e., all numbers that are 2nd members in pairs belonging to *F*) is called the *range* of *F* – briefly Ran *F*; the class of all pairs of numbers that are 1st members is referred to as the *domain* of *F* – briefly Dom *F*.

Hence *F* is the class of all pairs ((x, y), *F* (x, y)) for any (x, y) in Dom *F*. The function may be considered as the result of pairing a number with every element of its domain. The simple surface representing *F* is called the graph of *F* or the surface *F*. Conversely, the surface *P* mentioned in Exercise 1 is the graph of a function, the "product function" *P*, for which *P* (x, y) = x · y. Hence the following identifications:

number	point on a line with a scale;
ordered pair of numbers . . .	point on a plane with a Cartesian frame;
1-place-function	simple curve above a basic line with a scale;
2-place-function	simple surface above a basic plane with a Cartesian frame.

In the definition of *F*, x and y are two numerical variables which may be replaced by any non-identical letters. In particular, *F* might be defined as the class of all pairs ((a, b), *F* (a, b)) for any (a, b) in Dom *F;* of all pairs ((x, b), *F* (x, b)) for any (x, b) in Dom *F*; of all pairs ((y, x), *F* (y, x)) for any (y, x) in Dom *F*; . . . .

In the study of 1-place functions frequent references had to be made to the class of all numbers, which therefore was denoted by a special letter, U. In dealing with 2-place functions, a symbol for the class of all pairs of numbers is convenient. This class will be denoted by U^2. Thus, for any 2-place function, the domain is a subclass of U^2, the range is a subclass of U.

b. Constant Functions; *I* and *J*. In the classical literature, only few and complicated 2-place functions have designations. No traditional names or symbols exist for the most important ones: the constant functions and functions that are analogous to the identity function *j*.

A 2-place function is called *constant* if its range consists of a single number. The function with the domain U^2 whose range consists of the number 3 will be denoted by $3^{(2)}$ or, where the context leaves no doubt as to the 2-place character of the function, simply by *3*. Its graph is the horizontal plane of altitude 3 above the basic plane. The latter is the graph of the function $O^{(2)}$.

Of the utmost importance are 2-place functions which, hereinafter, will be denoted by *I* and *J* and referred to as the *selector* functions, since they select from any pair of numbers one member as their value – *I* the 1st member, *J* the 2nd member:

$$I(x, y) = x \quad \text{and} \quad J(x, y) = y \quad \text{for any } (x, y) .$$

In other words, *I* and *J* are the classes of all pairs ((x, y), x) and ((x, y), y), respectively. Their graphs are two planes inclined under $\pi/4$ against the basic plane: the plane *I* passes through the 2nd axis, the plane *J* through the 1st (see Fig. 43).

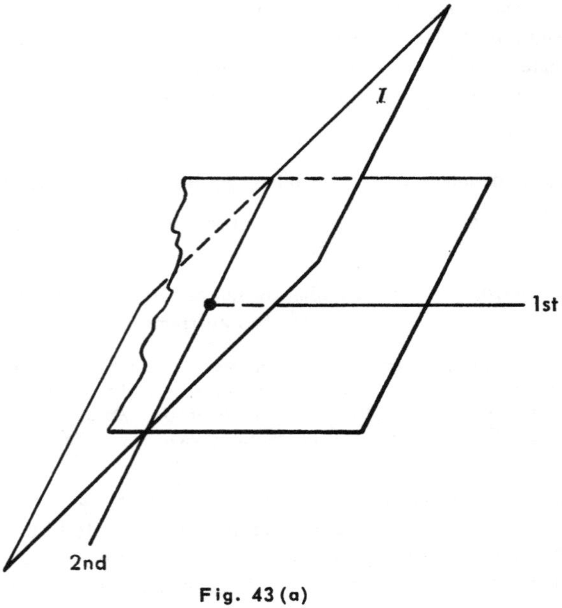

Fig. 43 (a)

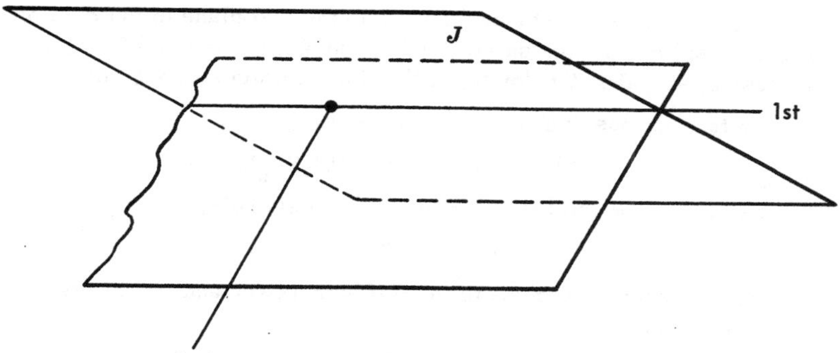

Fig. 43 (b)

Clearly, I and J might be defined as the functions such that

$$I(a, y) = a \quad \text{and} \quad J(a, y) = y \quad \text{for any} \quad (a, y) ;$$
$$I(y, x) = y \quad \text{and} \quad J(y, x) = x \quad \text{for any} \quad (y, x) .$$

c. Other Functions and Function Variables. In the following sections, by means of a few basic operations and with the use of 1-place functions, an enormous variety of 2-place functions will be derived from the constant functions, I and J. Permanent symbols for other 2-place functions will be unnecessary for the purposes of this book. (In fact, even S and P might be replaced by $I + J$ and $I \cdot J$, and in this way be dispensed with.) Only *ad hoc* designations of specific 2-place functions will be occasionally introduced – always italic capitals.

Many definitions and statements are valid for any 2-place function of a certain kind. They will be formulated in terms of a 2-place function *variable* – a symbol (usually, $F, G, H, \ldots$) that may be replaced by the designation of any 2-place function belonging to a certain class of such functions. An example is, of course, the definition of a 2-place function F on page 295; another, the following statement about equal functions.

d. Equality. Restrictions and Extensions. Two 2-place functions (i.e., two consistent classes of quantities whose objects are ordered pairs of numbers) are *equal* if any quantity belonging to either function belongs also to the other. In analogy to what has been proved about 1-place functions, the reader can easily verify that $F = G$ if and only if $\text{Dom}\, F = \text{Dom}\, G$ and $F(x, y) = G(x, y)$ for any pair (x, y) belonging to that common domain.

Restrictions and extensions of 2-place functions can be defined just as they have been in the realm of 1-place functions.

e. Classes of Pairs of Numbers. Just as classes of numbers or of points on a line can be described by relations between 1-place functions such as $\{ j = 1 \}$, $\{ 0 < j < 1 \}$, $\{ tan < exp \}$ and the like, classes of pairs of numbers or of points in a plane can be described by relations between

2-place functions. For instance, $\{I > O\}$ is the half-plane to the right of the 2nd axis; $\{P < O\}$ the sum of the 2nd and 4th quadrants; $\{I = J\}$ is the class of all pairs (t, t) for any t, that is, a certain straight line.

f. Limits. Suppose that, for some number c,

$$\text{if } (x, y) \underset{\text{suf}}{\sim} (a, b), \text{ then } F(x, y) \underset{\text{arb}}{\sim} c.$$

Then c is said to be the limit of F at (a, b); in symbols,

$$\lim_{(a,\, b)} F = c.$$

Here, (x, y) is said to be close to (a, b) if both $|x-a|$ and $|y-b|$ are small. For instance,

$$\lim_{(a,\, b)} P = a \cdot b.$$

Indeed, if $|x-a| < d$ and $|y-b| < d$ (where $d < 1$), then, as one readily proves, $|xy-ab| < (|a|+|b|+1)d$. Since $a \cdot b = P(a, b)$, it follows that $\lim_{(a,\, b)} P = P(a, b)$, which is expressed by saying that P is continuous at (a, b).

g. Direction Fields. The surface F is not the only geometric representation of a 2-place function, F. The latter may be represented within a plane with a Cartesian frame by a *direction field,* called the direction field F. "Within" the plane means that this representation makes no reference to space. To obtain the direction field I, draw through each point (a, b) the line whose slope is $I(a, b) = a$ – or rather, since in a drawing these lines would interfere with each other, draw a *short segment* of the line of slope a through the point (a, b); and since obviously even this cannot actually be done for *each* point (a, b), draw a large number of such segments. The following observation is helpful: Any line parallel to the 2nd axis is an *isocline* of the direction field I; that is to say, at any point of such a line the slope of the field is the same. After one has constructed one short segment, he therefore can draw parallel segments through many points on the isocline.

The direction field F is obtained by drawing through many points (a, b) short segments of the slope $F(a, b)$.

EXERCISES

4. Draw the graph as well as the direction field of the "Sum" function S for which $S(x, y) = x + y$. What is the *contour line* of altitude 3 on the surface S (consisting of all points at which S has the altitude 3 and corresponding to the isocline of slope 3 in the direction field S)?

5. Sketch the surfaces J and P with some contour lines and the direction fields J and P with some isoclines. Draw the direction field 2 corresponding to the constant function 2 and study the isocline situation.

6. Sketch various direction fields having the circles about (0, 0) as isoclines. What can be said about the corresponding surfaces?

7. Determine the classes $\{P = O\}$; $\{P < 1\}$; $\{S = -1\}$; $\{P < 1\}$; $\{S = P\}$. Is $\{P = I\} = \{J = 1\}$? In what relation is $\{S = 3\}$ *(a)* to the contour line of altitude 3 of the surface S? *(b)* to the isocline of slope 3 of the direction field S?

3. OPERATIONS ON TWO-PLACE FUNCTIONS

a. Addition and Multiplication. 2-place functions can be added and multiplied in the same way as 1-place functions. $F + G$ and $F \cdot G$ are the functions assuming the values

$$(F+G)(x, y) = F(x, y) + G(x, y) \quad \text{and} \quad (F \circ G)(x, y) = F(x, y) \cdot G(x, y)$$

for any (x, y) belonging to both Dom F and Dom G. Thus

$$\text{Dom}(F + G) = \text{Dom}(F \cdot G) = \text{Dom}\, F \cdot \text{Dom}\, G .$$

For instance, $P = I \cdot J$; and $I + J$ equals the function S mentioned in Exercise 4.

The surface $F + G$ can be constructed by the superposition of the surfaces F and G, the surface $F \cdot G$ by a graphical procedure similar to that by which the curve $f \cdot g$ was obtained (p. 83).

The sum of a 2-place function and a 1-place function or a number will remain undefined; so will the product of a 2-place function and a 1-place function. But, as in the case of 1-place functions, it is sometimes convenient to write:

$3F$ for $F + F + F$, which equals $3^{(2)} \cdot F$;

$4 + 3I + 2J$ instead of $4^{(2)} + 3^{(2)} \cdot I + 2^{(2)} \cdot J$;

F^3 for $F \cdot F \cdot F$ and to set $F^0 = 1^{(2)}$.

In view of this last convention, a 2-place polynomial can be defined as a function that is the sum of terms each of which is of the form $cI^m \cdot J^n$, where m and n are non-negative integers. The terms of such a polynomial may be arranged in various ways, e.g., according to increasing powers of J or increasing powers of I. For instance,

$$\begin{aligned} &4 - I + 3I \cdot J^2 + 2I^2 \cdot J^2 - I \cdot J^4 + J^5 \\ &= (4 - I) + (3I + 2I^2) \cdot J^2 - I \cdot J^4 + J^5 \\ &= (4 + J^5) + (-1 + 3J^2 - J^4) \cdot I + 2J^2 \cdot I^2 . \end{aligned}$$

b. Substitution of Two C.C.Q.'s. A 2-place function lends itself to the substitution of pairs of consistent classes of quantities having the same domain The result of substituting into the product function P gas pressure p and gas volume v is the variable quantity having the same domain as have p and v, namely, the class of all instantaneous gas samples, and associating with any such γ the value $P(p\gamma, v\gamma) = p\gamma \cdot v\gamma$. Thus $P(p, v) = p \cdot v$. Boyle's Law (p. 175) states the equality of $P(p, v)$ and $\mathrm{r}t$.

More generally, if F is any 2-place function and u and w are any two consistent classes of quantities with the same domain, A, then $F(u, w)$ is the class of all pairs

$$(a, F(ua, wa)) \text{ for any } a \text{ belonging to A and such that the pair } (ua, wa) \text{ belongs to Dom } F.$$

For instance, for the sum and product functions, S and P,

(1) $S(u,w) = u+w$ and $P(u, w) = u \cdot w$ for any two c.c.q.'s u and w.

Of particular importance is the substitution into a 2-place function of a pair of *functions* having the same domain: two 2-place functions or two 1-place functions. These operations will be called (2, 2)-substitution and (2, 1)-substitution, respectively. The substitution of a 1-place into a 1-place function will from now on be referred to as (1, 1)-substitution.

If it were tacitly understood that the two functions following F are to be *substituted* into F, then one could dispense with parentheses and write FGH and Fgh instead of $F(G, H)$ and $F(g, h)$, and $F\mathrm{xy}$ instead of $F(\mathrm{x}, \mathrm{y})$, for that matter. In what follows, parentheses will be retained to facilitate the reading.

c. (2, 2)-Substitution. The result of the (2, 2)-substitution, denoted by $F(G, H)$, is defined as the 2-place function assuming the value

$$F(G,H)(\mathrm{x}, \mathrm{y}) = F(G(\mathrm{x}, \mathrm{y}), H(\mathrm{x}, \mathrm{y})) \text{ for any } (\mathrm{x}, \mathrm{y}) \text{ belonging to Dom } G \cdot \text{Dom } H$$
$$\text{and such that } (G(\mathrm{x}, \mathrm{y}), H(\mathrm{x}, \mathrm{y})) \text{ belongs to Dom } F.$$

As an example, find $P(S, I)$. It is the function assuming for any (x, y) the value

$$P(S, I)(\mathrm{x}, \mathrm{y}) = P(S(\mathrm{x}, \mathrm{y}), I(\mathrm{x}, \mathrm{y})) = P(\mathrm{x}+\mathrm{y}, \mathrm{x}) = (\mathrm{x}+\mathrm{y})\,\mathrm{x} = \mathrm{x}^2 + \mathrm{xy}.$$

This is also the value of $I^2 + P$ for (x, y). Hence $P(S, I) = I^2 + P$. Moreover, in agreement with (1): $P(S, I) = S \cdot I$.

More generally, from (1) it follows that in the realm of 2-place functions, addition and multiplication might be dispensed with. They might be replaced by (2, 2)-substitution into the specific 2-place functions S and P. The same is, of course, true for other binary operations.

In the realm of 2-place functions, I and J play a role somewhat similar to that of j among 1-place functions.

(2) $F(I, J) = F$ for any 2-place function F.

Indeed, the value of $F(I, J)$ for (x, y) is $F(I(\mathrm{x}, \mathrm{y}), J(\mathrm{x}, \mathrm{y})) = F(\mathrm{x}, \mathrm{y})$.

(3) $I(G, H) = G$ and $J(G, H) = H$ for any two functions G and H.

Indeed, $I(G, H)(x, y) = I(G(x, y), H(x, y)) = G(x, y)$.

If C is a constant function, then $C(G, H)$ is the restriction of C to Dom $G \cdot$ Dom H.

d. (2, 1)-Substitution. The result of the (2, 1)-substitution, denoted by $F(g, h)$, is defined as the 1-place function assuming the value

$$F(g, h)\mathrm{a} = F(g\mathrm{a}, h\mathrm{a}) \text{ for any a in Dom } g \cdot \text{Dom } h \text{ such that}$$
$$(g\mathrm{a}, h\mathrm{a}) \text{ belongs to Dom } F.$$

Example 1. The value of $P(\sin, \cos)$ is $P(\sin \mathrm{a}, \cos \mathrm{a}) = \sin \mathrm{a} \cdot \cos \mathrm{a}$ for any a. Hence $P(\sin, \cos) = \frac{1}{2} \sin(2j)$.

Example 2. In analogy to (3) and as consequences of (1), one readily obtains

$$I(g, h) = g, \quad J(g, h) = h, \quad S(g, h) = g + h, \quad P(g, h) = g \cdot h.$$

Example 3. The constant 2-place and 1-place functions of the value a will be denoted by A or $a^{(2)}$ and a or $a^{(1)}$, respectively. Clearly,

$$A(g, h) = a \text{ (on Dom } g \cdot \text{Dom } h) \text{ or } a^{(2)}(g, h) = a^{(1)} \text{ (on Dom } g \cdot \text{Dom } h).$$

Example 4. If (a, b) belongs to Dom F, then $F(a, b)$ is the constant 1-place function of the value $F(\mathrm{a}, \mathrm{b})$; and $F(A, B)$ (the result of a (2, 2)-substitution) is the constant 2-place function of the value $F(\mathrm{a}, \mathrm{b})$.

Example 5. Substitute (g, h)

(a) into $F(A, J)$; using Examples 2, 3, 4 one finds
$F(A, J)(g, h) = F(A(g, h), J(g, h)) = F(a, h)$;

(b) into $F - F(A, J)$; the result is $F(g, h) - F(a, h)$;

(c) into $\dfrac{F - F(A, J)}{I - A}$; the result is $\dfrac{F(g, h) - F(a, h)}{g - a}$.

e. (1, 2)-Substitution. Being a consistent class of quantities, a 2-place function G may be substituted into a 1-place function f. The result of this operation, hereinafter referred to as (1, 2)-substitution, is the 2-place function fG assuming the value $f(G(x, y))$ for any pair (x, y) belonging to Dom G and such that $G(x, y)$ belongs to Dom f.

Example 6. $j^{1/2}(1 - I^2 - J^2)$ or $\sqrt{1 - I^2 - J^2}$ is the function assuming the value $\sqrt{1 - \mathrm{a}^2 - \mathrm{b}^2}$ for any (a, b) in $\{I^2 + J^2 < 1^{(2)}\}$ so that $1 - \mathrm{a}^2 - \mathrm{b}^2$ belongs to Dom $j^{1/2}$. The graph of $1 - I^2 - J^2$ is a paraboloid of revolution; that of its root is the upper hemisphere of radius 1 about (0, 0).

Example 7. fI is the function assuming for any (x, y) the value $f(I(x, y)) = f\mathrm{x}$. Its graph is a cylinder with generating lines parallel to the 2nd axis. The intersection of this cylinder with a vertical plane through the 1st axis is the simple curve f above the 1st axis. Similarly, fJ is a cylinder with generating lines parallel to the 1st axis.

Example 8. $\sin S = \sin(I + J) = \sin I \cdot \cos J + \cos I \cdot \sin J$. The graph of $\sin S$ is a corrugated plane. The last formula is related to the trigonometric law $\sin(x + y) = \sin x \cos y + \cos x \sin y$ for any x and any y, as is $j^2 - 1 = (j + 1)\cdot(j - 1)$ to $x^2 - 1 = (x + 1)(x - 1)$ for any x.

Example 9. If in Example 5(c), one replaces A by gA, in view of Example 3, he obtains

$$\left(\frac{F - F(gA, J)}{I - gA}\right)(g, h) = \frac{F(g, h) - F(ga, h)}{g - ga}.$$

SUMMARY OF THE FOUR TYPES OF SUBSTITUTION STUDIED ABOVE. An (m, n)-substitution is the substitution

of an m-tuple of n-place functions

into an m-place function

resulting in an n-place function.

A 2-place function F lends itself also to the substitution of a pair u, v of c.c.q.'s with not necessarily equal domains, the result being a c.c.q. that may be denoted by $F[u, v]$. Its domain is the class of all ordered pairs (α, β) for any α in Dom u and any β in Dom v. If u and v are functions of m and n places, respectively, then $F[u, v]$ is essentially an (m + n)–place function, i.e., a c.c.q. whose domain consists of (m + n)–tuples of numbers. If u and v are 1-place functions, g and h, then $F[g, h]$ is the 2-place function $F(gI, hJ)$. But unless m = n = 1, the operation transcends the realm of 2-place functions and will here not be studied. Only one interesting result concerning this substitution may be mentioned. The associativeness of multiplication can be expressed in the formula $P[j, P] = P[P, j]$.

EXERCISES

8. Construct the graph of $3 - 2I + 4J$. Show that for any three numbers a, b, c the graph of $a + bI + cJ$ is a non-vertical plane; and that every non-vertical plane is, for some three numbers a, b, c the graph of $a + bI + cJ$.

9. Sketch the graphs of $S(-3^{(2)}, 5^{(2)})$; $P(J, -2^{(2)})$; $S(S, 1^{(2)})$.

10. Show that if K, L, M are 2-place polynomials, then so is $K(L, M)$. Show that this implies that the sum and the product of L and M are polynomials.

11. A 2-place function is called symmetric if $F = F(J, I)$. What can be said about the graph of a symmetric function? Show that S and P are symmetric. Are I and J?

12. For the "quotient" function Q assuming the value y/x for any (x, y) such that $x \neq 0$, show that $Q(J, I) = 1/Q$ or $Q(J, I) = Q(Q, 1)$.

13. On top of a vertical bar magnet, put a horizontal cardboard with a Cartesian frame of reference whose origin is above the point of contact with the magnet. Iron splinters, poured over the cardboard, point toward the point (0, 0) when they come to rest. Show that the splinters materialize the direction field Q. What direction field would they materialize if the point (2, 3) were above the magnet?

14. Put a short horizontal bar magnet under a horizontal cardboard along the 2nd axis of a Cartesian frame with the point (0, 0) above the center of the bar. The laws of magnetic attraction imply that an iron splinter at the point (x, y) (not too close to the magnet) assumes the direction $M(x, y) = \frac{2}{3}\frac{y}{x} - \frac{1}{3}\frac{x}{y}$. Taking advantage of isoclines, draw the direction field of the function $M = \frac{1}{3}(2Q - 1/Q)$.

15. Show that $jG = G$, $j^{-1}G = 1/G$, and $j^2 G = G \cdot G$, for any G.

16. How are the contour lines of the surfaces G and fG related?

17. Translate $\log P = S(\log I, \log J)$ into a law about numbers expressed in terms of numerical variables. Then translate $e^{x+y} = e^x \cdot e^y$ into a formula connecting specific functions without reference to numerical variables. Find a function equal to $S(\sin I, \sin J)$ and translate the equality for $\cos x + \cos y$ into an equality of functions.

18. For the 2-place polynomial $4 - 3I \cdot J^2 + 2I^2 \cdot J + I \cdot J^4 + J^5$, *ad hoc* denoted by F, find 1-place polynomials $p_0, \ldots, p_5$ and q_0, q_1, q_2 such that $F = p_0 I \cdot + p_1 I \cdot J + p_2 I \cdot J^2 + \ldots + p_5 I \cdot J^5 = q_0 J + q_1 J \cdot I + q_2 J \cdot I^2$. Here, $p_k I$ and $q_k J$ denote results of (1, 2)-substitution.

19. Substitute (g, h) into $\dfrac{F - F(I, hB)}{J - hB}$.

20. Show that the substitution of polar coordinates (r, a) into $I \cdot \cos J$ yields the abscissa x relative to a Cartesian frame in a standard relation to the polar frame (p. 188). By what 2-place function is y connected with r and a? Into which functions should x and y be substituted to yield r and a?

21. Show that $\lim_{a} g = g\,a$, $\lim_{a} h = h\,a$, and $\lim_{(g\,a,\, h\,a)} E = c$ imply $\lim_{a} E(g, h) = c$.

★22. Show that two 1-place functions f and g such that $fI = gJ$ are equal and constant.

★23. Show that substitution of (G', H') into $F(G, H)$ and of $(G(G', H'), H(G', H'))$ into F have equal results. Illustrate this "associative" law of (2, 2)-substitution by simple examples. Using the symmetry of S and P, derive from it the commutative laws for addition and multiplication: $S(G, H) = S(H, G)$ and $P(G, H) = P(H, G)$. The associative law for addition can be formulated as follows: $S[S(F,G), H] = S[F, S(G,H)]$. Formulate the distributive law.

★24. Devise a construction for the point above (x, y) of the surface $F(G, H)$ if F, G, and H are given surfaces. Hint. Use the planes I and J in analogy to j in Fig. 27.

★25. Whereas $f + g$ and $f \cdot g$ are obtainable by (2, 1)-substitution of (f, g) into S and P, show that there is no 2-place function T such that $fg = T(f, g)$ *for any f and g*. Hint. Assume that there is such a T and derive a contradiction: $T(a, j) = a$ would imply $T = I$, which is impossible since $T(j, j^2) = j^2$ and $I(j, j^2) = j$.

4. PURE ANALYTIC GEOMETRY AND PURE KINEMATICS

a. The Pure Analytic Plane. In physical or postulational geometry, the abscissa x and the ordinate y relative to a Cartesian frame (p. 186) are consistent classes of quantities whose domain is the class of all points; and points are something like chalk dots or undefined elements satisfying certain assumptions, respectively. In pure analytic geometry, points are defined as ordered pairs of numbers. For instance, (0, 0), (1, 0), $\left(\frac{1}{2}, -\sqrt{3}\right)$ are called points. Consequently, *any c.c.q. whose domain consists of such points is a 2–place function.* For instance, the class of all quantities ((x, y), x) for any point (x, y) is the function I. Since x is called the abscissa, and y the ordinate, of the point (x, y), *the coordinates in pure analytic geometry are the selector functions.*

$\{2I + 3J = 5^{(2)}\}$, that is, the class of all pairs (x, y) such that 2x + 3y = 5, is called a straight line. Any straight line is defined as a class $\{L = 0^{(2)}\}$ for some linear 2-place-polynomial L. The circle of radius 1 about the point (0, 0) is defined as $\{I^2 + J^2 = 1^{(2)}\}$.

There is no reason why in pure analytic geometry the points (0, 0), (1, 0), and (0, 1) should be called "frame of reference," or the lines $\{J = 0^{(2)}\}$ and $\{I = 0^{(2)}\}$ "first and second axis of reference."

It is important to realize that x and y in definitions such as that of the line $\{2I + 3J = 5^{(2)}\}$ are numerical variables. They may be replaced, without change of the meaning, by any two non-identical letters; in particular, they may be interchanged. In contrast, in the definition of the corresponding line $\{2x + 3y = 5\}$

in a physical or postulational plane, the variable quantities x and y may not be interchanged. For reasons mentioned on page 186, there is a far-reaching parallelism between

x and y in a physical plane,

x and y in a postulational plane,

I and J in a pure analytical plane.

But *there is no parallelism whatever between consistent classes of quantities* such as x and y or I and J, on the one hand, *and numerical variables* such as x and y or a and b, on the other.

b. Transformations of the Plane. In pure analytic geometry, a transformation is defined as an ordered pair of 2-place-functions that has to be substituted into any consistent class of quantities, that is, into any 2-place-function. For instance, the transformation (G, H) replaces I by $I\,(G, H) = G$ and J by $J\,(G, H) = H$; it replaces F by $F\,(G, H)$; and $O^{(2)}$ by $O^{(2)}\,(G, H) = O^{(2)}$. Hence it replaces the class $\{F = O^{(2)}\}$ by the class $\{F\,(G, H) = O^{(2)}\}$.

According to Exercise 20, the transition from a pure Cartesian to a pure polar plane is defined by the transformation

$$(I \cdot \cos J,\ \ I \cdot \sin J).$$

The circle $\{I^2 + J^2 = 1^{(2)}\}$ is thereby replaced by

$\{I^2 (I \cdot \cos J,\, I \cdot \sin J) + J^2\,(I \cdot \cos J,\, I \cdot \sin J) = 1^{(2)}\} =$

$\{I^2 \cdot \cos^2 J + I^2 \cdot \sin^2 J = 1^{(2)}\} = \{I^2 = 1^{(2)}\}$;

the line $\{2\,I + 3\,J = 5^{(2)}\}$ by $\{I \cdot (2 \cos J + 3 \sin J) = 5\}$.

Indeed, in a physical plane

$\{x^2 + y^2 = 1\} = \{r^2 = 1\}$ and $\{2\,x + 3\,y = 5\} = \{r \cdot (2 \cos a + 3 \sin a) = 5\}$.

c. Pure Kinematics. A motion M in the plane is defined as a pair of 1-place-functions $(g,\ h)$ with the same domain. For any c belonging to that domain, the point $(g\ c,\ h\ c)$ is said to be a position in the course of M. The class of all positions is called the trajectory of M. For instance, the trajectories of the motions $(j,\ 2\,j)$ and $(j^3,\ 2\,j^3)$ are identical and contain those of $(j^2,\ 2\,j^2)$ and $(\cos,\ 2 \cos)$ as parts.

Geometrically, M may be represented by the (not necessarily different) graphs of g and h in what may be called the plane of time tables. (2, 1)—substitution of (g, h) into F may be represented by a graphical timetable of F along the motion in the plane of time tables. For instance, $\sqrt{I^2 + J^2}$ along $(g,\ h)$ is represented by the graph $\sqrt{g^2 + h^2}$.

SUMMARY OF DEFINITIONS IN PURE ANALYTICS AND PURE KINEMATICS

Point: a pair of numbers.

Plane: the class U^2 of all pairs of numbers.

Straight Line: a class $\{L = O^{(2)}\}$ for some linear 2-place-polynomial.

Unit Circle about (0, 0): $\{I^2 + J^2 = 1^{(2)}\}$.

Motion: a pair of 1-place-functions.

Transformation of the Plane: a pair of 2-place-functions.

EXERCISES

26. By what does the linear transformation

$$(a_{11} I + a_{12} J + b_1, a_{21} I + a_{22} J + b_2)$$

replace a) the line $\{p I + q J = r^{(2)}\}$, b) the circle $\{I^2 + J^2 = 1^{(2)}\}$? Study the replacing classes. Their nature depends upon whether or not $a_{11}, a_{12}, a_{21}, a_{22}$ satisfy certain conditions.

27. Apply the linear transformation of the preceding exercise (setting $b_1 = b_2 = 0$) to a quadratic curve $\{a I^2 + 2b I \cdot J + c J^2 = 1\}$ and study the replacing class.

5. PARTIAL DERIVATIVES

a. Profiles of Mountains. In the geometry of 2-place functions geographic terms are suggestive. One may regard the surface F as a *mountain* above a plane on sea level. The positive half of the 1st axis in a Cartesian frame, chosen in that basic plane, may be said to point *east*, that of the 2nd axis, *north*. Abscissa and ordinate of a point may then be referred to as the *longitude* and the *latitude* of that point as well as of the point of the mountain above it.

The curve $F(j, b)$ is not only a graphical timetable of F along a WE motion at the latitude b. It is also a vertical plane profile – briefly, a *profile*, of the mountain above the trajectory of that motion. Similarly, $F(a, j)$ is the SN profile at the longitude a. These two profiles will be called the *principal profiles* of F at (a, b), where they intersect.

Also $F(j^2, b)$ and any $F(g, b)$ are graphical timetables of F along WE motions at the latitude b. But, in general, unless $g = j$, they are distortions of the profile. Similarly, $F(j, j)$ is a distortion of the profile of F through (0, 0) from SW to NE. For instance, all profiles through (0, 0) of the hemisphere $\sqrt{1 - I^2 - J^2}$ are semicircles congruent with the WE profile $\sqrt{1 - j^2}$. But (2, 1)–substitution of (j, j) into the hemisphere results in $\sqrt{1 - 2j^2}$. The reader should prove that the plane profile of any surface F above the motion $(c\, j,\ d\, j)$ has the shape of (i.e., is congruent to) the curve

$$F\left(\frac{c}{\sqrt{c^2 + d^2}}\, j,\ \frac{d}{\sqrt{c^2 + d^2}}\, j\right).$$

The graphical timetable of F along the circular motion (cos, sin) is a flattened but undistorted circular panorama of the mountain above the unit circle about (0, 0).

Example 1. For the surface $\dfrac{I^2 - J^2}{I^2 + J^2}$, in this example denoted by G, all plane profiles through (0, 0) are horizontal lines, though of various altitudes. Indeed, $G\,(c\, j,\ d\, j)$ is the constant function of the value $\dfrac{c^2 - d^2}{c^2 + d^2}$. All these plane profiles meet above (0, 0) somewhat like the steps of a circular staircase meet at its axis. The surface G has indeed a ramp-like shape. Clearly, there is no $\lim\limits_{(0,0)} G$ even though $\lim\limits_{0} G(c\, j,\ d\, j)$ exists for every c and d, not both 0.

Example 2. For the surface $\dfrac{I \cdot J^2}{(I + J^2)^2}$, in this example denoted by G, one finds $G\,(c\, j,\ d\, j) = \dfrac{cd^2 j}{(c + d^2\, j)^2}$. Consequently, $\lim\limits_{0} G(c\, j,\ d\, j) = 0$, for any c and d; that is to say, each plane profile of G has the limit 0 at 0. Nevertheless, $\lim\limits_{(0,0)} G$ does not exist. As close to (0, 0) as one pleases, G has points at the altitude 1/4. Indeed, $G\,(j^2,\ j)$ is the constant function of the value 1/4. Thus a parabolic profile of the surface G has the constant altitude 1/4.

Example 3. Consider $\sqrt{I^2 + J^2}$ (on $\{ I \geq O$ and $J \geq O \}$), that is, a quarter of a cone with the vertex at (0, 0). The (2, 1)-substitution of (j, O) as well as of (O, j) yields j(on $\{ j \geq O \}$). These are lines of the slope 1. But for any b $\neq$ 0, the substitution of $(j,\ b)$ yields $\sqrt{j^2 + b^2}$ (on $\{ j \geq 0 \}$), that is, a hyperbola with the slope 0 at 0.

Example 4. Set, *ad hoc*, $G = \dfrac{\sqrt{I^2 + J^2} - \sqrt{J^2}}{I}$ (on $\{ I \geq O$ and $J \geq O \}$).

One easily verifies that $\lim\limits_{0} G\,(j, b) = 0$ if b $\neq$ 0. Clearly, $\lim\limits_{(0,\ 0)} G$ does not exist.

b. Partial Derivatives. In what follows, it will be assumed that all the profiles considered are smooth in the sense of having tangents at the points studied. The slopes of the two principal profiles of a surface F at (a, b) are called the *partial derivatives* of F at (a, b), More precisely,

$\mathbf{D}\, F\,(j, b)$ a is called the *1st-place derivative* of F at (a, b) and is denoted by $\mathbf{D}_1 F$(a, b);

$\mathbf{D}\, F\,(a, j)$ b is called the *2nd-place derivative* of F at (a,b) and is denoted by $\mathbf{D}_2\, F$(a, b).

Thus just as a curve has a slope at a point and a 1-place function f has a derivative $\mathbf{D}\, f$ a at a, a surface has two principal slopes at a point, and a 2-place function F has two partial derivatives, $\mathbf{D}_1\, F$ (a, b) and $\mathbf{D}_2\, F$ (a, b), at (a, b). Just as the pairing of $\mathbf{D}\, f$ a with a gives rise to a 1-place function $\mathbf{D}\, f$, the pairings of $\mathbf{D}_1\, F$ (a, b) with (a, b) and of $\mathbf{D}_2\, F$ (a, b) with (a, b) give rise to two 2-place functions $\mathbf{D}_1\, F$ and $\mathbf{D}_2\, F$.

For instance,

$\mathbf{D}_1\, I\,(\mathrm{a}, \mathrm{b}) = \mathbf{D} I(j, b)\, \mathrm{a} = \mathbf{D}\, j\, \mathrm{a} = 1$ and $\mathbf{D}_2\, I\,(\mathrm{a}, \mathrm{b}) = \mathbf{D}\, I\,(a, j)\, \mathrm{b} = \mathbf{D}\, a\, \mathrm{b} = 0.$

$\mathbf{D}_1 S(\mathrm{a}, \mathrm{b}) = \mathbf{D}\, S\,(j, b)\, \mathrm{a} = \mathbf{D}\,(j + b)\, \mathrm{a} = 1,\ \mathbf{D}_2 S(\mathrm{a}, \mathrm{b}) = \mathbf{D}\, S\,(a, j)\, \mathrm{b} = \mathbf{D}(a + j)\mathrm{b} = 1$

$\mathbf{D}_1\, P(\mathrm{a}, \mathrm{b}) = \mathbf{D}\, P(j, b)\mathrm{a} = \mathbf{D}\,(j \cdot b)\, \mathrm{a} = \mathrm{b}.$

Summarizing these and similar results one can say

(4) $$\mathbf{D}_1\, I = 1^{(2)},\ \mathbf{D}_2\, I = O^{(2)};\ \mathbf{D}_1\, J = O^{(2)},\ \mathbf{D}_2\, I = 1^{(2)};$$

$$\mathbf{D}_1 S = \mathbf{D}_2\, S = 1^{(2)};\ \mathbf{D}_1\, P = J,\ \mathbf{D}_2\, P = I\,.$$

Comparison of the last formulas with (4) yields special cases of Sum and Product Rules which can readily be reduced to those in Chapter VIII.

(5) $\mathbf{D}_i\,(F + G) = \mathbf{D}_i\, F + \mathbf{D}_i\, G;\ \mathbf{D}_i\,(F \cdot G) = \mathbf{D}_i\, F \cdot G + F \cdot \mathbf{D}_i\, G$ (i = 1 or 2).

c. Conventions concerning the Reach of $\mathbf{D}_1$ and $\mathbf{D}_2$. In accordance with what was said about $\mathbf{D}$ (p. 225), only the immediately following 2-place function will be considered to be within the reach of the symbols $\mathbf{D}_1$ and $\mathbf{D}_2$. For instance,

$\mathbf{D}_1\, F\,(G, H)$ stands for $(\mathbf{D}_1 F)\,(G, H)$ and not for $\mathbf{D}_1\,[F\,(G, H)]$.

Since partial derivatives are not defined for 1-place functions, $\mathbf{D}_1$ f is a meaningless symbol. Hence

$\mathbf{D}_1 f G$ stands for $\mathbf{D}_1\,(fG)$; it cannot mean anything else.

Similarly, derivatives are only defined for 1-place functions. $\mathbf{D}\, F$ is a meaningless symbol. Hence

$\mathbf{D}\, F(g, h)$ stands for $\mathbf{D}\,[F\,(g, h)]$; it cannot mean anything else. The last convention was used in defining $\mathbf{D}_1\, F$(a, b) as $\mathbf{D}\, F\,(j,\ b)$ a.

d. Derivation and (1, 2)–Substitution. What are the partial derivatives of the 2-place function $f\,G$? Determine the value $\mathbf{D}_1\, f\, G\,(\mathrm{a, b})$ of $\mathbf{D}_1\, f\, G$ (that is, $\mathbf{D}_1\,(fG)$) at (a, b). By the definition of $\mathbf{D}_1$, one has to find the derivative at a of the 1-place function $f\,G\,(j, b)$:

$\mathbf{D}_1\, f\, G\,(\mathrm{a,b}) = \mathbf{D}\,[f G(j, b)]\,\mathrm{a}$. (Here, the brackets cannot be omitted.) Since $fG(j, b)$ results from (1, 1)-substitution of $G\,(j, b)$ into f, the value at a of the derivative of $fG\,(j, b)$ is obtained by the (1, 1)-Substitution Rule of page 233:

$$\mathbf{D}\,[fG\,(j, b)]\;\mathrm{a} = [(\mathbf{D} f)\; G\,(j, b)\;\mathrm{a}] \cdot \mathbf{D}\; G\,(j, b)\;\mathrm{a}\,.$$

On the right side, the brackets and the first pair of parentheses might be omitted; but they facilitate the reading. The last factor equals $\mathbf{D}_1\; G\,(\mathrm{a, b})$ by the very definition of $\mathbf{D}_1$; the other factor is the value of $\mathbf{D}\, f$ for $G\,(\mathrm{a, b})$. Thus altogether,

$\mathbf{D}_1\; f\; G\,(\mathrm{a, b}) = (\mathbf{D}\; f)\; G\,(\mathrm{a, b}) \cdot \mathbf{D}_1\; G\,(\mathrm{a, b})$, and hence the

(1, 2)-SUBSTITUTION RULE. $\mathbf{D}_1\; f\, G = (\mathbf{D}\; f)\; G \cdot \mathbf{D}_1\; G\,.$

For instance, if f is replaced by sin or j^n, then
$\mathbf{D}_1\; sin\; G = cos\; G \cdot \mathbf{D}_1\; G$ and $\mathbf{D}_1\; G^n = \mathrm{n}\; G^{n-1} \cdot \mathbf{D}_1\; G$.

Of course, an analogous formula holds for $\mathbf{D}_2\; fG$. For instance,

$$\mathbf{D}_1\; exp\; P = exp\; P \cdot J \quad \text{and} \quad \mathbf{D}_2\; exp\; P = exp\; P \cdot I\,.$$

In view of (4), the (1, 2)-Substitution Rule implies

(6) $$\mathbf{D}_1\; fI = \mathbf{D}\; fI,\;\; \mathbf{D}_2\; fI = O^{(2)};\;\; \mathbf{D}_1\; fJ = O^{(2)},\;\; \mathbf{D}_2\; fJ = \mathbf{D}\; fJ.$$

e. Partial Derivatives as Limits. $\mathbf{D}_1\; F\,(\mathrm{a, b})$, being the value for a of the derivative of the 1-place function $F\,(j, b)$, is the limit at a of the difference quotient at a of $F\,(j, b)$; that is to say,

$$\mathbf{D}_1\; F\,(\mathrm{a, b}) = \lim_{\mathrm{a}} \frac{F\,(j, b) - F\,(a, b)}{j - a} \quad \text{and} \quad \mathbf{D}_2\; F\,(\mathrm{a, b}) = \lim_{\mathrm{b}} \frac{F\,(a, j) - F\,(a, b)}{j - b}\,,$$

where a and b denote the constant functions of the values a and b.

According to Example 5(c) on page 301, the difference quotient

$\dfrac{F\,(j, b) - F\,(a, b)}{j - a}$ is the result of the (2, 1)-substitution of (j, b) into

$$\frac{F - F\,(A, J)}{I - A} \quad \text{or, more symmetrically,} \quad \frac{F - F\,(A, J)}{I - I\,(A, J)}\,.$$

The preceding 2-place function unites, as it were, the difference quotients at a of all WE profiles of F at the various latitudes. It will be referred to as the *1st-place difference quotient* of F at a. The limit at (a, b) of this function, if it has one, will be denoted by

$$\lim_{(a,\,b)} \frac{F - F(A, J)}{I - I(A, J)} = \mathbf{D}_1^* F(a, b),$$

and will be called the *uniform 1st-place derivative* of F at (a, b). If there is such a number, it clearly is equal to $\mathbf{D}_1 F(a, b)$, But if the quarter of the cone in Example 3, page 307 is *ad hoc* denoted by F, then $\mathbf{D}_1 F(0, 0) = 1$, whereas there is no number $\mathbf{D}_1^* F(0, 0)$. Indeed, the 1st-place difference quotient of the cone F at 0 is the function G of Example 4 on page 307, which has no limit at (0, 0).

f. Derivation and (2, 1)-Substitution. What is the derivative of the 1-place function $F(g, h)$? The value for a of this function will be determined under two assumptions:

1. that there is a number $\mathbf{D}_1^* F(g\,a, h\,b)$,

2. that g and h are continuous at a; that is to say, that

$$\lim_a g = g\,a \text{ and } \lim_a h = h\,a.$$

Under these assumptions, $\mathbf{D}\, F(g, h)\, a$ can be determined by the Three-Step Rule (p. 132):

Step I. Write down the difference quotient of $F(g, h)$ at a:

$$\frac{F(g, h) - F(ga, ha)}{j - a} .$$

Step II. Transform it to make the computation of its limit at a feasible. Just as in proving the Product Rule (p. 228), add and subtract one and the same term in the numerator, namely, $F(ga, h)$, and split the quotient into two terms

$$\frac{F(g, h) - F(ga, h)}{j - a} + \frac{F(ga, h) - F(ga, ha)}{j - a} .$$

Just as in proving the (1, 1)-Substitution Rule (p. 233), multiply and divide each of these terms by one and the same expression, and regroup:

$$\frac{F(g, h) - F(ga, h)}{g - ga} \cdot \frac{g - ga}{j - a} + \frac{F(ga, h) - F(ga, ha)}{h - ha} \cdot \frac{h - ha}{j - a} .$$

Step III. Compute the limit at a of the difference quotient in the new form. It obviously equals the sum of two products of limits at a. Clearly, the limit at a of the 2nd factor is $\mathbf{D}\, g\, a$, that of the 4th, $\mathbf{D}\, h\, a$; that of the 3rd is easily seen to be $\mathbf{D}_2 F(g\,a, h\,a)$. Only the 1st factor presents a minor difficulty. This factor results from (2, 1)-substitution of (g, h) into $\frac{F - F(gA, J)}{I - gA}$ (see Example 9, p. 302). The latter function may *ad hoc* be denoted by E. By virtue of assumption *1*, E has a limit at $(g\ \mathrm{a}, h\ \mathrm{a})$, namely,

$$\lim_{(g\ \mathrm{a}, h\ \mathrm{a})} E = \mathbf{D}_1^* F(g\ \mathrm{a}, h\ \mathrm{a}).$$

In view of assumption *2*, it follows from Exercise 21 that

$$\lim_{(g\ \mathrm{a}, h\ \mathrm{a})} E = \lim_{\mathrm{a}} E(g, h).$$

Now $\lim_{\mathrm{a}} E(g, h)$ is just the 1st factor, which still had to be determined. The preceding two formulas imply that its value is $\mathbf{D}_1^* F(g\ \mathrm{a}, h\ \mathrm{a})$.

Thus, altogether, the following result has been obtained:

$$\mathbf{D}\, F(g, h)\ \mathrm{a} = \mathbf{D}_1^* F(g\ \mathrm{a}, h\ \mathrm{a}) \cdot \mathbf{D}\, g\ \mathrm{a} + \mathbf{D}_2 F(g\ \mathrm{a}, h\ \mathrm{a}) \cdot \mathbf{D}\, h\ \mathrm{a}.$$

Assuming the existence of a uniform 1st-place derivative of F along the entire motion (g, h), where both g and h are continuous, one thus arrives at the following

(2, 1)-SUBSTITUTION RULE. $\mathbf{D}\, F(g, h) = \mathbf{D}_1 F(g, h) \cdot \mathbf{D}\, g + \mathbf{D}_2 F(g, h) \cdot \mathbf{D} h.$

Example 5. Let g and h be the linear functions $g = \mathrm{a} + \mathrm{p}\ j$ and $h = \mathrm{b} + \mathrm{q}\ j$, for which $g\ 0 = \mathrm{a}$, $h\ 0 = \mathrm{b}$, $\mathbf{D}\, g\ 0 = \mathrm{p}$, $\mathbf{D}\, h\ 0 = \mathrm{q}$. The (2, 1)-Substitution Rule yields

(7) $$\mathbf{D}\, F(\mathrm{a} + \mathrm{p}\ j, \mathrm{b} + \mathrm{q}\ j)\ 0 = \mathrm{p}\ \mathbf{D}_1 F(\mathrm{a}, \mathrm{b}) + \mathrm{q}\ \mathbf{D}_2 F(\mathrm{a}, \mathrm{b}).$$

For any four numbers, a, b, p, q, $(\mathrm{a} + \mathrm{p}\ j, \mathrm{b} + \mathrm{q}\ j)$ is a motion with a straight trajectory and the position (a, b) at 0. Hence (7) shows that the slope at 0 of the graph of F along any such motion can be expressed in terms of p and q and the values at (a, b) of the two partial derivatives of F.

For instance, consider the linear polynomial

$$T = \mathrm{c} + \mathrm{r}(I - \mathrm{a}) + \mathrm{s}(J - \mathrm{b}),$$

for which $T(\mathrm{a}, \mathrm{b}) = \mathrm{c}$, $\mathbf{D}_1 T(\mathrm{a}, \mathrm{b}) = \mathrm{r}$, and $\mathbf{D}_2 T(\mathrm{a}, \mathrm{b}) = \mathrm{s}$. (In fact, $\mathbf{D}_1 T$ and $\mathbf{D}_2 T$ are constant functions of the values r and s.) Application of (7) to T yields

(8) $$\mathbf{D}\, T(\mathrm{a} + \mathrm{p}\ j, \mathrm{b} + \mathrm{q}\ j) = \mathrm{pr} + \mathrm{qs}.$$

An important result can be obtained from a comparison of the surfaces F and T (which is a plane). If, at (a, b), T and F have the same value, that is, if $\mathrm{c} = F(\mathrm{a}, \mathrm{b})$, then above (a, b) the plane T and the surface F

have a common point. If, at (a, b), the corresponding partial derivatives of T and F have the same values, that is, if

$$r = \mathbf{D}_1 F(a,b) \quad \text{and} \quad s = \mathbf{D}_2 F(a,b),$$

then, as a comparison of (7) and (8) proves, the profiles of T and F above any straight line through (a, b) have the same slope at (a, b). Thus the plane

$$(9) \qquad T = F(a,b) + \mathbf{D}_1 F(a,b) \cdot (I - a) + \mathbf{D}_2 F(a,b) \cdot (J - b)$$

contains the tangents at the point above (a, b) to all profiles of F. In other words, (9) is the tangential plane to F above (a, b), which on page 292 was denoted by T_F (a, b). One of the two basic problems concerning surfaces mentioned in Section 1 is thus solved, and notably in terms of derivatives.

If, in (9), one replaces F by $4 - I^2 - J^2$, whose graph is a paraboloid, he obtains

$$4 - a^2 - b^2 - 2a(I - a) - 2b(J - b)$$

as the tangential plane at (a, b) for any a and b. For instance, the tangential planes of the paraboloid at (2, 3), (0, 3), and (0, 0) are

$$-9 - 4(I - 2) - 6(J - 3), \ -5 - 6(J - 3), \ 4, \text{ respectively.}$$

Applying (9) to the hemisphere $\sqrt{4 - I^2 - J^2}$ one finds the tangential plane at (a, b)

$$\sqrt{4 - a^2 - b^2} - \frac{a}{\sqrt{4 - a^2 - b^2}}(I - a) - \frac{b}{\sqrt{4 - a^2 - b^2}}(J - b).$$

But one must keep in mind that the (2, 1)-Substitution Rule has been proved only under the assumption that F has a uniform partial derivative. If this condition is not fulfilled, then, even if the profiles of F at (a, b) are smooth so that $\mathbf{D}_1 F(a, b)$ and $\mathbf{D}_2 F(a, b)$ exist, there is no reason why the plane (9) should be tangential to F. In fact, for the quarter of the cone in Example 3 both partial derivatives assume the value 1 at (0, 0) and yet the plane (9), that is $I + J$, obviously is not tangential to the cone (which at (0, 0) has no tangential plane). Above the trajectory of (j, j) as one easily verifies, the cone and the plane $I + J$ have different slopes.

If only the numbers $\mathbf{D}_1 F(a, b)$ and $\mathbf{D}_2 F(a, b)$ are known, all that can be said is that *if the surface F has a tangential plane at* (a, b), *then it must be the plane* (9). But only if more is known, e.g., the existence of a uniform partial derivative (see Exercise 29), can one conclude that F actually has a tangential plane at (a, b), namely, the plane (9).

Example 6. A 2-place function F is called *homogeneous of degree* k if

(10) $F(\mathrm{a}\,j, \mathrm{b}\,j) = F(\mathrm{a},\mathrm{b}) \cdot j^{\mathrm{k}}$ for any a and b.

Equating the derivatives of the two 1-place functions, that on the left side being obtained by the (2, 1)-Substitution Rule, one obtains

$$\mathbf{D}_1 F(\mathrm{a}\,j, \mathrm{b}\,j) \cdot \mathrm{a} + \mathbf{D}_2 F(\mathrm{a}\,j, \mathrm{b}\,j) \cdot \mathrm{b} = \mathrm{k}\,F(\mathrm{a},\mathrm{b})\,j^{\mathrm{k}-1}.$$

Evaluation of the three 1-place functions for 1 yields

$\mathbf{D}_1 F(\mathrm{a},\mathrm{b}) \cdot \mathrm{a} + \mathbf{D}_2 F(\mathrm{a},\mathrm{b}) \cdot \mathrm{b} = \mathrm{k}\,F(\mathrm{a},\mathrm{b})$ for any a and b, or what is called

EULER'S HOMOGENEITY RELATION. $\mathbf{D}_1 F \cdot I + \mathbf{D}_2 F \cdot J = \mathrm{k}\,F$ for any F that is homogeneous of degree k.

Example 7. If f is a 1-place function, let F *ad hoc* denote the function $\int f$ assuming for any (a, b) the value $\int_{\mathrm{a}}^{\mathrm{b}} f$. The calculus solution of the area problem can be expressed as follows:

$$F = (\mathbf{D}^{-1} f)\,J - (\mathbf{D}^{-1} f)\,I,$$

where $\mathbf{D}^{-1}$ is any antiderivative of f, but the same both times. From (6) it follows that

$$\mathbf{D}_1 F = -fI \text{ and } \mathbf{D}_2 F = fJ.$$

Now consider $F(g, h)$, that is, the 1-place function assuming for any c the value $F(g\,\mathrm{c}, h\,\mathrm{c}) = \int_{g\,\mathrm{c}}^{h\,\mathrm{c}} f$. If that 1-place function is symbolized by $\int_g^h f$, then the (2, 1)-Substitution Rule implies

$$\mathbf{D}\left(\int_g^h f\right) = fh \cdot \mathbf{D}\,h - fg \cdot \mathbf{D}\,g.$$

g. Derivation and (2, 2)-Substitution. The determination of $\mathbf{D}_1[F(G, H)]$ can be reduced to the (2, 1)-Substitution Rule. Indeed, $\mathbf{D}_1[F(G, H)](\mathrm{a},\mathrm{b}) = \mathbf{D}[F(G, H)(j, b)]\,\mathrm{a} = \mathbf{D}[F(G(j, b), H(j, b))]\,\mathrm{a}$, and the last derivative can be obtained by applying the (2, 1)-Rule to $g = G(j, b)$ and $h = H(j, b)$. The result is

$$(\mathbf{D}_1 F)(G(\mathrm{a},\mathrm{b}), H(\mathrm{a},\mathrm{b})) \cdot \mathbf{D}\,G(j, b)\,\mathrm{a} + (\mathbf{D}_2 F)(G(\mathrm{a},\mathrm{b}), H(\mathrm{a},\mathrm{b})) \cdot \mathbf{D}H(j, b)\mathrm{a}$$
$$= (\mathbf{D}_1 F)(G(\mathrm{a},\mathrm{b}), H(\mathrm{a},\mathrm{b})) \cdot \mathbf{D}_1 G(\mathrm{a},\mathrm{b}) + (\mathbf{D}_2 F)(G(\mathrm{a},\mathrm{b}), H(\mathrm{a},\mathrm{b})) \cdot \mathbf{D}_1 H(\mathrm{a},\mathrm{b}),$$

under the assumption that a uniform partial derivative of F exists.

(2, 2)-SUBSTITUTION RULES.

$$\mathbf{D}_1\, F(G,H) = (\mathbf{D}_1\, F)(G,H)\cdot \mathbf{D}_1\, G + (\mathbf{D}_2\, F)(G,H)\cdot \mathbf{D}_1\, H$$

$$\mathbf{D}_2\, F(G,H) = (\mathbf{D}_1\, F)(G,H)\cdot \mathbf{D}_2\, G + (\mathbf{D}_2\, F)(G,H)\cdot \mathbf{D}_2\, H.$$

The reader should determine

$$\mathbf{D}_i[F(a_{11}\, I + a_{12}\, J,\ a_{21}\, I + a_{22}\, J)] \text{ and } \mathbf{D}_i\, [F(I\cdot \cos J,\ I\cdot \sin J)]\,.$$

h. Higher Partial Derivatives. The 2-place functions $\mathbf{D}_1 F$ and $\mathbf{D}_2\, F$, may have partial derivatives. If they do, set

$$\mathbf{D}_i\, (\mathbf{D}_k\, F)(x,y) = \mathbf{D}_{ik}\, F\,(x,y) \text{ and } \mathbf{D}_i\,(\mathbf{D}_k\, F) = \mathbf{D}_{ik}\, F, \text{ for i, k} = 1 \text{ or } 2.$$

Thus $\mathbf{D}_1\ (\mathbf{D}_1\ F)$ is abbreviated to $\mathbf{D}_{11}\, F$ and not to $\mathbf{D}_1^2\ F$, whereas for 1-place functions one sets $\mathbf{D}\,(\mathbf{D}\, f) = \mathbf{D}^2 f$. The four functions

$$\mathbf{D}_{11}\, F,\ \mathbf{D}_{21}\, F,\ \mathbf{D}_{12}\, F,\ \mathbf{D}_{22}\, F$$

are called the *second order partial derivatives* of F; the two in the middle are often referred to as the *mixed* derivatives of second order.

As an example, consider the function $I\cdot J\cdot \dfrac{I^2 - J^2}{I^2 + J^2}$, which is defined on $U^2 - \{(0,\ 0)\}$, that is, for any pair of numbers except (0, 0), where it has the limit 0. Let F denote, *ad hoc*, the (continuous) extension of this function, for which $F\,(0,\ 0) = 0$. A simple computation shows that, on $U^2 - \{(0,\ 0)\}$,

$$\mathbf{D}_1\, F = \frac{J\cdot (I^4 + 4\, I^2\cdot J^2 - J^4)}{(I^2 + J^2)^2},\ \ \mathbf{D}_2\, F = \frac{I\cdot (I^4 - 4\, I^2\cdot J^2 - J^4)}{(I^2 + J^2)^2}$$

$$\mathbf{D}_{21}\, F = \frac{I^8 + 10\, I^6\cdot J^2 - 10\, I^2\cdot J^6 - J^8}{(I^2 + J^2)^4} = \mathbf{D}_{12}\, F.$$

What are the principal profiles of the surfaces F, $\mathbf{D}_1 F$, and $\mathbf{D}_2\, F$ at (0, 0)?

$$F\ (j,\ O) = O \text{ and } F\ (O,\ j) = O;$$

$$\mathbf{D}_1\, F\ (j,\ O) = O\cdot \frac{j^4}{(j^2)^2} = O \text{ and } \mathbf{D}_1\ F\ (O,\ j) = j\cdot \frac{-j^4}{(j^2)^2} = -j;$$

$$\mathbf{D}_2\ F\ (j,\ O) = j\cdot \frac{j^4}{(j^2)^2} = j \text{ and } \mathbf{D}_2\ F\ (O,\ j) = O.$$

Hence the values at (0, 0) of the partial derivatives of F are

$$\mathbf{D}_1\ F\ (0,\ 0) = \mathbf{D}\ [F\ (j,\ O)]\ 0 = \mathbf{D}\ O\ 0 = 0 \text{ and } \mathbf{D}_2\ F\ (0,\ 0) = 0;$$

$$\mathbf{D}_{11}\, F\,(0,\ 0) = \mathbf{D}[\mathbf{D}_1\ F\ (j,\ O)]\ 0 = \mathbf{D}\ O\ 0 = 0;$$

$\mathbf{D}_{21} F(0, 0) = \mathbf{D}[\mathbf{D}_1 F(O, j)]\, 0 = \mathbf{D}(-j)\, 0 = -1;$
$\mathbf{D}_{12} F(0, 0) = \mathbf{D}[\mathbf{D}_2 F(j, O)]\, 0 = \mathbf{D}\, j\, 0 = 1;$
$\mathbf{D}_{22} F(0, 0) = \mathbf{D}[\mathbf{D}_2 F(O, j)]\, 0 = \mathbf{D}\, O\, 0 = 0.$

Thus, while $\mathbf{D}_{12}F = \mathbf{D}_{21} F$ on $U^2 - \{(0, 0)\}$, $\mathbf{D}_{21} F(0, 0) \neq \mathbf{D}_{12} F(0,0)$.

For any function F, it can be shown that $\mathbf{D}_{21}F = \mathbf{D}_{12}F$ at any place where the mixed partial derivatives are continuous. For the function F in the preceding example, $\mathbf{D}_{21}F$ and $\mathbf{D}_{12} F$ are continuous on $U - \{(0, 0)\}$, where indeed $\mathbf{D}_{21} F = \mathbf{D}_{12}F$. At (0, 0), where the mixed derivatives are unequal, they are not continuous.

Applying the (2, 1)-Substitution Rule to $\mathbf{D}_1F$ and (g, h), one obtains

$$\mathbf{D}(\mathbf{D}_1 F)(g, h) = \mathbf{D}_{11} F(g, h) \cdot \mathbf{D}\, g + \mathbf{D}_{21} F(g, h) \cdot \mathbf{D}\, h$$

and a similar result for $\mathbf{D}(\mathbf{D}_2 F)(g, h)$. These results yield an expression for the second derivative of the 1-place function $F(g, h)$:

$$\mathbf{D}^2 F(g, h) = \mathbf{D}[\mathbf{D} F(g, h)] = \mathbf{D}[\mathbf{D}_1 F(g, h) \cdot \mathbf{D}\, g + \mathbf{D}_2 F(g, h) \cdot \mathbf{D}\, h].$$

Applying to either term the Product Rule, and to the derivatives of the first factors the two preceding results, one obtains in cases where $\mathbf{D}_{21} F = \mathbf{D}_{12} F$ the important formula

$$\mathbf{D}^2 F(g, h) = \mathbf{D}_{11}F(g,h) \cdot (\mathbf{D}\, g)^2 + 2\mathbf{D}_{12}F(g,h) \cdot \mathbf{D}\, g \cdot \mathbf{D}\, h + \mathbf{D}_{22}F(g,h) \cdot (\mathbf{D}\, h)^2 + \\ + \mathbf{D}_1 F(g, h) \cdot \mathbf{D}^2 g + \mathbf{D}_2 F(g, h) \cdot \mathbf{D}^2 h.$$

If g and h are linear functions, say, $g = a + p\, j$ and $h = b + q\, j$, then $\mathbf{D}\, g = p$ and $\mathbf{D}\, h = q$, while $\mathbf{D}^2 g = \mathbf{D}^2 h = O$. Hence

$$\mathbf{D}^2 F(g, h) = p^2\, \mathbf{D}_{11} F(g, h) + 2pq\, \mathbf{D}_{12} F(g, h) + q^2 \mathbf{D}_{22} F(g, h).$$

EXERCISES

28. Determine the partial derivatives of *(a)* *arctan* Q, where $Q = J/I$; *(b)* *arctan* (G/F); (c) $j^{1/2}(I^2 + J^2)$; (d) $j^{1/2}(F^2 + G^2)$.

29. Find the tangential planes to (a) $\sqrt{I^2 + J^2}$ at (1, 2); (b) $\mathrm{a}I^2 + 2\mathrm{b}\, I \cdot J + \mathrm{c}J^2 + \mathrm{m}I + \mathrm{n}J + \mathrm{p}$ at (x, y); (c) $(I^4 + J^4)/(I^2 + J^2)$ and $I \cdot J/\sqrt{I^2 + J^2}$ at $(\mathrm{a}, \mathrm{b}) \neq (0, 0)$. What about the last two surfaces at (0, 0) if the values of the function at (0, 0) are defined to be 0?

30. If F stands, *ad hoc*, for $I^2 + J^2$, then for $I^2 + J^2 - J^3$, and finally for any 2-place function, for what motions (g, h) is $\mathbf{D}\, F(g, h) = O$? In the first case, for what motions is $\mathbf{D}\, F(g, h) = 1$?

31. Reconstruct the proof of the Product Rule of page 228 (which was not quite symmetric in f and g) by applying the proof of the (2, 1)-Substitution Rule to $P(f, g)$.

32. Obtain the Quotient Rule of page 230 by studying $Q(f, g)$.

33. Prove the (2, 1)-Rule under the assumption that $\mathbf{D}_2^*\, F(g\,\mathrm{a}, h\,\mathrm{a})$ exists. *Show that the assumption of the existence of either uniform partial derivative can be replaced by the assumption of the continuity of either $\mathbf{D}_1\, F$ or $\mathbf{D}_2\, F$ at $(g\,\mathrm{a}, h\,\mathrm{a})$.

6. IMPLICIT FUNCTIONS

So far in this chapter, cross sections of a surface F by vertical planes (called *profiles*) have been studied. Now we consider its intersections with horizontal planes, called *contour lines* (although in many cases they are not lines). If F is a mountain whose highest peak is 10,000, then the intersection of F with the horizontal plane at the altitude 10,001 is empty; that with the plane at 10,000 contains exactly one point; that with a lower plane may consist of a system of complicated curves and include even plateaus.

If c is a number, then $\{F = \mathrm{c}^{(2)}\}$ is the projection of the contour line at the altitude c onto the basic plane. No general method for a closer description of this class of points or pairs of numbers is known. This is not surprising considering that there is no general method for the description of the class of numbers $\{g = \mathrm{c}\}$, that is, for finding

$$\mathrm{x} \text{ such that } g\,\mathrm{x} = \mathrm{c}.$$

In some cases (that is, for some F and some c), one can find a pair

(g, h) such that $F(g, h) = c$. For instance, if $F = 3I^2 + 4J^2$ and $c = 5$, one readily sees that each of the three motions

$$(\sqrt{5/3}\ cos,\ \sqrt{5}/2\ sin);\ (\sqrt{5/3}\ sin,\ \sqrt{5}/2\ cos);$$
$$\left(\sqrt{5/3}\ \frac{j^2-1}{j^2+1},\ \sqrt{5}\ \frac{j}{j^2+1}\right)$$

substituted into F yields 5.

If $\mathbf{D}_1^* F$ and $\mathbf{D}_2 F$ exist along the trajectory of (g, h), then, by virtue of the (2, 1)-Substitution Rule, $F(g, h) = c$ implies

$$\mathbf{D}_1 F(g, h)\, \mathbf{D} g + \mathbf{D}_2 F(g, h)\, \mathbf{D} h = O.$$

The reader should verify this formula in the preceding examples.

A 1-place function f is said to be *implicitly defined* by the 2-place function G if

$$G(j, f) = O. \tag{11}$$

The graph of f belongs to the contour line of G at the altitude 0. If G is the elliptical paraboloid $4I^2 + J^2 - 9$, then (11) amounts to $4j^2 + f^2 - 9 = O$, and the only continuous functions implicitly defined by G are

$$f_1 = \sqrt{9 - 4j^2} \quad \text{and} \quad f_2 = -\sqrt{9 - 4j^2}. \tag{12}$$

Clearly, the surfaces $(4I^2 + J^2 - 9)^3$ and $exp(4I^2 + J^2 - 9)$ have the same contour lines and define the same implicit functions.

In view of the (2, 1)-Substitution Rule and $\mathbf{D} j = 1$, (11) implies $\mathbf{D}_1 G(j, f)\mathrm{x} + \mathbf{D}_2 G(j, f) \cdot \mathbf{D} f \mathrm{x} = 0$, that is,

$$\mathbf{D}_1\ G(\mathrm{x}, f\mathrm{x}) + \mathbf{D}_2\ G(\mathrm{x}, f\mathrm{x}) \cdot \mathbf{D} f\mathrm{x} = 0$$

for any x in Dom $\mathbf{D} f$. This relation is linear in $\mathbf{D} f$ x and therefore can easily be solved:

$$\mathbf{D} f\mathrm{x} = -\frac{\mathbf{D}_1 G(\mathrm{x}, f\mathrm{x})}{\mathbf{D}_2 G(\mathrm{x}, f\mathrm{x})} \quad \text{or} \quad \mathbf{D} f = -\frac{\mathbf{D}_1 G(j, f)}{\mathbf{D}_2 G(j, f)} \quad \text{on } \{\mathbf{D}_2 G(j, f) \neq O\}. \tag{13}$$

In the example of $4I^2 + J^2 - 9$, (13) implies

$$\mathbf{D} f = -\frac{4j}{f}, \quad \text{thus} \quad \mathbf{D} f_1 = -\frac{4j}{f_1} \quad \text{and} \quad \mathbf{D} f_2 = -\frac{4j}{f_2}.$$

In this way, without a direct application of the rules of Chapter VIII, $\mathbf{D} f_1$ and $\mathbf{D} f_2$ have been determined in the sense that they are expressed in terms of f_1 and f_2. Using (12) one finds

$$\mathbf{D} f_1 = -\frac{4j}{\sqrt{9 - 4j^2}} \quad \text{and} \quad \mathbf{D} f_2 = -\frac{4j}{-\sqrt{9 - 4j^2}}$$

in agreement with the results of Chapter VIII concerning $\mathbf{D}\sqrt{9 - 4j^2}$.

In general, (13) expresses $\mathbf{D} f$ in terms of f. What then is the use of (13) in the numerous cases where f cannot be determined? The answer is that *for any* x *such that* $f\,x$ *can be guessed or determined,* (13) *makes it possible to determine* $\mathbf{D}\,f\,x$ – *accurately or approximately, according to whether* $f\,x$ *has been determined accurately or approximately.*

As an example, consider the functions f determined by

(14) $\quad -2 + I \cdot J + J^5$, that is, such that $-2 + j \cdot f + f^5 = O$.

It can be proved that f is not an elementary function.

(14) implies that

for		satisfies the equation	having the roots
any x	$f\,x$	$-2 + x \cdot f\,x + (f\,x)^5 = 0$	
0	$f0$	$-2 + (f0)^5 = 0$	$2^{1/5}$ and no other root
1	$f1$	$-2 + f1 + (f1)^5 = 0$	1 " " " "
2	$f2$	$-2 + 2f2 + (f2)^5 = 0$	approximately, .9

Proceeding in this way, with sufficient patience one can determine values of f at as many more places as one pleases. The equation

$$-2 - 2\,f(-2) + f(-2)^5 = 0$$

can readily be shown to have three solutions. Hence (14) implicitly defines at least three functions whose domains include -2. On the other hand, it can be shown that for no x has $-2 + x \cdot f\,x + (f\,x)^5 = 0$ more than three solutions.

Unless one notices a short cut (see Exercise 41), the computation of the values of f, that is, the determination of *points* on the curves f, by approximately solving 5th degree equations is very tedious. In contrast, $\mathbf{D}\,f$, that is, the *slope* of the curve f at any of its known points, can be found very easily. According to (13),

$$\mathbf{D}\,f = -\frac{f}{j + 5f^4} \quad \text{or} \quad \mathbf{D}\,f\,x = -\frac{f\,x}{x + 5(f\,x)^4} \quad \text{if } x + 5(f\,x)^4 \neq 0.$$

For instance, in view of the results tabulated above,

$$\mathbf{D}\,f0 = \frac{-2^{1/5}}{5 \cdot 2^{4/5}} = -1/5 \cdot 2^{-3/5},\ \mathbf{D}\,f1 = \frac{-1}{1+5} = -1/6,\ \mathbf{D}\,f2 \sim \frac{-.9}{2+(.9)^4}.$$

Functions implicitly defined by any 2-place polynomial, P, are called *algebraic.* The terms of P can be arranged according to increasing powers of J (p. 299), say,

$$P = p_0 I + p_1 I \cdot J + p_2 I \cdot J^2 + \ldots + p_n I \cdot J^n,$$

where $p_n \neq O$. Then $P(j, f) = O$ amounts to

(15) $\quad p_0 + p_1 \cdot f + p_2 \cdot f^2 + \ldots + p_n \cdot f^n = O.$

Here, $p_0, \ldots, p_n$ are 1-place polynomials.

EXERCISES

34. Study the contour lines of $I^2 + J^2 - J^3$. Show that the shore line (i.e., contour line at the altitude 0) of this mountain has an isolated point due to a submarine rock which, at one point, reaches sea level (See Exercise 10c, p. 66). What is the 0-contour line of $(I - 1)\cdot(I^2 + J^2 - 1)$? What are the graphs of the functions implicitly defined by the last function?

35. Find the slope at 1 of each of the three functions having 1 in their domain that are implicitly defined by $-1 + j^5 + (1 - 3j)\cdot f + 2f^7 = 0$.

36. If $-1 + j + (-1 + j^2)\cdot f + (-3 + 2j)\cdot f^5 = O$, find the values of f and $\mathbf{D}f$ for 1, -1, and 2.

37. Find $\mathbf{D}f$ for the algebraic function (15). Show that a 1-place function is rational if and only if it can be implicitly defined by a linear 2-place polynomial. Determine the algebraic functions implicitly defined by a quadratic 2-place polynomial.

38. A function f, implicitly defined by G, is said to be *singular* at a if $\mathbf{D}_1 G(a, fa) = 0$ and $\mathbf{D}_2 G(a, fa) = 0$. Examples at 0: the functions defined by $j^2 - f^3 = O$, $j^3 - f^2 = O$, $j^2 - j^3 + f^2 = O$. Give further examples.

39. If f is implicitly defined by G, show that *(a)* the tangent at a to the curve f in the plane of timetables is the graph of

$$fa - \frac{\mathbf{D}_1 G(a, fa)}{\mathbf{D}_2 G(a, fa)}(j - a) \quad \text{provided} \quad \mathbf{D}_2 G(a, fa) \neq 0;$$

(b) the tangent to the trajectory of the motion (j, f) at the instant a is

$$\{\mathbf{D}_1 G(a, fa)\cdot(I - a) + \mathbf{D}_2 G(a, fa)\cdot(J - fa) = O^{(2)}\}.$$

Simplify this equation in the case of the paraboloid $4I^2 + J^2 - 9$ by taking into account that $4a^2 + (fa)^2 - 9 = 0$. Set up a simple equation for the tangent at (a, fa) of the 0-contour line f implicitly defined by $aI^2 + 2bI\cdot J + cJ^2 + pI + qJ + r$.

7. PARTIAL INTEGRALS AND THE VOLUME PROBLEM

Just as the tangential planes to the surface F are connected with the slopes of the principal profiles of F, the volume under F (see p. 292) is connected with the areas under those profiles. The number comparable to $\frac{F(\mathrm{b}, \mathrm{c}) - F(\mathrm{a}, \mathrm{c})}{\mathrm{b} - \mathrm{a}}$, the slope of the chord between a and b of the WE profile of F at the latitude c, is $\int_{\mathrm{a}}^{\mathrm{b}} [F(j, \mathrm{c})]$, the area from a to b under that profile.

The 1-place function assuming for any c the value $\int_{\mathrm{a}}^{\mathrm{b}} [F(j, \mathrm{c})]$ will be denoted by $\int_{1\ \mathrm{a}}^{\mathrm{b}} F$ and will be called the *1st-place-area* under F from a to b. This function unites, as it were, all areas from a to b at various latitudes, just as the 1-place-function $\frac{F(b, j) - F(a, j)}{\mathrm{b} - \mathrm{a}}$ assuming the value $\frac{F(\mathrm{b}, \mathrm{c}) - F(\mathrm{a}, \mathrm{c})}{\mathrm{b} - \mathrm{a}}$ unites all slopes between a and b at various latitudes.

The 2-place-function assuming for any (b, c) the value $\int_{\mathrm{a}}^{\mathrm{b}} [F(j, \mathrm{c})]$ will be denoted by $\int_{1\ \mathrm{a}} F$ and will be called the *1st-place a-integral* of F. This function unites the a-integrals under all WE profiles of F at the various latitudes, just as the 2-place-function $\frac{F - F(A, J)}{I - I(A, J)}$ assuming for (b, c) the value $\frac{F(\mathrm{b}, \mathrm{c}) - F(\mathrm{a}, \mathrm{c})}{\mathrm{b} - \mathrm{a}}$ unites the chord functions at a of the various WE profiles of F.

Since the integral from a to b is defined only for 1-place-functions, $\int_{\mathrm{a}}^{\mathrm{b}} F$ is meaningless just as is $\mathbf{D} F$. Hence the brackets in $\int_{\mathrm{a}}^{\mathrm{b}} [F(j, \mathrm{c})]$ can be dispensed with, just as they are superfluous in $\mathbf{D}[F(j, \mathrm{c})]$. In the sequel that number will be denoted by $\int_{\mathrm{a}}^{\mathrm{b}} F(j, \mathrm{c})$.

Example 1. $\int_a^b I(j,c) = \int_a^b j = \frac{1}{2}(b^2 - a^2)$. Thus ${}_1\int_a^b I$ is the 1-place function assuming the value $\frac{1}{2}(b^2 - a^2)$ for any c, that is the constant function of value $\frac{1}{2}(b^2 - a^2)$. In a formula, ${}_1\int_a^b I = \frac{1}{2}(b^2 - a^2)$.

Similarly, ${}_1\int_a^b I$, being the 2-place-function that assumes the value $\frac{1}{2}(b^2 - a^2)$ for any (b, c), is the function $\frac{1}{2}(I^2 - A^2)$, if A denotes the constant 2-place-function of value a.

Example 2. $\int_a^b J(j,c) = \int_a^b c = c\cdot(b-a)$; ${}_1\int_a^b J = (b-a)\cdot j$;

${}_1\int_a J = (I - A)\cdot J$.

Example 3. For $P = I\cdot J$, one finds

$\int_a^b P(j,c) = \int_a^b cj = \frac{1}{2}c\cdot(b^2 - a^2)$; ${}_1\int_a^b P = \frac{1}{2}(b^2 - a^2)\cdot j$;

${}_1\int_a P = \frac{1}{2}(I^2 - A^2).\ J$.

Analogous concepts can be defined with regard to SN profiles at various longitudes:

${}_2\int_c^d F$ is the 1-place-function assuming for any a the value $\int_c^d F(a,j)$;

${}_2\int_c F$ is the 2-place-function assuming that value for any (d, a).

Example 4. $\int_c^d I(a,j) = \int_c^d a = a\cdot(d-c)$. Hence

${}_2\int_c^d I = (d-c)\cdot j$ and ${}_2\int_c I = (I-C)\cdot J$.

In all preceding definitions and examples, the letters a, b, c, d are numerical variables that may be replaced, without any change of the meaning, by other distinct letters. For instance, one might interchange a and b, or (a, b) and (c, d).

Just as $\mathbf{D}_1 F$ and $\mathbf{D}_2 F$ are called partial derivatives, $\int_{1\ p}^{q} F$ and $\int_{2\ p}^{q} F$ will be referred to as *partial integrals* from p to q of F, and $\int_{1\ p} F$ and $\int_{2\ p} F$ as *partial* p-*integrals* of F, more specifically, as *1st place* and *2nd place* partial integrals, respectively.

If the constant 2-place function of value m is denoted by M, the last formula in Example 4 reads $\int_{2\ m} I = (I - M) \cdot J$. Hence, according to Example 2, $\int_{2\ m} I = \int_{1\ m} J$. This is a special case of a general law relating the two partial integrals of a 2-place-function F, analogous to the law $\mathbf{D}_2 F = \mathbf{D}_1 [F(J, I)]$ relating its partial derivatives. (Here and in what follows, the brackets are, of course, indispensable.)

Connection between the Partial Integrals.

$$\int_{2\ m} F = \int_{1\ m} [F(J,I)] \quad \text{and} \quad \int_{1\ m} F = \int_{2\ m} [F(J,I)].$$

Indeed, $\int_{1\ m} [F(J,I)]$ is the function assuming for (p, q) the value

$\int_{m}^{p} [F(J,I)(j,q)] = \int_{m}^{p} F(q,j)$, which is also the value for (p, q) of $\int_{2\ m} F$.

Example 5. The Laplace Transform of a 1-place function f (see p. 271) is defined as the function $\mathbf{L} f$ assuming for a the value $\int_{0}^{\infty} \frac{f}{\exp(\mathrm{a}j)}$. Thus

$$\mathbf{L} f = \int_{1\ 0}^{\infty} \frac{fJ}{\exp P} = \int_{2\ 0}^{\infty} \frac{fI}{\exp P}.$$

Another important law extends to 2-place-functions the formula $\int_{a}^{b} (cf) = c \int_{a}^{b} f$, which is a consequence of the Constant Factor

Rule for Antiderivatives. In a 1st-place-partial integral, one can factor out a function gJ; in a 2nd-place integral, a function gI.

(16) $$\,_2\!\int_m^p (gI\cdot F) = g\cdot \,_2\!\int_m^p F \quad\text{and}\quad \,_1\!\int_m^p (gJ\cdot F) = g\cdot \,_1\!\int_m^p F .$$

Indeed, $\,_2\!\int_m^p (gI\cdot F)$ is the function assuming for any q the value

$$\int_m^p [(gI\cdot F)(q,j)] = \int_m^p [gq\cdot F(q,j)] = gq\cdot \int_m^p F(q,j) .$$

Thus, for any q, the value of $\,_2\!\int_m^p (gI\cdot F)$ is gq times that of $\,_2\!\int_m^p F$.

Reading (16) from right to left one sees that a 1-place-factor g may be transferred into the integrand of a 2nd-place partial integral in the form gI, and of a 1st-place integral in the form gJ.

Example 6. Integration of a 1-place-function f by the substitution of aj yields

$$\int_0^\infty f = \int_0^\infty [\mathrm{a}f(\mathrm{a}j)] = \mathrm{a}\int_0^\infty f(\mathrm{a}j) \quad\text{for any a.}$$

It follows that the 1-place-function $\,_2\!\int_0^\infty (I\cdot fP)$ which, according to (16) equals $j\cdot \,_2\!\int_0^\infty (fP)$, is constant and has the value $\int_0^\infty f$. For instance, let p denote the number $\int_0^\infty [exp(-j^2)]$ (which is of fundamental importance for probability theory). If p is the constant 1-place-function of the value p, then what was said implies

$$\,_2\!\int_0^\infty [I\cdot exp(-P^2)] = j\cdot \,_2\!\int_0^\infty [exp(-P^2)] = p .$$

Of course, also $$\,_1\!\int_0^\infty [J\cdot exp(-P^2)] = j\cdot \,_1\!\int_0^\infty [exp(-P^2)] = p .$$

Partial integrals will now be utilized in determining $\int_{(a,\,c)}^{(b,\,d)} F$, the volume under the portion of the surface F that lies above the rectangle with the corners (a, c) and (b, d). According to page 293, this volume is close to that under the nth regular step surface $R_n^{(F)}$ adapted to F. This step surface consists of n^2 rectangles each of which has its center point in common with the surface F. If one sets $s = b - a$ and $t = d - c$, then the centers of these rectangles are at

$$\left(a + \frac{2i-1}{2n}\, s,\ c + \frac{2k-1}{2n}\, t\right),\quad \text{briefly,}\quad (s_i, t_k) \text{ for } i, k = 1, 2, \ldots, n.$$

The WE profile of $R_n^{(F)}$ at the latitude t_k is the nth regular step line adapted to the WE profile $F(j, t_k)$ of F at that latitude. Clearly, the area $\int_a^b R_n^{(F)}$ under that step-profile, multiplied by t/n equals the volume of the layer under $R_n^{(F)}$ between the latitudes $y_k - t/2n$ and $y_k + t/2n$. The sum of the volumes of the n layers so obtained equals the volume under $R_n^{(F)}$. Thus

$$\left[\int_a^b R_n^{(F)}(j, y_1) + \int_a^b R_n^{(F)}(j, y_2) + \ldots + \int_a^b R_n^{(F)}(j, y_n)\right] \cdot \frac{t}{n} = \int_{(a,c)}^{(b,d)} R_n^{(F)}.$$

The area under the profile of $R_n^{(F)}$ is close to that under the profile of F at the same latitude. Hence

$$\left[\int_a^b F(j, y_1) + \int_a^b F(j, y_2) + \ldots + \int_a^b F(j, y_n)\right] \cdot \frac{t}{n} \sim \int_{(a,c)}^{(b,d)} R_n^{(F)}.$$

Since $\int_a^b F(j, y_k)$ is the value of the 1-place-function ${}_1\!\int_a^b F$ for y_k, the sum on the left side of the approximate equality is the product of t and the average of the values that the 1-place-function ${}_1\!\int_a^b F$ assumes at $y_1, y_2, \ldots, y_n$. Hence, for large n, the expression on the left side is close to $\int_c^d {}_1\!\int_a^b F$. It follows that

$$\int_c^d {}_1\!\int_a^b F \sim \int_{(a,c)}^{(b,d)} R_n^{(F)}.$$

Studying SN profiles, one obtains in a similar way

$$\left[\int_c^d F(x_1, j) + \int_c^d F(x_2, j) + \ldots + \int_c^d F(x_n, j)\right] \cdot \frac{s}{n} \sim \int_{(a,c)}^{(b,d)} R_n^{(F)}.$$

In fact, one can show that if n is sufficiently large, then $\int_{(a,c)}^{(b,d)} R_n^{(F)}$ is as close as he pleases to

$$\int_{(a,c)}^{(b,d)} F \quad \text{as well as to} \quad \int_c^d {}_1\!\int_a^b F \quad \text{and} \quad \int_a^b {}_2\!\int_c^d F,$$

which implies the equality of the last three numbers:

$$\int_{(a,c)}^{(b,d)} F = \int_a^b {}_2\!\int_c^d F = \int_c^d {}_1\!\int_a^b F.$$

In this way, the volume problem has been reduced to integrations. The volume under F can be found by two successive integrations (the first of which is a partial integration) and this procedure can be performed in one of two ways.

Example 7. For $\int_{(a,c)}^{(b,d)} I$, one finds

$$\int_c^d {}_1\!\int_a^b I = \int_c^d [\tfrac{1}{2}(b^2-a^2)] = [\tfrac{1}{2}(b^2-a^2)j]_c^d = \tfrac{1}{2}(b^2-a^2)\cdot(d-c);$$

$$\int_a^b {}_2\!\int_c^d I = \int_a^b (d-c)\,j = [\tfrac{1}{2}(d-c]\,j^2]_a^b = \tfrac{1}{2}(d-c)\cdot(b^2-a^2).$$

Example 8. Using (16) one can compute the important number p mentioned in Example 6. In the last formula in Example 6, multiply both sides by the 1-place-function $2\,exp(-j^2)$:

$$2\,p\,exp(-j^2) = 2\,j\cdot exp(-j^2)\cdot {}_1\!\int_0^\infty [exp(-P^2)].$$

Applying (16), one can write the right side in the form

$${}_1\!\int_0^\infty [2\,J\cdot exp(-J^2)\cdot exp(-P^2)], \quad \text{which equals}$$

$${}_1\!\int_0^\infty [2\,J\cdot exp(-(1+I^2)\cdot J^2)].$$

Hence

$$2p\,exp(-j^2) = {}_1\!\int_0^\infty [2\,J\cdot exp(-1+I^2)\cdot J^2)].$$

Equating the integrals from 0 to ∞ of these two 1-place-functions, one finds

$$2p^2 = \int_0^\infty {}_1\!\int_0^\infty [2\,J\cdot exp(-(1+I^2)\cdot J^2)].$$

Applying the Exponential Rule on p. 255 to the right side, one obtains

$$2p^2 = \int_0^\infty {}_1\!\int_0^\infty [2J\cdot exp(-(1+I^2)\cdot J^2)]$$

$$= \int_0^\infty \frac{1}{1+j^2} = [arctan]_0^\infty = \pi/2$$

and hence

$$p = \int_0^\infty [exp(-j^2)] = \sqrt{\pi/2}\,.$$

(In this simplicity, the computation is due to Sandham, 1946).

The a-integral $\int_a f$ of a 1-place function f is the function assuming for any b the value $\int_a^b f$. Its graph is the a-area curve of the curve f.

Analogously, one can introduce the (a, c)-integral $\int_{(a,c)} F$ of a 2-place function F as the 2-place-function assuming for (b,d) the value $\int_{(a,c)}^{(b,d)} F$.

This 2-place-function can be represented by what might be called the (a,c)-*volume surface* of F – a surface whose altitude at any point (c,d) equals the volume under the portion of F above the rectangle from (a,c) to (b,d). For instance, the results of Example 7 can be expressed in the formula

$$\int_{(a,c)} I = \tfrac{1}{2}(I^2 - A^2)\cdot(J - C). \quad \text{Similarly,}$$

$$\int_{(a,c)} P = \tfrac{1}{4}(I^2 - A^2)\cdot(J^2 - C^2).$$

In conclusion, it may be mentioned that also the reciprocity laws have analogues for partial derivatives and partial integrals. For instance, it is easy to show that, in analogy to the First Reciprocity Law,

$$\mathsf{D}_1\left({}_1\int_a F\right) = F,\quad \mathsf{D}_2\left({}_2\int_c F\right) = F,\quad \text{and}\quad \mathsf{D}_{12}\left(\int_{(a,\,c)} F\right) = F.$$

If $\mathsf{D}_{12}^{-1} F$ is any (1,2)-antiderivative of F, that is, any function G such that $\mathsf{D}_{12} G = F$, then

$$\int_{(a,\,c)}^{(b,\,d)} F = \left[\mathsf{D}_{12}^{-1} F\right]_{(a,\,c)}^{(b,\,d)}$$

The symbol on the right side denotes the rise of $\mathsf{D}_{12}^{-1} F$ from (a,c) to (b,d). The rise of a 2-place function G from (a,c) to (b,d) is defined as

$$[G]_{(a,\,c)}^{(b,\,d)} = G(a,c) - G(b,c) - G(a,d) + G(b,d).$$

EXERCISES

40. Determine ${}_1\int_a^b (I \cdot J^2)$, ${}_1\int_m (I \cdot J^2)$, ${}_2\int_m (I \cdot J^2)$, $\int_{(a,\,c)}^{(b,\,c)} (I \cdot J^2)$, and $\int_{(a,\,c)} (I \cdot J^2)$. Then compute $\mathsf{D}_1\left[{}_1\int_m (I \cdot J^2)\right]$ and $\mathsf{D}_{12}\left[\int_{(ac)} (I \cdot J^2)\right]$.

41. For any two numbers p and q, determine the (a,c)-volume surfaces of $pI + qJ$ and $pI^2 + qJ^2$.

42. Find $\int_{(0,\,0)} \sqrt{4 - I^2 - J^2}$.

43. By a modification of the procedure applied to rectangular domains, find the volume under the surface F above

(a) the triangle with the corners (a,c), (a,d), and (b,c);

(b) the part of the plane bounded by parts of the straight lines $\{x = a\}$ and $\{y = b\}$ and the curves $\{y = g(x)\}$ and $\{y = h(x)\}$.

Check the result *(a)* for the pyramid under the plane $1 - \frac{1}{b} I - \frac{1}{d} J$ above the triangle with the corners (0,0), (b,0), (0,d).

Then modify the procedure to obtain the volume under the hemisphere $\sqrt{r^2 - I^2 - J^2}$ above the circle of radius r about (0,0).

44. Determine the following antiderivatives of the function $I \cdot J^2$: D_1^{-1}, D_2^{-1}, D_{11}^{-1}, D_{12}^{-1}, D_{21}^{-1}, D_{22}^{-1}.

45. Prove that $\mathsf{D}_1 G = \mathsf{D}_1 H$ implies $G - H = fJ$ for some 1-place function f.

8. THE MEAN VALUE THEOREM AND THE TAYLOR EXPANSION

If the surface F has a tangential plane at (a, b), then

(17) $F \sim F(a,b) + \mathbf{D}_1 F(a,b) \cdot (I - a) + \mathbf{D}_2 F(a,b) \cdot (J - b)$ near (a, b);

that is to say, F can be approximated by the tangential plane, or

(17′) $F(c,d) \sim F(a,b) + \mathbf{D}_1 F(a,b) \cdot (c-a) + \mathbf{D}_2 F(a,b) \cdot (d-b)$ for $(c,d) \sim (a,b)$.

These vague statements can be made precise by studying F above the segment joining (a, b) and (c, d), that is, by considering the auxiliary one-place function

$$f = F(a + (c-a)\,j,\ b + (d-b)\,j),$$

which assumes the value $f\,t = F(a + (c-a)t,\ b + (d-b)t)$ for t. Thus

$$f\,0 = F(a, b) \text{ and } f\,1 = F(c, d),$$

while, for $0 < t < 1$, $f\,t$ equals the value of F at a point between (a, b) and (c, d) (i.e., on the segment joining these two points). According to the (2, 1)-Substitution Rule and Exercise 33, if $\mathbf{D}_1 F$ and $\mathbf{D}_2 F$ are continuous, $\mathbf{D} f$ exists between 0 and 1, and

$$\mathbf{D} f = \mathbf{D}_1 F(a+(c-a)\,j,\ b+(d-b)\,j) \cdot (c-a) + \mathbf{D}_2 F(a+(c-a)j,\ b+(d-b)\,j) \cdot (d-b).$$

Applying the Mean Value Theorem to f and the interval from 0 to 1 one obtains

$$f\,1 = f\,0 + \mathbf{D} f\,t^* \cdot (1-0) = f\,0 + \mathbf{D} f\,t^* \text{ for some } t^* \text{ between 0 and 1.}$$

Setting $x^* = a + (c-a)\,t^*$ and $y^* = b + (d-b)\,t^*$ one finds the following

INDETERMINATE EQUALITY.

$F(c,d) = F(a,b) + \mathbf{D}_1 F(x^*, y^*) \cdot (c-a) + \mathbf{D}_2 F(x^*, y^*) \cdot (d-b)$
for some point (x^*, y^*) *between* (a, b) *and* (c, d).

Since here both partial derivatives of F are evaluated at the same point (x^*, y^*) between (a, b) and (c, d), the preceding equality is often referred to as the *Mean Value Theorem* for 2-place functions. From it, as in the case of one-place functions, one deduces the

DETERMINATE INEQUALITY.

$$F(a, b) + \mathit{Min}\,[\mathbf{D}_1 F \cdot (c-a)] + \mathit{Min}\,[\mathbf{D}_2 F \cdot (d-b)] \leq F(c, d)$$
$$\leq F(a, b) + \mathit{Max}\,[\mathbf{D}_1 F \cdot (c-a)] + \mathit{Max}\,[\mathbf{D}_2 F \cdot (d-b)].$$

Here, the maxima and minima (or, in amended inequalities, upper and lower bounds) refer to all points between (a, b) and (c, d). For instance, $\mathit{Max}\,[\mathbf{D}_1 F \cdot (c-a)]$ is the largest of the numbers $\mathbf{D}_1 F(x^*, y^*) \cdot (c-a)$ for all points (x^*, y^*) between (a, b) and (c, d). Hence

$$Max\,[\mathbf{D}_1 F \cdot (c-a)] = \begin{array}{l} Max\ \mathbf{D}_1 F \cdot (c-a) \text{ if } c \geq a \\ Min\ \mathbf{D}_1 F \cdot (c-a) \text{ if } c \leq a \end{array}.$$

The reader should set up the expressions for the upper and lower bounds distinguishing the four cases

$$c \geq a \text{ and } d \geq b;\ c \leq a \text{ and } d \geq b;\ c \leq a \text{ and } d \leq b;\ c \geq a \text{ and } d \leq b,$$

which correspond to the quadrants with regard to (a, b) in which (c, d) may lie.

If $\mathbf{D}^2 f$ exists between 0 and 1, then the preceding results can be improved by applying to f the Taylor Expansion with quadratic remainder:

$$f1 = f0 + \mathbf{D}f0 \cdot (1-0) + \frac{1}{2}\mathbf{D}^2 f t^{**} \cdot (1-0)^2 = f0 + \mathbf{D}f0 + \frac{1}{2}\mathbf{D}^2 f\, t^{**}$$

for some t^{**} between 0 and 1.

$\mathbf{D}^2 f$ can be obtained from the formula for $\mathbf{D}^2 F\,(g, h)$ on p. 315 if g is replaced by $a + (c-a)\,j$ and h by $b + (d-b)\,j$. It can be shown that $\mathbf{D}^2 f$ exists if F has continuous second order partial derivatives. If the point between (a, b) and (c, d) that corresponds to t^{**} is denoted by (x^{**}, y^{**}), then the Taylor Expansion for f yields the following

TAYLOR EXPANSION WITH QUADRATIC REMAINDER FOR F.

$$F(c,d) = F(a,b) + \mathbf{D}_1 F(a,b) \cdot (c-a) + \mathbf{D}_2 F(a,b) \cdot (d-b) +$$
$$\frac{1}{2}[\mathbf{D}_{11} F(x^{**}, y^{**}) \cdot (c-a)^2 + 2\,\mathbf{D}_{12} F(x^{**}, y^{**}) \cdot (c-a) \cdot (d-b) + \mathbf{D}_{22} F(x^{**}, y^{**}) \cdot (d-b)^2]$$

Again this refined indeterminate equality may be transformed into determinate inequalities. The latter permit the approximate computation of F near a place where the values of F, $\mathbf{D}_1 F$, and $\mathbf{D}_2 F$ are known, provided that upper and lower bounds for the values of the second order derivatives of F near this place are known.

Suppose, for instance, that

$$F(2, 3) = 4,\quad \mathbf{D}_1 F(2, 3) = 1/2,\quad \mathbf{D}_2 F(2, 3) = 3$$

and $.25 \leq \mathbf{D}_{ik} F \leq 1.5$ (i, k = 1, 2) within a circle of radius .5 about (2,3).

Even if this is all one knows about F, he can compute good approximations to the values of F near (2, 3); e.g., one finds

$$F(2.1, 3.2) \leq 4 + \frac{1}{2} \cdot (.1) + 3 \cdot (.2) + \frac{1}{2} \cdot 1.5 \cdot (.1^2 + 2 \cdot (.1) \cdot (.2) + .2^2) = 4.7175.$$

Similarly, $F(2.1, 3.2) \geq 4.66425$. Thus one has confined $F(2.1, 3.2)$ to an interval of a length < .05. The reader can easily confine $F(2.01, 3.02)$ to a much shorter interval. He then should find bounds for

$$F(1.9, 3.2),\ F(1.9, 2.8),\ F(2.1, 2.8)\,.$$

On the other hand, no bounds whatever can be derived from the available information for F (2.4, 3.4), since the point (2.4, 3.4) lies outside the circle of radius .5 about (2, 3).

If $\mathbf{D}_1 F$ and $\mathbf{D}_2 F$ are continuous at (a, b), then, for (c, d) close to (a, b), everywhere between these two points the values of $\mathbf{D}_1 F$ and $\mathbf{D}_2 F$ differ little from $\mathbf{D}_1 F$ (a, b) and $\mathbf{D}_2 F$ (a, b). Hence the true value of F (c, d), as given by the Indeterminate Equality, differs from the approximate value on the right side of (17′) by a number that is small *in comparison with* c – a *and* d – b. In other words, if $\mathbf{D}_1 F$ and $\mathbf{D}_2 F$ are continuous, then (17′) is a good approximation provided (c, d) is close to (a, b).

In (17′), one frequently replaces the numerical variables a and b by x and y, and c – a and d – b by Δx and Δy. (The letter Δ in Δx is not a factor. It is a symbolic reminder of the fact that one is dealing with a difference in which the subtrahend is x.) After these replacements, (17′) reads

(17″) $F(x + \Delta x, y + \Delta y) - F(x, y) \sim \mathbf{D}_1 F(x, y) \cdot \Delta x + \mathbf{D}_2 F(x, y) \cdot \Delta y.$

The difference on the left side of (17″) is often denoted by $\Delta F(x, y)$ although, by itself, this symbol is incomplete. Even in the context of the formula

$$\Delta F(x, y) \sim \mathbf{D}_1 F(x, y) \cdot \Delta x + \mathbf{D}_2 F(x, y) \cdot \Delta y,$$

$\Delta F(x, y)$ on the left side is comprehensible only with the tacit understanding that the minuend of this difference is the value of F at $(x + \Delta x, y + \Delta y)$ for whatever Δx and Δy are on the right side. If $\mathbf{D}_1 F$ and $\mathbf{D}_2 F$ are continuous, then (17″) is a good approximation provided Δx and Δy are small.

This last fact is even nowadays frequently expressed by formulas such as

(17*) $$dF(x, y) = \mathbf{D}_1 F(x, y)\, dx + \mathbf{D}_2 F(x, y)\, dy.$$

with traditionally italicized *x* and *y*.

Really, (17*) expresses the 18th century idea that, if Δx and Δy are "infinitely small" (which was indicated by calling them dx and dy), then (17″) is a *precise* equality. The ("infinitely small") difference ΔF (x, y) corresponding to *dx* and *dy* was denoted by $dF(x, y)$ and called *"the differential of F (x, y)."* It is important for the reader to realize that, no matter how he interprets (17*), this formula yields no conclusions concerning finite numbers that cannot be inferred from (17″). In fact, if (17*) is to yield *any* inferences concerning finite numbers, (17*) must be considered as rendering (17″); in other words, (17*) is (17″) or (17′) in a symbolic disguise. True improvements of (17′) are the refined equalities and inequalities.

EXERCISES

46. For $F = I^2 \cdot J^3$ determine upper and lower bounds for F (c, d) – F (2, 3) using the inequalities that result (a) from the Mean Value Theorem; (b) from the Taylor Expansion with quadratic remainder. Then compute all bounds for F (2.1, 3.2).

47. Assuming a) $F(3, 2) = 4$, $\mathbf{D}_1 F(3, 2) = -2$, $\mathbf{D}_2 F(3, 2) = 3$; and b) $3 \le \mathbf{D}_{11} F \le 4$; $0 \le \mathbf{D}_{12} F \le 2$; $-3 \le \mathbf{D}_{22} F \le -2$ in a circle of radius .4 about (3, 2), what can one say about

$F(3.2, 2.1)$, $F(3.2, 1.8)$, $F(2.9, 3.01)$, $F(2.99, 1.98)$, $F(3, 2.1)$?

48. If height and radius of a circular cylinder (a circular cone) are increased by Δh and Δr, what are the approximate increases of the volume, the lateral surface, and the entire surface according to (17″)? Then answer these questions by the inequalities resulting (a) from the Mean Value Theorem; (b) from Taylor's Expansion with quadratic remainder.

49. Show that for every continuous function F

$\Delta F(x, y) \sim 3\Delta x + 2\Delta y$ (even if F has no partial derivatives), and

$\Delta F(x, y) \sim 2\mathbf{D}_1 F(x, y) \cdot \Delta x - 3\mathbf{D}_2 F(x, y) \cdot \Delta y$ (if partial derivatives exist).

Then show by examples that the inaccuracies in the preceding approximate equalities may be large *in comparison with* Δx *and* Δy. Show that, at the vertex of a quarter of a cone, even the inaccuracy in (17″) may be appreciable in comparison with Δx and Δy, no matter how small the latter are. (Only under the assumption of *continuous* $\mathbf{D}_1 F$ and $\mathbf{D}_2 F$ has (17″) been proved to be a good approximation for small Δx and Δy.)

★50. Assuming that even the second-order partial derivatives of F are continuous and that in a circle about (x, y) the absolute values of none of these second-order derivatives exceed the number m, prove that the inaccuracy in (17″) does not exceed $m[(\Delta x)^2 + (\Delta y)^2]$. Hint. Use the Taylor Expansion with quadratic remainder and apply to the middle term the so called

Schwarz Inequality: $2\,\Delta x \cdot \Delta y \le (\Delta x)^2 + (\Delta y)^2$,

which follows from the fact that $(\Delta x - \Delta y)^2 \ge 0$ (since the square of a real number is non-negative).

51. Interpret the old-fashioned formula $df(x) = \mathbf{D}f(x)\,dx$. For what function f is $df(x) = \frac{1}{x}\,dx$ (usually written $df(x) = \frac{dx}{x}$)? Apply the result to Daniel Bernoulli's hypothesis that a man owning a fortune of x dollars attributes to an increase of Δx dollars a value that is inversely proportional to x and, for small Δx, roughly proportional to Δx.

9. PARTIAL RATES OF CHANGE

For an ideal gas, if proper units are chosen, the volume is connected with pressure and temperature by the quotient function $Q = J/I$ (see Exercise 12); that is to say, $v = Q(p, t) = t/p$. For two instantaneous gas samples, γ_0 and γ_1, of the same temperature, consider

$$\frac{v\,\gamma_1 - v\,\gamma_0}{p\,\gamma_1 - p\,\gamma_0}.$$

If $p\,\gamma_1$ is close to $p\,\gamma_0$, then this quotient is close to the value for γ_0 of a consistent class of quantities, called *a partial rate of change of v*, more specifically, *the partial rate of change of v with regard to p keeping t constant*. This c.c.q. is denoted by $\left(\frac{\partial v}{\partial p}\right)_t$ and is connected with p and t by the function $\mathbf{D}_1\, Q = -J/I^2$; that is to say,

$$v = Q\,(p, t) \text{ implies } \left(\frac{\partial v}{\partial p}\right)_t = \mathbf{D}_1\, Q\,(p, t) = -t/p^2.$$

Physical chemists refer to $-\frac{1}{v}\left(\frac{\partial v}{\partial p}\right)_t$ as isothermic compression; to $\frac{1}{v}\left(\frac{\partial v}{\partial t}\right)_p$, as the isobaric expansion. Here, $\left(\frac{\partial v}{\partial t}\right)_p$ is a c.c.q. connected with p and t by the function $\mathbf{D}_2\, Q = 1/I$, wherefore $\left(\frac{\partial v}{\partial t}\right)_p = 1/p$.

More generally, one has the following

SCHEME FOR THE APPLICATION OF PARTIAL DERIVATIVES TO PARTIAL RATES OF CHANGE. *If* $w = F\,(u, v)$, *then* $\left(\frac{\partial w}{\partial u}\right)_v = \mathbf{D}_1\, F\,(u, v)$ *and* $\left(\frac{\partial w}{\partial v}\right)_u = \mathbf{D}_2\, F\,(u, v)$, *for any three consistent classes of quantities*, u, v, w.

In the following examples extensive use is made of the formulas

(18) $\qquad \mathbf{D}_1\, I = \mathbf{D}_2\, J = 1 \text{ and } \mathbf{D}_2\, I = \mathbf{D}_1\, J = O.$

All 2-place functions are assumed to have continuous partial derivatives, so that the three (i, k)-Substitution Rules are applicable.

Example 1. Since $w = J\,(u, w)$ for any u and w, it follows that

$$\left(\frac{\partial w}{\partial w}\right)_u = \mathbf{D}_2\, J\,(u, w) \text{ and } \left(\frac{\partial w}{\partial u}\right)_w = \mathbf{D}_1\, J\,(u, w)\,.$$

Hence

(19) $\qquad \left(\frac{\partial w}{\partial w}\right)_u = 1 \text{ and } \left(\frac{\partial w}{\partial u}\right)_w = O$ for any u and w.

Example 2. Suppose that u, v, and w are three c.c.q.'s such that

(20) w is a function of u and v; and v, a function of u and w.

In view of $u = I(u, w)$, from $w = F(u, v)$ and $v = G(u, w)$ it follows that $w = F(I(u, w), G(u, w))$; that is to say, w is connected with u and w by the function $F(I, G)$. On the other hand, J connects w with u and w. Hence $F(I, G) = J$, wherefore

$$\mathbf{D}_1 [F(I, G)] = \mathbf{D}_1 J = O \text{ and } \mathbf{D}_2 [F(I, G)] = \mathbf{D}_2 J = 1.$$

Applying to the left sides the (2, 2)-Substitution Rule one finds

$$\mathbf{D}_1 F(I,G) \cdot \mathbf{D}_1 I + \mathbf{D}_2 F(I,G) \cdot \mathbf{D}_1 G = O \text{ and } \mathbf{D}_1 F(I,G) \cdot \mathbf{D}_2 I + \mathbf{D}_2 F(I,G) \cdot \mathbf{D}_2 G = 1;$$

thus, in view of (18),

(21a) $\mathbf{D}_1 F(I,G) + \mathbf{D}_2 F(I,G) \cdot \mathbf{D}_1 G = O$ and (21b) $\mathbf{D}_2 F(I,G) \cdot \mathbf{D}_2 G = 1.$

Formulas (21) relate the partial derivatives of the functions which, according to assumption (20), connect u, v, and w. What can be concluded about the partial rates of change of these c.c.q.'s? According to the basic scheme,

$\left(\frac{\partial w}{\partial v}\right)_u = \mathbf{D}_2 F(u, v)$, which is equal to $\mathbf{D}_2 F(I(u,w), G(u, w))$.

Hence $\mathbf{D}_2 F(I,G)$ connects $\left(\frac{\partial w}{\partial v}\right)_u$ with u and w. Thus, if u and w are substituted into (21b), the first factor becomes $\left(\frac{\partial w}{\partial v}\right)_u$, while the second, according to the basic scheme, equals $\left(\frac{\partial v}{\partial w}\right)_u$. Hence, remembering assumption (20), one deduces from (21b):

(22) $\left(\frac{\partial w}{\partial v}\right)_u \cdot \left(\frac{\partial v}{\partial w}\right)_u = 1$ if w is a function of u and v, and v a function of u and w.

Interchanging u and v one obtains

(22′) $\left(\frac{\partial w}{\partial u}\right)_v \cdot \left(\frac{\partial u}{\partial w}\right)_v = 1$ if w is a function of v and u, and u a function of v and w.

Similarly, (21a) yields

(23) $\left(\frac{\partial w}{\partial u}\right)_v + \left(\frac{\partial w}{\partial v}\right)_u \cdot \left(\frac{\partial v}{\partial u}\right)_w = O$ for any u, v, w, satisfying (20).

Hence, except where $\left(\frac{\partial v}{\partial u}\right)_w$ or $\left(\frac{\partial w}{\partial v}\right)_u$ assume the value 0,

(23′) $\left(\frac{\partial w}{\partial v}\right)_v = -\left(\frac{\partial w}{\partial u}\right)_v \Big/ \left(\frac{\partial v}{\partial u}\right)_w$ and $\left(\frac{\partial v}{\partial u}\right)_w = -\left(\frac{\partial w}{\partial u}\right)_v \Big/ \left(\frac{\partial w}{\partial v}\right)_u$.

Multiplying (23) by $\left(\frac{\partial u}{\partial w}\right)_v$ and using (22′), which presupposes, of course, the combination of the assumptions made in (23) and (22′), one finds

$$1 + \left(\frac{\partial w}{\partial v}\right)_u \cdot \left(\frac{\partial v}{\partial u}\right)_w \cdot \left(\frac{\partial u}{\partial w}\right)_v = O \text{ or}$$

(24) $\left(\frac{\partial w}{\partial v}\right)_u \cdot \left(\frac{\partial v}{\partial u}\right)_w \cdot \left(\frac{\partial u}{\partial w}\right)_v = -1$ for any three c.c.q.'s each of which is a function of the other two.

Example 3. Suppose that $w = F\,(u, v)$ and that u and v are functions of some fourth c.c.q., say, $u = g\,(t)$ and $v = h\,(t)$ – briefly, $u = gt$ and $v = ht$. Then $w = F\,(g\,t, h\,t)$. Thus w is connected with t by the function $F\,(g, h)$, and $\mathbf{D}\,F\,(g, h)$ connects $\frac{\mathrm{d}\,w}{\mathrm{d}\,t}$ with t. Applying the (2, 1)-Substitution Rule one sees that $\frac{\mathrm{d}\,w}{\mathrm{d}\,t}$ is connected with t by the function

$\mathbf{D}_1\,F(g, h) \cdot \mathbf{D}\,g + \mathbf{D}_2\,F\,(g, h) \cdot \mathbf{D}\,h$. Since $\left(\frac{\partial w}{\partial u}\right)_v = \mathbf{D}_1 F(u,v) = \mathbf{D}_1 F(gt,ht)$, it follows that $\mathbf{D}_1\,F(g, h)$ connects $\left(\frac{\partial w}{\partial u}\right)_v$ with t. Hence

(25) $$\frac{\mathrm{d}\,w}{\mathrm{d}\,t} = \left(\frac{\partial w}{\partial u}\right)_v \cdot \frac{\mathrm{d}\,u}{\mathrm{d}\,t} + \left(\frac{\partial w}{\partial v}\right)_u \cdot \frac{\mathrm{d}\,v}{\mathrm{d}\,t}\,.$$

This relation between rates and partial rates of change of functionally connected variable quantities is much used in science. If for a gas $v = F\,(p, T)$, where T is the temperature, then one finds that the rates of change with regard to the time t are related as follows:

$$\frac{\mathrm{d}\,v}{\mathrm{d}\,t} = \left(\frac{\partial v}{\partial p}\right)_T \cdot \frac{\mathrm{d}\,p}{\mathrm{d}\,t} + \left(\frac{\partial v}{\partial T}\right)_p \cdot \frac{\mathrm{d}\,T}{\mathrm{d}\,t}; \text{if } F = Q = J/I, \text{then } \frac{\mathrm{d}\,v}{\mathrm{d}\,t} = -\frac{T}{p^2} \cdot \frac{\mathrm{d}\,p}{\mathrm{d}\,t} + \frac{1}{p} \cdot \frac{\mathrm{d}\,T}{\mathrm{d}\,t}.$$

Example 4. Suppose that $w = F\,(u, v)$ and $w = g\,u$. Then, clearly, $F = g\,I$ and, by (1, 2)-Substitution, $\mathbf{D}_1\,F = \mathbf{D}\,g\,I$ and $\mathbf{D}_2\,F = O$. Hence

$$\left(\frac{\partial w}{\partial u}\right)_v = \mathbf{D}_1\,F\,(u, v) = \mathbf{D}\,g\,I\,(u, v) = \mathbf{D}\,g\,u \text{ and} \left(\frac{\partial w}{\partial v}\right)_v = O.$$

As a corollary, one sees that

$w = F(u, v)$, $w = g\,u$ and $w = h\,v$ imply $F = g\,I = h\,J$.

Hence (see p. 303, Ex. 22) $g = h = c$ for some constant function, and $\left(\frac{\partial w}{\partial u}\right)_v = \left(\frac{\partial w}{\partial v}\right)_u = O.$

This corollary explains why one cannot obtain significant results about partial rates of change of 1-place functions. Suppose, for instance, that $t = j$, $u = j^2$, $v = j^3$, and $w = j^5$ in (25). Then w is connected with u and v by the product function P since $j^5 = j^2 \cdot j^3 = P(j^2, j^3)$. Yet $\left(\frac{\partial j^5}{\partial j^3}\right)_{j^2} = \left(\frac{\partial j^5}{\partial j^2}\right)_{j^3} = O$ since $j^5 = j^{5/3}\, j^3$, while j^5 is connected with j^2 by the function $j^{5/2}$ on P, and by $-j^{5/2}$ on N.

Example 5. Denote the values of p, v, and t for two instantaneous gas samples γ_0 and γ_1 by p_0, v_0, t_0 and p_1, v_1, t_1, respectively, and suppose that t is connected with p and v by a function F. Since $t_0 = F(p_0, v_0)$ and $t_1 = F(p_1, v_1)$, it follows from (17′) that

$$(26)\quad t_1 - t_0 \sim \mathbf{D}_1 F(p_0, v_0) \cdot (p_1 - p_0) + \mathbf{D}_2 F(p_0, v_0) \cdot (v_1 - v_0) \text{ or}$$
$$t_1 - t_0 \sim \left(\frac{\partial t}{\partial p}\right)_v (\gamma_0) \cdot (p_1 - p_0) + \left(\frac{\partial t}{\partial v}\right)_p (\gamma_0) \cdot (v_1 - v_0),$$

where γ_0 indicates that one has to consider the values of the partial rates of change for the gas sample γ_0. For an ideal gas, F is Boyle's product function P, for which $\mathbf{D}_1 P = J$ and $\mathbf{D}_2 P = I$, wherefore $\mathbf{D}_1 P(p_0, v_0) = v_0$ and $\mathbf{D}_2 P(p_0, v_0) = p_0$. Consequently,

$$t_1 - t_0 \sim v_0 \cdot (p_1 - p_0) + p_0 \cdot (v_1 - v_0).$$

The most important application of (26), however, is to cases where one has reason to assume that t is a function of p and v, and that the connecting function has continuous partial derivatives but where this function is *unknown*. In such cases, if for one gas sample γ_0 the values of p, v, t, $\left(\frac{\partial t}{\partial p}\right)_v$, and $\left(\frac{\partial t}{\partial v}\right)_p$ are experimentally determined, formula (26) permits to compute approximately the temperature of any gas sample γ_1 whose pressure and volume are known and differ little from those of γ_0.

This result is often expressed in the formula

$$(26^*)\quad dt = \left(\frac{\partial t}{\partial p}\right)_v dp + \left(\frac{\partial t}{\partial v}\right)_p dv; \text{ for an ideal gas, } dt = v\,dp + p\,dv.$$

This and similar 18th century relics are harmless as long as one keeps in mind

a) that (26*) is (26) in a symbolic disguise and nothing more;

b) that the validity of (26), and therefore of its travesty (26*), is contingent to the assumptions that t is a function of p and v, and that the connecting function has continuous partial derivatives.

What nonsense may result from overlooking these points can be seen in the case of the formula

(27*) $$dq = du + p\,dv,$$

which expresses an important special case of the celebrated First Law of thermodynamics. Here, p and v are gas pressure and volume, u is the internal energy of the gas, and q is the quantity of heat added to the gas (since a certain stage of the process). If one interprets (27*) as a symbolic disguise of a formula

(27) $$q_1 - q_0 \sim u_1 - u_0 + p_0 \cdot (v_1 - v_0),$$

then he arrives at the conclusion that, knowing u_0, v_0, q_0, and p_0, he can approximately compute q_1 for any stage of the process at which u_1 and v_1 are known and differ little from u_0 and v_0. But really, on the basis of this information, one cannot compute q_1 at all. Even one and the same process may pass through two stages for which both $u_1 - u_0$ and $v_1 - v_0$ are very small, and yet $q_1 - q_0$ is very large. In fact, $q_1 - q_0$ may be very large even though $u_1 - u_0$ and $v_1 - v_0$ are precisely 0; that is to say, there are processes in the course of which, after considerable addition of heat, the gas reassumes its initial energy and volume. The quantity of heat that one has to put into a gas in order to change (u_0, v_0) to (u_1, v_1) depends upon the nature of the process by means of which he achieves the change. Thus q is not a function of u and v, and hence (27*) cannot be interpreted as (27). To forestall the danger of such a misinterpretation, many physicists and chemists write

(27′) $$đq = du + p\,dv$$

instead of (27*), where the bar through the letter d on the left side (or in some texts the symbols $\dot{d}$ or δ) is to warn the reader against confusing (27′) with a formula such as (26*) or, as it is often put, to indicate that $đq$ in (27′) is "an incomplete differential" and not a "complete differential" as is dt in (26*). Other texts refer to $đq$ as an "inexact differential" and stress that it would be incorrect to treat it like an "exact differential," i.e., to replace (27′) by (27).

Since the most essential question concerning formulas such as (27′) (which are of paramount importance in thermodynamics and in mechanics) is what they *do* mean (and not what they do *not* mean), two *correct* interpretations may be mentioned. If the variable quantities q, u, p, and v are connected with the time t by functions having derivatives, then (27′) interrelates their time rates of change:

(27″) $$\frac{dq}{dt} = \frac{du}{dt} + p \cdot \frac{dv}{dt} \quad \text{or (in Newton's notation)} \quad \dot{q} = \dot{u} + p \cdot \dot{v}\,.$$

More generally, (27′) may be interpreted in terms of cumulations:

$$\int_{\gamma_0}^{\gamma_1} \mathrm{d}\,q = \int_{\gamma_0}^{\gamma_1} \mathrm{d}\,u + \int_{\gamma_0}^{\gamma_1} p\,\mathrm{d}\,v \text{ or } q_1 - q_0 = u_1 - u_0 + \int_{\gamma_0}^{\gamma_1} p\,\mathrm{d}\,v.$$

Example 6. u, v, and w may well be 2-place functions, i.e., c.c.q.'s whose domains consist of pairs of numbers. For instance, if $u = I$ and $v = J$, then, if w is any function F, it follows from $F = F(I, J)$ that $\left(\frac{\partial w}{\partial u}\right)_v = \mathbf{D}_1 F(I, J) = \mathbf{D}_1 F$; thus

(28) $$\left(\frac{\partial F}{\partial I}\right)_J = \mathbf{D}_1 F \text{ and, similarly, } \left(\frac{\partial F}{\partial J}\right)_I = \mathbf{D}_2 F.$$

Example 7. Suppose that u and v are two c.c.q.'s which, relative to some pairing of their domains, and for some 2-place function G, are connected by

(29) $$G(u, v) = O.$$

Let f be a function implicitly defined by $G(j, f) = O$. Then $v = f(u)$, briefly, $v = fu$. If f has a derivative, $\mathbf{D}f$ connects $\frac{\mathrm{d}\,v}{\mathrm{d}\,u}$ with u. According to (13) on p. 317,

(30) $$\frac{\mathrm{d}\,v}{\mathrm{d}\,u} = \mathbf{D}\,f\,u = -\frac{\mathbf{D}_1 G(u, fu)}{\mathbf{D}_2 G(u, fu)} = -\frac{\mathbf{D}_1 G(u, v)}{\mathbf{D}_2 G(u, v)}.$$

If w is a consistent class of quantities such that $w = G(u, v)$, then

$$\mathbf{D}_1 G(u, v) = \left(\frac{\partial w}{\partial u}\right)_v \text{ and } \mathbf{D}_2 G(u, v) = \left(\frac{\partial w}{\partial v}\right)_u,$$

wherefore (30) can be written in the form

(30′) $$\frac{\mathrm{d}\,v}{\mathrm{d}\,u} = -\left(\frac{\partial w}{\partial u}\right)_v \Big/ \left(\frac{\partial w}{\partial v}\right)_u \text{ or } \frac{\mathrm{d}v}{\mathrm{d}u} = -\left(\frac{\partial[G(u, v)]}{\partial u}\right)_v \Big/ \left(\frac{\partial[G(u, v)]}{\partial v}\right)_u.$$

The first formula (30′) and the second formula (23′) must be clearly distinguished, in spite of their similarity. They have been proved under quite different assumptions.

As an example of (29), consider pressure and volume during an isothermic process at the temperature t_0. For an ideal gas, they are related by $P(p, v) - t_0 = O$, where P is the product function and t_0 and O are the constant c.c.q.'s of the values t_0 and 0, defined for all gas samples. From $G = P - t_0$ it follows that $\mathbf{D}_1 G = \mathbf{D}_1 P = J$ and $\mathbf{D}_2 G = \mathbf{D}_2 P = I$. Thus (30) and (30′) read

(30″) $$\frac{\mathrm{d}\,v}{\mathrm{d}\,p} = -\frac{J(p, v)}{I(p, v)} = -\frac{v}{p} \text{ and } \frac{\mathrm{d}\,v}{\mathrm{d}\,p} = -\left(\frac{\partial[P(p, v)]}{\partial p}\right)_v \Big/ \left(\frac{\partial P(p, v)}{\partial v}\right)_p$$

If x and y are the Cartesian coordinates along a hyperbola $\{x \cdot y = c\}$, then (30) determines the slope $\frac{dy}{dx}$ along that curve in a similar way.

Utter nonsense would result if in (30) or (30′) the references to the fluent variables u and v were omitted or if one wrote $\left(\frac{\partial P}{\partial p}\right)_v$ in (30″). What is the rate of change of the product function with regard to the gas pressure keeping the volume constant? This makes as little sense as the words "the rate of change of the logarithm with regard to the temperature."

EXERCISES

52. According to van der Waals, for each gas there are three numbers a, b, and r such that

$$(p - \frac{a}{v^2}) \cdot (v - b) = r\,T. \quad \text{(If a = b = 0, the gas is ideal.)}$$

a) Find the rate of change of each of the variable quantities p, v, T with either of the other two keeping the third constant, and verify (22), (23), and (24).

b) Interrelate the time rates of change of p, v, and T.

c) For an isothermic process, say at the temperature t_0, determine $\frac{dp}{dv}$ by the method of the preceding section. Then express p as a 1-place function of v and determine the rate of change in this way. Compare the answers. ⋆Then do the same for $\frac{d^2 p}{dv^2}$.

53. For any number $c \geq 0$, set up the equation in Cartesian coordinates of the curve consisting of all points for which the product of their distances from the points (–1, 0) and (0, 1) equals c. For each of these so called Cassini curves determine the slope at any of its points. If c = 1, the curve is called *lemniscate*. Why has it no slope at (0,0)?

54. Assuming that s, t, u, v, w are any five c.c.q.'s such that w is a function of u and v, both of which are functions of s and t, find the rates of change of w with s keeping t constant, and with t keeping s constant.

55. From (27″) prove that, if u_0, v_0, p_0, q_0 refer to a gas at the instant t_0, and u_1, v_1, q_1 to that gas at the instant t_1, then

$$q_1 - q_0 \sim u_1 - u_0 + p_0 \cdot (v_1 - v_0) \text{ if } t_1 \sim t_0 .$$

Reconcile this result with what has been said about (27). To clarify this important question completely, consider the following four functions:

$$f = -\frac{1}{6}(2j-1)^3,\ g_1 = j^2 - j,\ g_2 = j^2 - j\ (= g_1) \text{ and } h = -2j.$$

Verify that $\mathbf{D} f = \mathbf{D} g_1 + h \cdot \mathbf{D} g_2$, and prove that

$$f\,x'' - f\,x' \sim g_1\,x'' - g_1\,x' + h\,x' \cdot (g_2\,x'' - g_2\,x') \text{ for } x'' \sim x'.$$

Yet one cannot claim that $f\,x - f\,x_0$ must be small if both $g_1\,x - g_1\,x_0$ and $g_2\,x - g_2\,x_0$ are small. For instance, set $x_0 = 0$ and $x = 1$.

10. REMARKS CONCERNING PARTIAL DERIVATIVES AND RATES OF CHANGE IN THE LITERATURE

In 1840, Cauchy, who referred to a 2-place function F as "the function $F(x, y)$," introduced the symbols $\mathbf{D}_x F(x, y)$ and $\mathbf{D}_y F(x, y)$ for the 1st and 2nd place partial derivatives, herein denoted by $\mathbf{D}_1 F$ and $\mathbf{D}_2 F$. Sometimes, $\mathbf{D}_x F(x, y)$ designates also the value of $\mathbf{D}_1 F$ for (x, y). Other symbols for partial derivatives of F used in the literature are

$$F_x \text{ and } F_y;\ F'_x \text{ and } F'_y;\ \text{and, occasionally, } D_1 F \text{ and } D_2 F.$$

The typographical distinction made in this book between 1-place and 2-place functions by reserving italic capitals for the latter is not universally maintained, so that $D_x f$ and f'_y sometimes designate partial derivatives of 2-place functions, while $F'(x)$ may denote the derivative of a 1-place function called F and referred to as "the function $F(x)$".

However, the symbols for partial derivatives most frequently used in the literature are those introduced by Jacobi (1841): $\dfrac{\partial F}{\partial x}$ and $\dfrac{\partial F}{\partial y}$.

Correspondingly, Jacobi denoted $\displaystyle {}_1\!\int_a^b F$ and $\displaystyle {}_2\!\int_a^b F$ by

$$\int_a^b F(x, y)\,\partial x \text{ and } \int_a^b F(x, y)\,\partial y.$$

In science, $\dfrac{\partial}{\partial}$ is used for the partial rate of change of one variable quantity with regard to another. Especially physical chemists indicate the v.q. (or, in more complicated cases, the v.q.'s) kept constant by subscripts, as in $\left(\dfrac{\partial t}{\partial p}\right)_v$. But one finds also $\dfrac{\partial t(p, v)}{\partial p}$ as a notation for this partial rate of change.

Frequently, a variable quantity such as t, which is a function of p and v, is confused with the connecting 2-place function. Especially rates of change of one physical v.q. with regard to another (which themselves are physical v.q.'s) and functions are sometimes equated.

Following Jacobi, one reserves the symbol $\frac{\mathbf{d}}{\mathbf{d}\,x} F(x, y)$ for cases where y is replaced by a 1-place function, say h, referred to as "the function $h(x)$," while x designates the identity function j; that is to say, the aforementioned so-called "*total* derivative of $F(x, y)$ with regard to x" designates $\mathbf{D}\,[F(j, h)]$ briefly, $\mathbf{D}\,F(j, h)$, while $\frac{\partial}{\partial x} F(x, y)$ stands for $\mathbf{D}_1\, F$.

Various ambiguities obscure the traditional treatment of implicit functions, which is based on a formula such as

$$G(x, y) = 0, \tag{31}$$

while the derivative of the implicit function is denoted by $\frac{\mathbf{d}\,y}{\mathbf{d}\,x}$. Here, the letters x and y are capable of four totally different interpretations.

(a) Some books denote the value of the implicitly defined function for x (traditionally italicized) by $y(x)$. In this case

$$G(x, y(x)) = 0 \text{ for any } x,$$

where x is a numerical variable, y a function variable. Instead we would denote the implicit function by f and would write

$$G(x, f(x)) = 0 \text{ for any x in Dom } f.$$

In this case, $\frac{\mathbf{d}\,y}{\mathbf{d}\,x}$ is $\mathbf{D}f$, its value for x is $\mathbf{D}f\,x$.

(b) Some books consider x as the identity function and y as a function variable. We express this point of view by $G(j, f) = O$. In this case, $\frac{dy}{dx}$ is $\frac{\mathbf{d}\,f}{\mathbf{d}\,j} = \mathbf{D}\,f$.

(c) Frequently, (31) is considered as a relation between Cartesian coordinates. We would express this idea by $\{G = O\}$ in pure analytic geometry, and by $\{G(x, y) = O\}$ in a physical plane. In the latter case, $\frac{\mathbf{d}\,y}{\mathbf{d}\,x}$ is the slope of the curve, and, if $y = fx$, then $\frac{\mathbf{d}\,y}{\mathbf{d}\,x} = \mathbf{D}\,f\,x$ and not $\frac{\mathbf{d}\,y}{\mathbf{d}\,x} = \mathbf{D}\,f$. According to (30) and (30′),

$$\frac{\mathbf{d}\,y}{\mathbf{d}\,x} = -\frac{\mathbf{D}_1\, G(x, y)}{\mathbf{D}_2\, G(x, y)} = -\left(\frac{\partial\,[G(x, y)]}{\partial\,x}\right)_y \Big/ \left(\frac{\partial\,[G(x, y)]}{\partial\,y}\right)_x .$$

(d) x and y in (31) are sometimes considered as fluent variables.

As such we have been using the letters $u, v, w, \ldots$, having reserved x and y for specific fluents, namely, Cartesian coordinates. But if x and y are fluent variables, then $\frac{dy}{dx}$ is determined by (30) and (30').

Again, if $y = fx$, then $\frac{dy}{dx} = \mathbf{D}fx$ and not $= \mathbf{D}f$.

Without a clarification of the meaning of x and y in (31) and in $\frac{dy}{dx}$ it is thus indeed hard to understand a treatment of implicit functions. Moreover, those who consider the (1, 1)-Substitution Rule in Leibniz' notation (p. 238)

$$\frac{dw}{du} = \frac{dw}{dv} \cdot \frac{dv}{du}$$

as evident, are bound to regard

$$\frac{dy}{dx} = \frac{\partial G}{\partial x} \Big/ \frac{\partial G}{\partial y}$$

as equally obvious. They are puzzled when reading

(32) $$\frac{dy}{dx} = -\frac{\partial G}{\partial x} \Big/ \frac{\partial G}{\partial y}$$

in traditional texts.

So is, for quite different reasons, the mature reader. Since he cannot associate any meaning with the rate of change of a 2-place function with regard to any fluent (e. g., of the product with regard to pressure, see p. 338), he is forced to conclude that, in (32), meaning has been sacrificed to brevity by omitting on the right side references to x and y in the numerators. Why then, he wonders, do those same texts never shed x in

$$\mathbf{D} \sin x = \cos x,$$

where brevity could be achieved without making the formula nonsensical?

WHAT ARE x AND y?

In: ALGEBRA	ANALYTIC GEOMETRY	CALCULUS
$\frac{x^2-1}{x-1} = x+1$ for $x \neq 1$; $x^2 - 9y^2 = (x+3y)(x-3y)$ for any x and y; $y^2 - 9x^2 = (y+3x)(y-3x)$ is equivalent.		$(\log x)' = 1/x$ for $x > 0$ $(\sin 2x)' = 2 \cos 2x$ If $F(x, y) = x^4 y^3$ for any x an then $\frac{\partial F}{\partial y}(x, y) = 3x^4y^2$ for any x and y.
	The class of all pairs (x, y) such that $2x + 3y = 5$ is called a straight line; it can also be defined as the class of all (y, x) such that $2y + 3x = 5$.	
Find x and y	Find the point (x, y) common to the lines which are the classes of all (x, y).	
such that $2x + 3y = 5$ and $x + y = 2$.		
		$x + 1$ is an extension of eith of the (non-identical) functic $\frac{x^2-1}{x-1}$ and $\frac{x^2-x-2}{x-2}$; the derivative of $x + 1$ is 1. $\frac{df}{dx}, \int_0^1 f\,dx$.
	$2x + 3y = 5$ and $2y + 3x = 5$ are different lines.	If a function of x equals a function of y, then the functions are equal and constant $\frac{\partial y}{\partial x} = 0$.
ALGEBRA OF FORMS $\frac{x^2-1}{x-1} = x + 1 = \frac{x^2-x-2}{x-2}$		
		$\int_0^1 \cos x\,dx = \int_0^1 \cos y\,dy$
	PHYSICAL PLANE The line $2x + 3y = 5$ does not pass through the origin.	$\frac{dy}{dx} = \frac{d f(x)}{dx}$ is the slope of the curve $y = f(x)$.
[Traditionally, x and y are printed in italics.]		$y = \sin x$ implies $\frac{dy}{dx} = \cos$ for any two variables x and The class of all functions y such that $\sin y = x$ is infinite. DIFFERENTIAL EQUATIO Find y such that $yy' = 1$.

d *y* are (and, in Cases II and IV, *x* is):

UMERICAL VARIABLES (p. 60) Herein, letters in roman type serve as numerical variables, as in

a) **INDICATIVELY** (p. 63)

$$x^2 - 9y^2 = (x + 3y)(x - 3y)$$

In CALCULUS, numerical variables may be dispensed with as in

$D\ log = j^{-1}$ (on P);

$D\ sin\ (2j) = 2\ cos\ (2j)$;

$D_2\ (I^4 \cdot J^3) = 3\ I^4 \cdot J^2$.

b) **CONJUNCTIVELY** (p. 63)

$$\{(x, y) \mid 2x + 3y = 5\} = \{(y, x) \mid 2y + 3x = 5\}$$

c) **IMPERATIVELY** (p. 64) Find x and y such that

$$2x + 3y = 5 \text{ and } x + y = 2$$

HE IDENTITY FUNCTION (p. 74), herein denoted by *j*, as in

$$\frac{j^2 - 1}{j - 1} = j + 1 \text{ (on U} - \{1\}), \quad \frac{j^2 - j - 2}{j - 2} = j + 1 \text{ (on U} - \{2\})$$

$D\ (j + 1) = 1$,

$\frac{df}{dj}$ (= D *f*), $\int_0^1 f\ dj = \int_0^1 f$.

THE SELECTOR FUNCTIONS (p. 296), herein denoted by *I* and *J*, as in

$\{2I + 3J = 5\} \neq \{2J + 3I = 5\}$;

If $fI = gJ$, then there exists a constant function c such that $f = g = c$;

$D_1\ J = O$.

AN INDETERMINATE (p. 104). Herein, $*$, $*^2$, ... serve as indicators for the places of coefficients in forms, as in $\frac{*^2 - 1}{* - 1} = * + 1 = \frac{*^2 - * - 2}{* - 2}$.

UMMIES (p. 154), herein shed, as in $\int_0^1 cos$.

SPECIFIC FLUENTS (CARTESIAN COORDINATES p. 186), herein denoted by *x* and *y*, as in $\{2x + 3y = 5\}$ (that is, the class of all points σ such that $2\,x\sigma + 3\,y\sigma = 5$) does not pass through the point *o*.

FLUENT VARIABLES (p. 173). Herein, the letters *u*, *v*, *w*, ... serve as fluent variables, *f*, *g*, *h*, ... as function variables.

a) **INDICATIVELY**

w = *sin u* implies $\frac{dw}{du}$ = *cos u* for any two fluents *u* and *w*.

b) **CONJUNCTIVELY**

$\{f \mid sin\ f = j\}$ is a class containing infinitely many functions.

c) **IMPERATIVELY**

Find *f* such that $f \cdot D\ f = 1$.

REMARKS

1. Trigonometry abounds in formulas such as

$$\sin(x + y) = \sin x \cos y + \cos x \sin y \text{ for any } x \text{ and } y$$

containing numerical variables. So does arithmetic, where in many formulas the scope of the numerical variables is the class Y of all positive integers as in

$$1 + 3 + \ldots + (2x - 1) = x^2 \text{ for any positive integer } x.$$

2. Also outside of calculus, formulas containing numerical variables may be replaced by statements about functions; e.g., those in Remark 1 $\sin(I + J) = \sin I \cdot \cos J + \cos I \cdot \sin J$ and $\sum_1 (2j - 1) = j^2$ (on Y).

3. Some expressions are tabulated in several places, e.g., "the line $2x + 3y = 5$." For this traditional expression means the class $\{2I + 3J = 5\}$ in pure analytic geometry, and the class of all points σ such that $2x\sigma + 3y\sigma = 5$ with regard to a physical plane.

4. In some formulas, x and y belong to different categories; e.g., in the differential equation $yy' = x$ (herein rendered by $f \cdot \mathbf{D}f = j$) y is a function variable, and x is the identity function.

5. In some cases, x has different meanings even at different places of one and the same traditional expression; e.g., in $\frac{\partial F}{\partial x}(x, y) = 4x^3y^3$, the x in the "denominator" is not a numerical variable, but rather part of an operator, herein denoted by $\mathbf{D}_1$.

6. In certain branches of analysis, especially in the theories of complex functions and of operators, x and y are used in various meanings not included in the table.

The English language originally "lacked an adequate system of pronouns and ambiguities were multiplied in Middle English when *hē* 'he', *hēo* 'she', and Anglian *hēo* 'they' ..., became identical in pronounciation... The listener or reader [had to] gather the meaning from the context," says S. Potter in *"Our Language"*; and he continues: "That is why Middle English adopted and adapted these structural words [*they, them, their,* and perhaps *she*] from Scandinavian to supply its needs. Then, as now, intelligibility was a strong determining factor." "But although the [new] forms must ... be reckoned a great advantage to the language, it took a long time before the old forms were finally displaced," says O. Jespersen in *"Growth and Structure of the English Language."* Eventually, however, the new forms did prevail. For, as Potter puts it, "when men find that their words are imperfectly apprehended they naturally modify their speech and they deliberately prefer the unambiguous form."

The tabulated ambiguities in the traditional mathematical symbolism are at the root of an often deplored fact: Even excellent teachers of mathematics "find that their words are imperfectly apprehended" by many beginners. What else can be expected considering that x (as a numerical and as a fluent variable) plays the roles of two mathematical pronouns as different from one another as are 'he' and 'they', while, in other usages, x is something like a specific mathematical noun or a mathematical verb or a mathematical adverb or void of any meaning? In view of all that, teachers of mathematics perform miracles.

The preceding table also exhibits natural modifications of that symbolism which obviate the need to gather the meaning of symbols from the context. Unambiguous forms are listed whose selection has been determined by one and only one factor – intelligibility. For it is the author's conviction that in the language of mathematics, and expecially of calculus, everything should be intelligible: every paragraph, every sentence, every clause, and every word.

BIBLIOGRAPHY

(Other Publications by the Author on the Modern Approach to Calculus)

1. ALGEBRA OF ANALYSIS. Notre Dame Mathematical Lectures, vol. 3, 1944 (out of print). In this booklet of 50 pages, symbols for the identity function and the n-th power function (j and n-po) were introduced into analysis. All laws of the calculus (including integration by substitution and by parts) were formulated without the use of numerical variables. Functions of "higher rank" (the analogue of functions of several places) and their partial derivatives were studied in the same spirit.

2. TRI-OPERATIONAL ALGEBRA. Reports of a Mathematical Colloquium, 2nd ser., issue 5-6, 1945, pp. 3-10. A postulational approach to analysis. A study of a class of undefined elements (called functions) for which addition, multiplication, and substitution are defined; in particular, of classes containing only a finite number of functions – analogues of planes containing only a finite number of points.

3. GENERAL ALGEBRA OF ANALYSIS. Above Reports, issue 7, 1946, pp. 46-60. An extension of the theory to functions of several places, into which several functions can be substituted. – Numerous papers on tri-operational algebra by collaborators and students of the author are published in the same issue of the Reports.

4. ANALYSIS WITHOUT VARIABLES. Journal of Symbolic Logic, 11, 1946, pp. 30/31. Abstract of an address before the Association of Symbolic Logic, February 23, 1946.

5. METHODS OF PRESENTING e AND π. American Mathematical Monthly, 52, 1945, pp. 28-33.

6. ON THE TEACHING OF DIFFERENTIAL EQUATIONS. Ibid., 51, 1944, pp. 392/395.

7. THE CONCEPT OF A FUNCTION. Edited by Burton Fried. 20 pp., 1948. (mimeographed). Second Edition. 22 pp. 1949.

8. ARE VARIABLES NECESSARY IN CALCULUS? American Mathematical Monthly, 56, 1949, pp. 609-620. An outline of calculus without numerical variables.

9. THE MATHEMATICS OF ELEMENTARY THERMODYNAMICS. American Journal of Physics, 18, 1950, pp. 89-103 (and 19, 1951, p. 476). Formulations of the laws of thermodynamics without the use of so-called incomplete differentials – either in terms of time rates of physical fluents or by means of Stieltjes Integrals.

10. INTRODUCTION TO CALCULUS (edited by H. S. Levin), a pamphlet of 64 pages published by the Bookstore, I.I.T., 1950.

11. UNDERSTANDING CALCULUS. Part I (pp. 1-120), Part II (pp. 121-192), published by the Bookstore, I.I.T., 1951. Graphical and numerical methods leading to the central facts of calculus – the Reciprocity Laws.

12. CALCULUS. A MODERN APPROACH. xxv + 255 pp. (mimeographed), published by the Bookstore, I.I.T., 1952.

13. THE IDEAS OF VARIABLE AND FUNCTION. Proceedings, National Academy of Sciences, 39, 1953, pp. 956-961.

14. CALCULUS. A MODERN APPROACH. Second, enlarged Edition. xxiii + 303 pp. (Mimeographed) published by the Bookstore I.I.T., 1953. This edition contains an extensive treatment of variable quantities and fluent variables (called "general variable quantities").

15. ON VARIABLES IN MATHEMATICS AND IN NATURAL SCIENCE. British Journal for the Philosophy of Science, 5, 1954, pp. 134-142.

16. IS CALCULUS A PERFECT TOOL? Journal of Engineering Education, 45, 1954, pp. 261-264.

17. TOSSING A COIN. American Mathematical Monthly, 61, 1954, pp. 634-636.

18. A SIMPLE DEFINITION OF ANALYTIC FUNCTIONS AND GENERAL MULTI-FUNCTIONS. Proceedings, National Academy of Sciences, 40, 1954, pp. 819-821. The so-called multi-valued functions studied in the theory of complex functions and generalizations are defined as classes of pairs of complex numbers.

19. VARIABLES DE DIVERSES NATURES. Bulletin des Sciences Mathématiques 78, 1954, pp. 229-234.

20. RANDOM VARIABLES AND THE GENERAL THEORY OF VARIABLES. To be published in the Proceedings of the 3rd Berkeley Symposium on Math. Statistics and Probability. Univ. of California Press, 1955.

21. CALCULUS 1950 – GEOMETRY 1880. To be published in Scripta Mathematica, 1956. The traditional presentation of calculus is compared with the development of geometry in Euclid's "Elements" which, for 2000 years, was universally regarded as a flawless deductive theory. Nowadays geometers (while, of course, still ranking the "Elements" as one of the great books of all time) feel that only comparatively recently geometry was transformed into a strictly deductive theory in the modern sense of the word. It was the work of Pasch in the 1880's that started a development in the course of which geometry was completely automatized.

TOPICAL INDEX

(* indicates that the reference is to an exercise.)

INDEX OF SYMBOLS

(Numbers refer to pages.)